Aotus: The Owl Monkey

Aotus: The Owl Monkey

Edited by

Janet F. Baer
California Institute of Technology
Division of Biology
Pasadena, California

Richard E. Weller
Batelle
Pacific Northwest Laboratories
Richland, Washington

Ibulaimu Kakoma
College of Veterinary Medicine
University of Illinois
Urbana, Illinois

Academic Press
San Diego New York Boston
London Sydney Tokyo Toronto

This book is printed on acid-free paper. ∞

Academic Press, Inc.
A Division of Harcourt Brace & Company
525 B Street, Suite 1900, San Diego, California 92101-4495

United Kingdom Edition published by
Academic Press Limited
24-28 Oval Road, London NW1 7DX

Library of Congress Cataloging-in-Publication Data

Aotus : the owl monkey / edited by Janet F. Baer, Richard E. Weller, and Ibulaimu Kakoma.
p. cm.
Includes bibliographical references and index.
ISBN 0-12-072405-7
1. Aotus. I. Baer, Janet F. II. Weller, Richard E. III. Kakoma, Ibulaimu.
QL737.925A94 1994
599.8'2--dc20 94-7468
CIP

PRINTED IN THE UNITED STATES OF AMERICA
94 95 96 97 98 99 EB 9 8 7 6 5 4 3 2 1

Contents

6 Handreating the Owl Monkey

Carlos A. Malaga

7 Infectious and Noninfectious Diseases of Owl Monkeys

Richard E. Weller

10 Ophthalmologic Research in the Owl Monkey
Thomas E. Ogden

11 The Functional Organization of Visual Cortex in Owl Monkeys
John Allman, Richard Jeo, and Martin Sereno

12 The Organization of Sensory and Motor Cortex in Owl Monkeys
Jon H. Kaas

13 Parasites of the *Aotus* Monkey
Manuel Tantaleán and Alfonso Gozalo

Contributors

Numbers in parentheses indicate the pages on which the authors' contributions begin.

John Allman (287), Division of Biology, California Institute of Technology, Pasadena, California 91125

Rolando Aquino (59), Centro de Reproducción y Conservación de Primates del Instituto Veterinario de Investigaciones Tropicales y de Altura, Universidad Nacional Mayor de San Marcos, Iquitos, Perú

Janet F. Baer (133), California Institute of Technology, Division of Biology, Pasadena, California 91125

William E. Collins (217), Centers for Disease Control, Atlanta, Georgia 30333

Allen F. Dixson (113), Medical Research Council Group on the Development and Integration of Behavior, Sub-Department of Animal Behavior, University of Cambridge, Madingley Cambridge CB3 8AA, United Kingdom

Filomeno Encarnación (59), Centro de Reproducción y Conservación de Primates del Instituto Veterinario de Investigaciones Tropicales y de Altura, Universidad Nacional Mayor de San Marcos, Iquitos, Perú

Susan M. Ford (1), Department of Anthropology, Southern Illinois University, Carbondale, Illinois 62901

Alfonso Gozalo (353), Nonhuman Primate Reproduction and Conservation Center, Tropical and High Altitude Research Veterinary Institute, Universidad Nacional Mayor de San Marcos, Casilla 621, Iquitos, Perú

Richard Jeo (287), Division of Biology, California Institute of Technology, Pasadena, California 91125

Jon H. Kaas (321), Department of Psychology, Vanderbilt University, Nashville, Tennessee 37240

Norval W. King (245), New England Regional Primate Research Center, One Pine Hill Drive, Southborough, Massachusetts 01772

Carlos A. Malaga (165), Animal Resources Center, Medical Sciences Campus, University of Puerto Rico, San Juan, Puerto Rico 00936

Thomas E. Ogden (263), Doheny Eye Institute, University of Southern California, School of Medicine, Los Angeles, California 90033
Martin Sereno (287), Department of Cognitive Science, University of California at San Diego, La Jolla, California 92093
Manuel Tantaleán (353), Tropical Medicine Institute "Daniel Alcides Carrión", Universidad Nacional Mayor de San Marcos, Casilla 2670, Lima 100, Perú
Richard E. Weller (177), Battelle, Pacific Northwest Laboratories, Richland, Washington 99352
Patricia C. Wright (97), Department of Anthropology, State University of New York at Stony Brook, Stony Brook, New York 11794

Foreword

Aotus, the owl monkey, has become an important nonhuman primate model for biomedical research, not only for developing a human malaria vaccine but also for understanding viral pathogenesis and the visual system *inter alia*. The popularity of the owl monkey as a laboratory animal model is recent. Owl monkeys are among important primate species kept in zoological gardens and primate centers. While their natural distribution is known to occur from Panama to Argentina, their taxonomy was not completely understood until the early 1980s. Moreover, the embargo imposed by source countries in the 1970s, which were alarmed by the uncontrolled trade, destruction of natural habitats, and the lack of national and international supervision during capture and transport of neotropical primates, made them inaccessible.

Because nonhuman primates are indispensable for modern biomedical research, as well as for production of biologics and testing of compounds for toxicity, the embargo polarized the issues of conservation and use. These issues are viewed to be diametrically opposed, but are indeed complementary, since the policies of integrating conservation and rational use have been shown to be beneficial to the preservation of nonhuman primate species and the protection of their native habitat.

Recognizing that conservation and rational use of neotropical primates are inextricably linked, the Pan American Health Organization, which is also the Regional Office of the World Health Organization for the Americas, developed a program of technical cooperation with its member governments based on a comprehensive approach consisting of integrated activities focused on conservation, reproduction in the natural habitat and in captivity, and rational use of nonhuman primates in 1973.

In 1976, the Government of Peru was the first to recognize that total embargo was not the appropriate solution to conservation. Instead it promulgated a national policy for the rational management of its wildlife fauna, and its Primate Project was one of its lines of action. It considered its wild population of nonhuman primates as a renewable natural resource that could be managed for the mutual benefit of man, animals, and the environment. The results of the census and population dynamics and allied studies conducted at the Peruvian Primatological Project are contained in this volume. The Project stands as a paradigm to programs designed to conserve wild nonhuman primates as a renewable resource.

There is no doubt that *Aotus: The Owl Monkey* will prove invaluable to the research community. Until now, information regarding *Aotus* was scattered in the literature and not readily available to those interested in, or working with members of this genus. This text is intended to consolidate much of that information and increase awareness of the owl monkey. The editors and chapter authors have certainly attained that goal. The materials contained herein represent the multidisciplinary efforts of biologists, primatologists, veterinarians, geneticists, and biomedical researchers who have integrated their expertise to produce the first comprehensive text on the owl monkey.

I therefore urge each and every user of this book to recommend its value to their colleagues, and to support those programs and activities that will benefit the conservation and rational use of this unique primate species.

Primo Arambulo III
Program Coordinator, Veterinary Public Health
Pan American Health Organization/World Health Organization
Washington, D.C.

Preface

The impetus for this volume resulted from a meeting of a small group of scientists in Washington, D.C., in 1989. Although the meeting participants had diverse backgrounds they were all linked through their interest in the genus *Aotus* and all agreed that a comprehensive volume would be a valuable aid to the scientific community. *Aotus,* also referred to as owl monkeys, night monkeys, or douroucouli, have been maintained in captivity for more than 30 years and are well represented in the scientific literature. However, published information has been scattered in a number of disparate, specialized textbooks and journals. The general description of the genus in textbooks has invariably tended to be brief, leaving the reader with only a minimal knowledge of these unique, nocturnal, neotropical primates.

Providing the reader with a comprehensive overview of the genus *Aotus* dictated the selection of the diverse group of contributors and the contents of their respective chapters in this volume. The volume begins with a thorough review by Susan Ford of the historical distribution of the owl monkey and a discussion on taxonomy based on extensive body measurements, and chromosomal and blood protein heterogeneity. As a result of complex analysis, applying standard criteria such as the species recognition concept, vicariant biogeographic patterns, and karyotic polymorphisms, she concludes that the number of species of *Aotus* remains putative as seven or probably five.

Next, features of the natural forest habitat and complexities of sleeping site selection are characterized and analyzed by Rolando Aquino and Filomeno Encarnación. Data on group size, group composition, and breeding season are presented and utilized to devise a population census and harvest strategy based on the calculated impact that cropping may have on population structure. Patricia Wright continues the discussion of environmental influences on the evolutionary development of the genus in Chapter 3 in which she describes the complex evolutionary struggle and adaptation of the owl monkey as it relates to its behavioral repertoire. In Chapter 4, Allen Dixson focuses his discussion on the reproductive biology of the genus. Testicular function, ovarian cycles, and sexual behavior are described in relation to the fertility and fecundity problems of owl monkeys. The paucity of data on the influence of seasonal factors on reproduction in the wild, the problem of low sperm counts and the limited data on endocrine changes in pregnancy are discussed. All of these authors emphasize the need to balance the human demand for biomedical research with the welfare of the species in meeting natural and artificial ecological challenges.

The theme of the book then shifts to the owl monkey in captivity, starting with husbandry and medical management in Chapter 5. In this chapter, Janet Baer systematically presents the husbandry needs, dietary requirements, social grouping parameters, and veterinary procedures for the genus. Additionally, she presents a wide range of normative physiologic data that will certainly be helpful to the clinician and other investigators. She emphasizes the need to create an enriched environment that contains key critical components of the natural habitat and is consistent with the requirements of various regulatory agencies. Basic guidelines that promote an enriched captive environment for these animals are provided.

In Chapter 6, Carlos Malaga complements chapters by Dixson and Baer presenting handrearing techniques for the owl monkey, concentrating on infant care and natural and foster parenting. A useful catalog of the various equipment, supplies, and formulas available for infant care is provided.

In Chapter 7, Richard Weller highlights the infectious and noninfectious diseases of owl monkeys, ranging from classical infectious to metabolic, neurologic, gastrointestinal, and neoplastic disease. In that treatise, he systematically describes clinical and laboratory diagnostic criteria for a wide range of diseases and makes numerous useful suggestions on prevention and management of a wide array of diseases in both wild and captive animals. Although he primarily addresses veterinarians, his chapter, including comments on potential zoonotic diseases, will be useful to animal handlers at all levels.

Subsequent chapters deal with the use of the owl monkey as an animal model for basic and applied investigation. A comprehensive review of the use of these animals in malaria research is presented by William Collins in Chapter 8. Collins, emphasing the unique role of the owl monkey in malaria research, provides a detailed description of natural and experimental malaria infections in owl monkeys followed by a review of the use of these animals in malaria vaccine trials. In Chapter 13, Manuel Tantaleán and Alfonso Gozalo complement Collins' work in their review of the miscellaneous endo- and ectoparasites of owl monkeys. Emphasizing nonmalaria parasites, these authors present detailed morphologic descriptions of a number of potential and known owl monkey parasites and include useful information on methods of detection, control, and eradication of these organisms.

A thorough account of the use of the owl monkey as an animal model for the study of viral oncogenesis is presented in Chapter 9 by Norval King.

Susceptiblility to infection, clinical signs of disease, pathologic features, and the etiology and pathogenesis of oncogenic herpesviruses and polyomaviruses are discussed in a comprehensive and systematic manner.

In a similar treatise, Thomas Ogden reviews the use of the owl monkey as an animal model for ophthalmologic research in Chapter 10. He sets the stage for his review of the experimental use of the owl monkey in ocular research by first providing a detailed description of the structure and function of the owl monkey eye.

The remaining chapters focus on the use of the owl monkey for studies of neocortical organization and function. These chapters emphasize the value of the owl monkey in neurobiological research because of its relatively small sized and less fissured brain as compared to Old World primates, and it has a uniquely adapted nocturnal visual system. In Chapter 11, John Allman and his colleagues describe the multiple maps in the owl monkey visual cortex and the relationship of these maps to perceptual functions. Several of the cortical visual areas now known for many species were first discovered in the owl monkey. They also describe a new procedure for recording from single neurons in awake, behaving owl monkeys performing perceptual and memory-related tasks. In Chapter 12, Jon Kaas describes in detail the intricate anatomical organization and function of the auditory, visual, somatosensory, and motor systems of the owl monkey. Readers interested in pathology and genetics will find useful information and references in chapters throughout the volume.

This volume is the result of countless hours of effort on the part of the individual chapter authors. The editors gratefully acknowledge this effort and extend their appreciation to each contributor for the enthusiasm, support, and patience demonstrated during the development and production process. The editors are also indebted to the staff at Academic Press, especially Charles Crumly, for their unstinting patience, guidance, and editorial assistance. The editors are also grateful to the staff of Battelle Northwest Laboratories, the University of Illinois, and the California Institute of Technology for various contributions in time, data, and oganizational support that made this volume possible.

In conclusion, the editors hope that the availability of this collection of authoritative articles in a single comprehensive volume will represent a useful resource for the scientific community who share our interest in this small nocturnal primate. It is our hope that this volume will ultimately promote both human and owl monkey health.

Janet Baer
Richard Weller
Ibulaimu Kakoma

1

Taxonomy and Distribution of the Owl Monkey

Susan M. Ford
Department of Anthropology
Southern Illinois University
Carbondale, Illinois 62901

I. INTRODUCTION

The South American owl monkey, *Aotus*, a somewhat small, monogamous, generally quadrupedal platyrrhine lacking a prehensile tail, is unique among anthropoids in its nocturnal habits. Owl monkeys, or douroucoulis, average around 1 kg in weight (Ford and Davis, 1992), with no sexual dimorphism in size or noticeably in morphology. They share fairly short hair, with a generally grayish/agouti dorsum, often with a slightly more reddened stripe down the back. The central face is brown, surrounded by white, with large white patches over the eyes. There are three black stripes extending back onto the forehead (one central and one from the lateral side of each eye); these vary in width, in darkness, and in whether or not they converge posteriorly. Slight black stripes extend toward each cheek as well. The ears are small. The sides of the neck vary from gray (generally north of the Amazon) to red (generally south of the Amazon). The ventrum is generally yellow-tan or buffy, with populations varying in the presence of orange-to-red pigmentation expanding broadly on the ventrum and onto the inner sides of the arms and legs. The tail is nonprehensile and agouti-brown, with varying amounts of black-tipped or totally black hairs extending from the distal end up the tail, and in one part of the range, gray hairs on the proximal end. The eyes are large and are generally assumed to lack a tapetum; however, while Hershkovitz (1983) found eyeshine to be absent, Emmons (1990, p. 110) described it as present and "bright, reddish orange."

Owl monkeys are widely distributed from Panama to Argentina, and from the mouth of the Amazon (on its southern bank) west into its headwaters in Peru and Ecuador, covering most of tropical South America and extending from low to fairly high altitudes (up to 3200 m). Although they have been reported in Surinam, Guyana, and French Guiana in the past (Cabrera, 1957; de Cruz Lima, 1945; Tate, 1939), no newer studies suggest their presence there (Emmons, 1990; Fleagle and Mittermeier, 1980; Hershkovitz, 1983; Mittermeier *et al.*, 1988; Thorington and Vorek, 1976).

Despite this wide distribution and the common tendency in the 1800s and early 1900s to establish many local varieties as separate species or subspecies, until recently there has generally been assumed to be only one species, *Aotus trivirgatus*, following Hershkovitz (1949) and Cabrera (1957; see, e.g., Honacki *et al.*, 1982; Mittermeier and Coimbra-Filho, 1981; Moynihan, 1976; Napier, 1976; Rathbun and Gache, 1980; Wright, 1981). However, there has been little attempt to carefully assess this assumption. One study, that of Thorington and Vorek (1976) based on a quantitative multivariate analysis of cranial data and a discussion of some pelage features, appeared to reinforce the general assumption that differences in geographical populations of *Aotus* are entirely clinal, and that only one true species is present today. On the other hand, Hershkovitz (1983) suggested that there are nine valid species, two with two valid subspecies each (Fig. 1). His conclusions are based primarily on coloration of the side of the neck and karyotype, supported by other features of pelage coloration and malaria susceptibility.

FIGURE 1 Distribution of nine species and four subspecies recognized by Hershkovitz (1983). Modified from Hershkovitz's Fig. 2, with the distribution of *A. nancymae* in particular altered some in accord with Aquino and Encarnación (1988); see text.

His suggestions follow on and draw from the growing body of data on karyotypes and chromosome structure in *Aotus*, which has long led Brumback (1973, 1974) to suggest that there are more than one species of *Aotus*. Recently, several populations in the highlands of Colombia and Peru have been suggested to represent valid species; as yet, none of these has been formally named or described (see Mittermeier *et al.*, 1988). Hall (1981) named an additional species from Panama, *A. bipunctatus*, although he indicated that it was most likely simply a local variant of the more widespread Central American and northern South American species.

This study takes the validity of the species and subspecies categories proposed by Hershkovitz (1983) as a null hypothesis to be tested; therefore, his taxonomy will be used in the body of the paper. Hershkovitz finds a very basic division between populations from north of the Rio Amazonas, characterized by gray-agouti hairs on the sides of the neck (including behind and below the ears) and susceptibility to malaria, and populations south of the river, with orange (light to dark reddish-orange) hairs on the side of the neck (including behind and below the ears), which are resistant to malaria. These comprise his gray-neck and red-neck species groups, respectively.

Four species north of the Amazon make up Hershkovitz's gray-neck group, which he considers the ancestral condition for the genus. As can be seen in Fig. 1, his species include *A. lemurinus* I. Geoffroy 1843 from northern Colombia and Venezuela, with *A. l. lemurinus* including the Panamanian and western and southern Colombian specimens as well as those from a narrow band in the Andes on the eastern portion of the range, and *A. l. griseimembra* Elliot 1912 extending from the north coast of Colombia east into extreme northwestern Venezuela, and possibly in a narrow tongue south along the valley of the Rio Magdalena, bisecting the range of its sister subspecies. An enclave on the eastern slopes of the Andes around Villavicencio and Fusagasuga, Colombia, possibly east to the Venezuelan border, forms the species *A. brumbacki* Hershkovitz 1983. *Aotus trivirgatus* Humboldt 1811 extends from the Rio Trombetas west to the Rio Negro, and from the north bank of the Rio Amazonas into the area of the Rio Caura/Rio Orinoco confluence in Venezuela, apparently restricted to the south bank of the Rio Orinoco. *Aotus vociferans* Spix 1823 borders *A. trivirgatus* to the west, extending from the west bank of the Rio Negro west to the Rio Cenepa in the Andean highlands of Ecuador and Peru, and from the north bank of the Amazonas north for an undetermined distance. There is also an enclave, possibly of recent origin, south of the Rio Amazonas along the Rio Purus.

South of the Amazon are the remaining five species, his red-necked group. Hershkovitz suggests that the red-neck trait linking these populations developed as a founder effect in a population, possibly in northern Peru, which became separated from the north-bank parent population by isolation on the south bank of the Rio Amazonas. The species include *A. infulatus* Kuhl 1820 on the east, extending from the mouth of the Amazon west to the Rio Tapajos and Rio Juruena and into Rondonia (Pieczarka *et al.*, 1993), and south in the Mato Grosso of Brazil to the Rio Paraguai. The extent of its distribution east of the Rio Tocantins is uncertain, but it does exist on the east bank (Sampaio *et al.*, 1991; Schneider *et al.*, 1989). This is bordered to the west by *A. nigriceps* Dollman 1909, west to the Rio Huallaga and south from the Amazon possibly to the Rio Guapore and Rio Madre de Dios. *Aotus nancymae* Hershkovitz 1983 is found in northern Peru and western Brazil, in a slightly larger range than that originally described by Hershkovitz. [Note: I here use the feminine spelling for Hershkovitz's *A. nancymae*, following Aquino and Encarnación (1988; see also Groves, 1989).] According to Aquino and Encarnación (1988), *A. nancymae* expands south from the Rio Amazonas/Marañon to about 7 degrees south latitude, bordered there by *A. nigriceps*, and from west of the Rio Huallaga (possibly up into the eastern slopes of the Andes) east to the Rio Yavari (Aquino and Encarnación, 1988) or on to the Rio Jutai (Hershkovitz, 1983). There is also a small enclave north of the Rio Marañon, from the lower Rio Tigre west to the lower Rio Pastaza (see Fig. 1). Specimens from west and south of the Rio Huallaga to about 10 degrees south on the eastern slope of the Andean highlands are *A. miconax* Thomas 1927. The southernmost members of *Aotus* are placed in *A. azarae* Humboldt 1811, extending south from the Rio Madre de Dios and Rio Guapore in Bolivia into Paraguay and northern Argentina south of the Rio Bermejo, and east from the Rio Paraguai to the Andes in the north of its range. This species is divided by an arc extending roughly from Lago Uberaba on the east to Sucre, Bolivia, on the west into a northern subspecies, *A. a. boliviensis* Elliot 1907, and a southern one, *A. a. azarae*.

Use of the species and subspecies designations of Hershkovitz throughout the text, tables, and figures does not imply any necessary acceptance of these groups, since examining the nature of geographic variation in *Aotus* is the subject of this study. However, use of these categories should serve to simplify and clarify discussion.

II. MATERIALS AND METHODS

Two separate data sets were examined. The first consisted of a series of 15 cranial and dental measurements taken by R. W. Thorington, Jr. on 193 adult male and female *Aotus* skulls from a number of institutions, and originally analyzed in a somewhat different manner from that used here (Thorington and Vorek, 1976). The measurements taken are described in Appendix A. The second set consisted of qualitative scores on 16 aspects of pelage coloration on 105 adult male and female specimens of *Aotus* housed in the National Museum of Natural History and the Museu de Zoologia da Universidade de São Paulo (data collected by the author). Pelage variables and scoring are given in Appendix B and follow guidelines given by Thorpe (1976, 1979; Thorpe and McCarthy, 1978) for scoring coloration patterns and other qualitative variables for use in multivariate analysis.

Sample sizes for several of Hershkovitz's geographic species and subspecies are quite large, others are very small or not represented. In particular, many of the skins examined were incomplete (especially affected by missing tails, arms, or legs), and thus were not able to be used in the multivariate aspects of this study. Sample sizes by species/subspecies are given in Table I. Obviously, for those taxa not represented or present in small numbers, conclusions must be extremely tentative.

t-Tests were performed for all variables in both data sets to determine whether there were any significant differences between the sexes, using the t-test procedure of SAS (SAS Institute, Inc., 1990). This was performed on the total data set, as sample sizes within many of the groups were so small. None of the metric cranial variables showed any significant differences between males (n = 84) and females (n = 89). Only TUFT and WHORL in the pelage data set indicated significant differences between the sexes ($p \leq 0.01$, n = 35–37 males, n = 43–46 females); however, for

TABLE I Sample Size by Hershkovitz's (1983) Taxonomy

Taxon	*N* Cranial data	*N* Pelage data[a]	
		Total	Sample 1
A. azarae azarae	2	—	—
A. azarae boliviensis	31	1	1
A. brumbacki	4	9	6
A. infulatus	1	23	18
A. lemurinus griseimembra	21	11	9
A. lemurinus lemurinus	41	22	16
A. miconax	8	9	7
A. nancymae	1	1	1
A. nigriceps	6	8	8
A. trivirgatus	19	11	8
A. vociferans	26	3	2
Barranquilla origin	33	3	3
Unsure of taxon	—	4	3
Total	193	105	82

[a] Sample size 1 is that actually used in most multivariate analyses, after deletion of incomplete specimens and of two head stripe variables. The total pelage sample was used in the univariate analyses (see text).

both of these variables, only a very small portion of the specimens exhibited the trait, and this may have biased the results of the t-tests. As discussed in the following, these two variables were ultimately deleted from the multivariate analyses for other reasons. Although testing for sexual dimorphism was not done within groups, given the absence of any significant differences between males and females for the features examined coupled with the fact that *Aotus* in general exhibits little to no sexual dimorphism in most traits (Ford and Davis, 1992; Hershkovitz, 1983; Moynihan, 1976; Wright, 1981), the sexes were pooled in all subsequent analysis.

The protocol chosen for the multivariate study of population structure within the genus *Aotus* closely follows that developed by Thorpe (1976, 1979, 1980a,b, 1983, 1984; see also Thorpe and McCarthy, 1978; Benton, 1980) in his studies of the ringed garter snake, *Natrix natrix*, and other reptiles.

In Thorpe's studies, large samples from specific localities were grouped prior to further analysis by a cluster technique. In this study, samples from each relatively confined geographic locality were generally quite small, often single specimens. Therefore, specimens were grouped into 11 categories (variable "HERSH"—see Appendix C) based on geographic locality, these categories being the 11 species and subspecies proposed by Hershkovitz (1983). All localities (taken from the museum specimen tags) were located by latitude and longitude when possible, primarily through the use of published gazeteers (especially Fooden, 1963; Hershkovitz, 1977; Paynter, 1982; Paynter and Traylor, 1977, 1981; Paynter *et al.*, 1975; Stephens and Traylor, 1983), reference to collectors' notes in museum collections, and by location on maps of South America (gazeteer available upon request). In some cases, although taxon identification was secure, latitude and longitude had to be estimated. This was particularly true if the locality information consisted solely of a state or department, where the coordinates of the center of the department were used, or simply a river with no indication of location along the course of the river, where the coordinates of the mouth of the river were used (following the policy of the ornithological gazeteers published by Harvard University). A number of specimens come from Barranquilla, northern Colombia, a coastal city at the mouth of the Rio Magdalena. Most if not all of these specimens were undoubtedly collected at various unspecified localities inland. On the assumption that they are from northwestern Colombia, possibly along the Rio Magdalena itself, they are often placed in *A. lemurinus griseimembra,* with the latitude and longitude of Barranquilla. However, it is recognized that this assumption may be in error [see, e.g., Ma *et al.* (1985) for a likely specimen of *A. vociferans* imported from Barranquilla]. Because of this, all analyses were done with the Barranquilla specimens treated as a distinct group ("Q"); all analyses were then repeated with the Barranquilla specimens excluded from the analysis. A small number of specimens included in the analysis either could not be identified as to "HERSH" group or could not be assigned latitude and longitude values because of the vagueness of the locality information (e.g., "Amazon" or "Neotropics"). These were assigned to a thirteenth category of "HERSH", "U" for unknown population group.

Univariate frequencies, expressed as percentages of each morphology (Appendix B) for each taxon, were determined for the qualitative features comprising the pelage data set. These were examined both in chart form and displayed in histograms. This allowed better interpretation of the results of the multivariate analyses, emphasizing the amount of overlap between taxa and serving as a caution in cases where only a few or a single specimen was examined. In addition, many specimens were not complete skins, and because not all variables could be scored for these individuals, they were of necessity deleted from the multivariate analyses (see Table I). Univariate analysis allowed some consideration of the entire sample of 105 skins.

Variables were deleted from further consideration if significant geographic variation was not apparent, based on a one-way analysis of variance (AOV) for each variable across groups (see Benton, 1980; Thorpe, 1980b, 1984). None of the variables in the cranial data was deleted as a result of this step. One-way AOV on the pelage data indicated that there was insignificant geographic variation between groups in SIDENECK, TUFT, and WHORL. When a second one-way AOV was performed on the pelage data set excluding these three variables, there was not significant variation in the additional variables HEADCHK and HEADWID. It is interesting to note that one of these pelage variables, SIDENECK, is the basis of the primary division of *Aotus* into subcategories by Hershkovitz (1983), and that two others, TUFT and WHORL, figure prominently in his classification. As mentioned earlier, interscapular tufts and whorls were found on only a very small number of the 105 skins examined, and this fact may have caused them not to exhibit any significant geographic variation in this study. Since the color of the side of the neck forms the basis of Hershkovitz's classification, which is used as the original groupings in this analysis, "Red Neck" is either present in all members of a group or absent in all. Therefore, it is to be expected that it would not exhibit significant values and would not be allowed in an analysis based on predefined groups. SIDENECK was excluded from further consideration in the canonical variates analysis described in the following, which directly uses and tests the predetermined groups, as were the other 4 variables (TUFT, WHORL, HEADCHK, and HEADWID), leaving 11 pelage characteristics. Because of the apparent importance of SIDENECK, at least in Hershkovitz's earlier assessment of geographic variation in *Aotus*, it was included in the principal components analyses (the following), which was therefore based on 12 pelage characteristics.

Only variables that contribute significant information independent of information gained from other variables should be included, as indicated by a low pooled within-group correlation and a lack of logical correlation (Thorpe, 1980b; also Thorpe, 1984; Benton, 1980). This method is particularly effective with "all things being equal," that is, no size effect driving overall correlation (R. S. Thorpe, personal communication). As will be seen, size is a factor in some of the population differences. However, these monkeys do not differ markedly in overall size; rather, portions of the cranium may be slightly larger in one population and other portions in a different population. This is apparent in the fact that only four pairs of cranial variables had a greater than .65 pooled within-group correlation, and of these, only two pairs were greater than .7: TRL and MLL (.72) and ANH and TMJ (.88) (the other two pairs are CBL and ANL, .66, and CBH and PSL, .67). ANH was dropped from further consideration after this step, because of the high correlation and logical connection in angular height (ANH) and temperomandibular joint height (TMJ) (note that in a study of forms differing more widely in diet, e.g., different genera, such a high correlation would not be expected). Only one pair of pelage variables had a pooled within-group correlation higher than .5 (ARM and LEG), and that pair's correlation was only .65. Therefore, all remaining variables (excepting those listed here) were included in further analyses.

These two data sets were then analyzed using two different multivariate techniques. The first was a principal components analysis, with the data first standardized to a mean of 0 and a standard deviation of 1 to control for differences in measurement scale (Thorpe, 1980a). This was done using the STANDARD procedure followed by the FACTOR procedure of SAS. The results were then examined in a number of ways. The first three factors were considered to express any primary pattern of variation. In fact, the first three factors explained 49, 13, and 9% (respectively, 71% cumulative) of the variation seen in the cranial data, and 30, 15, and 12% (respectively, 57% cumulative) of the variation seen in the pelage data. These factors were plotted against one another in scatterplots to look for any significant clusters of specimens that might form geographic populations.

The primary factors were then each plotted against both latitude and longitude. By visually examining the distribution along east–west and north–south transects, it is much easier to determine if significant variation results from gradual, clinal shifts or reflects abrupt transitions across small geographic distances, indicating more discrete population shifts [perhaps subspecies or species differences, Thorpe's "categorical" or "stepwise" shifts (as in Thorpe, 1979)]. In all cases, the third factor was found to contribute no discernible additional separation based on Hershkovitz's categories and will not be discussed further.

The second multivariate technique used was a canonical variates analysis. Since canonical variates analysis determines distance between the centroids of predetermined groups (in this case, Hershkovitz's species and subspecies) using the Mahalanobis D^2 statistic, which is not affected by differences in scale of variables, it is not necessary to standardize the data first (Thorpe, 1980a). The CANDISC procedure of SAS was used. The results were then examined as in the principal components analysis: the first three canonical variates (which explained 53, 17, and 9%, respectively, 79% cumulative, of the variation seen in the cranial data, and 49, 28, and 9%, respectively, 87% cumulative, of the variation seen in the pelage data) were plotted against one another in scatterplots, and the primary canonical variates were each plotted against both latitude and longitude. Although several of the species/subspecies had very small samples (even as low as 1), they were included in these analyses. For canonical variates analysis, it is more important that one have a sufficient number of groups, many with sufficiently large and reasonably balanced sample sizes. The few groups with small samples should not unduly bias the results of the overall analysis; however, conclusions based on the placement of those taxa with samples of 1–3 should be made with extreme care and not weighted too heavily (P. Woehlke, personal communication).

Because the canonical variates analysis is based on predetermined groups, and Hershkovitz (1983) included two species divided into subspecies as well as suggesting that three other species may in fact be conspecific, this analysis was repeated with the groups in question combined. This second set of canonical variates analyses did not significantly alter the results attained, except to obscure what appears to be a separation between some of the western Amazon Basin populations by lumping them together (see following discussion). Therefore, the analyses with lumped groups will not be discussed further.

Both multivariate analyses were also performed on subsets of the data, to seek clearer discrimination of groups in close geographic contact, in particular, the northern species *A. lemurinus*, *A. brumbacki*, *A. vociferans*, and *A. trivirgatus*. In this analysis of the northern species, several additional pelage features had to be dropped because of failure to demonstrate significant variation in a one-way AOV; the variables dropped were LEG, NECK, TAILRED, HEADWID, and RUMP, leaving seven variables. All of these analyses were performed twice, once including the group of specimens from Barranquilla and once excluding them; the results for the other groups were not significantly altered. Therefore, only the analyses that included the Barranquilla specimens will be discussed.

As a final note, Thorpe (1979, 1980a,b, 1983) found that canonical variates analysis yielded more clearly defined patterns than other approaches, including principal components analysis. When there is a large size effect, in particular, size should either be controlled for by some mechanism in using principal components analysis or one should focus on canonical variates analysis (Thorpe, 1988). In this study, only 49% of the variation in the metric (cranial) data was explained by the first principal component, indicating a strong but not overwhelming size effect. For this reason, principal components analysis was performed on the cranial data, although the results of the canonical variates analysis are given greater weight. On the other hand, canonical variates analysis can tend to allow single characters to have particularly strong effects in the analysis of pelage data coded as here, because of the heteroscedasticity of these data (R. S. Thorpe, personal communication); the results of the principal components analysis are thus given somewhat more weight for the pelage data set, although the two analyses largely concur.

III. RESULTS

A. Cranial Data

The principal components analysis on the cranial data does not clearly isolate any single subset of the data. However, Factor 1 does serve to separate *A. a. boliviensis* specimens fairly well from all other specimens (Fig. 2). Based on the factor pattern, it appears that this discrimination is primarily due to very high values in *A. a. boliviensis* for CBL, ANL, and BMW, and also to higher values for MSW, BZW, ANH, TMJ, MLL, and TRL (Table II). The two specimens of *A. a. azarae* fall at the end of the *A. a. boliviensis* spectrum closest to the other *Aotus* species, despite the fact that they are farther removed geographically from other *Aotus*.

The canonical variates analysis more clearly indicates that at least two populations of *Aotus* are distinct from the others, and that this distinction is not due to gradual clinal change. Canonical variate one separates the southernmost population, *A. a. boliviensis*, from other *Aotus*, reinforcing the results of the principal components analysis (Table III). As one tracks south in latitude (Fig. 3) or east to west (not shown), *A. a. boliviensis* jumps out as a distinct population. The transition is abrupt, not gradual or clinal, although there is some overlap. The single specimens of both *A. infulatus* and *A. nancymae* fall within the *A. a. boliviensis* group, and the two specimens of southern neighboring *A. a. azarae* overlap the low end of variation of this population. The distinctness of *A. a. boliviensis* is due primarily to loadings for large values for MSW, CBH, and BZW, in particular, and also for CBL and BMW.

Canonical variate two (Fig. 4) separates the eastern population north of the Amazon, *A. trivirgatus*, from others, especially from its neighbor, *A. vociferans*. There is in fact no overlap between these two populations on this variate. Again, the separation of *A. trivirgatus* appears to be a morphological "leap" rather than the result of a clinal gradient of change. This distinctiveness of *A. trivirgatus* is even more marked when only the northernmost groups are analyzed, as seen in the leap on canonical variate one as one "travels" south on Fig. 5. The separation of *A. trivirgatus* appears to be

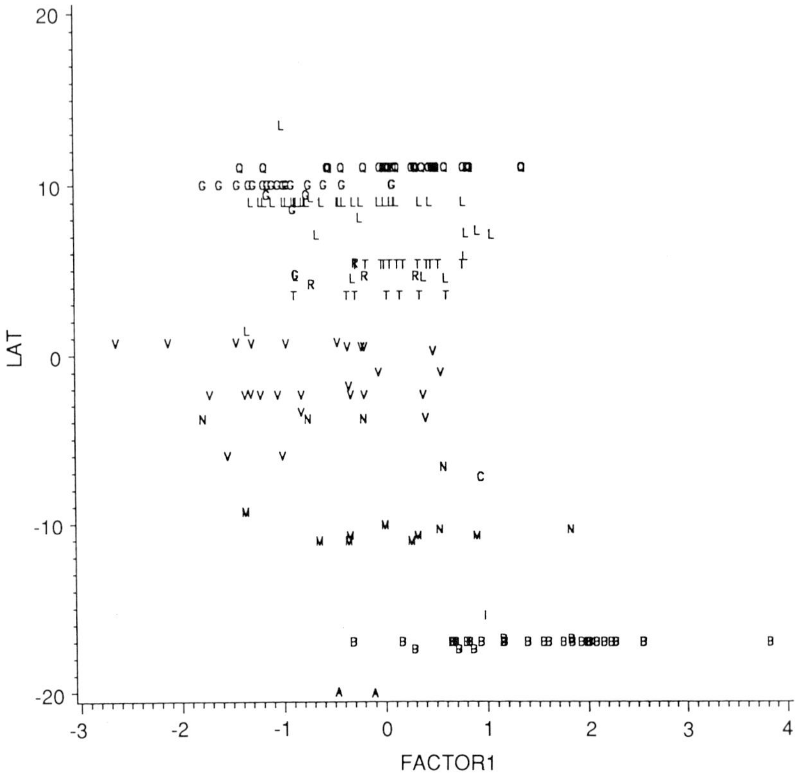

FIGURE 2 Principal components analysis of craniodental morphometric data, showing Factor 1 scores plotted against latitude (on Y axis). Note that as one proceeds south, population B (*A. azarae boliviensis*) demonstrates a marked jump in morphology. Letters indicate species and subspecies recognized by Hershkovitz (1983): A = *Aotus azarae azarae*, B = *A. azarae boliviensis*, C = *A. nancymae*, G = *A. lemurinus griseimembra*, I = *A. infulatus*, L = *A. lemurinus lemurinus*, M = *A. miconax*, N = *A. nigriceps*, Q = Barranquilla, Colombia, source, R = *A. brumbacki*, T = *A. trivirgatus*, V = *A. vociferans*.

due primarily to its large values for MLL, and also to large values for BCW, and ANL and TRL.

Both the principal components and canonical variates analyses indicate mixed variation across much of the rest of the range of *Aotus*, especially in the northern populations identified as *A. l. lemurinus*, *A. l. griseimembra*, *A. brumbacki*, and *A. vociferans* (see Figs. 2–5). These geographic populations do not appear to vary in a stepwise (or categorical) manner in craniodental quantitative features. There is some suggestion in, for example, the distribution seen in Fig. 5 of a clinal gradient within these other northern, gray-necked populations, with gradually decreased dimensions, particularly dentally, as one proceeds south.

TABLE II Means and Standard Deviations for Cranial Data (in Millimeters)[a]

Variable	"Species" N:	A 2	B 31	C 1	G 21	I 1	L 41	M 8	N 6	Q 33	R 4	T 19	U —	V 26
CBL														
Mean		49.00	51.56	50.60	46.70	50.10	48.43	48.70	48.83	48.75	48.98	48.14	—	47.41
S.D.		1.27	1.33	—	0.95	—	1.41	1.59	2.26	1.48	1.64	0.87	—	1.54
BOW														
Mean		42.45	43.40	44.10	42.04	43.60	42.88	42.91	43.30	43.47	42.63	42.80	—	42.33
S.D.		0.35	1.36	—	1.04	—	0.84	1.11	2.47	1.01	0.83	1.05	—	1.33
POW														
Mean		30.85	32.36	32.10	31.89	33.20	31.30	32.20	32.08	32.76	31.13	31.78	—	31.55
S.D.		0.35	1.32	—	0.82	—	1.04	1.31	1.60	0.95	0.85	0.92	—	1.11
BZW														
Mean		40.90	40.65	41.50	38.90	41.10	39.10	38.88	39.08	40.67	40.20	39.74	—	38.81
S.D.		0.00	1.51	—	1.06	—	1.21	1.66	1.01	1.29	1.21	0.92	—	1.57
MSW														
Mean		36.65	37.34	37.80	34.54	39.00	34.10	34.98	35.33	35.06	34.38	35.13	—	33.46
S.D.		0.07	1.44	—	1.01	—	1.18	0.91	1.63	1.08	0.50	0.96	—	1.10
ANL														
Mean		60.10	64.20	64.10	60.07	62.30	61.58	61.54	62.63	61.83	60.90	61.93	—	60.39
S.D.		1.98	1.59	—	1.15	—	1.58	1.41	2.50	1.36	0.70	0.89	—	1.64
ABL														
Mean		44.00	47.75	47.70	46.74	48.60	46.56	48.25	47.23	47.34	45.38	47.23	—	46.46
S.D.		0.42	1.56	—	1.52	—	2.00	1.02	1.88	1.75	1.36	1.09	—	1.99
PSL														
Mean		30.55	29.87	29.30	26.05	28.10	27.62	27.55	29.28	27.48	27.38	27.71	—	27.76
S.D.		0.07	1.77	—	1.89	—	1.59	2.56	2.06	1.58	0.84	1.45	—	1.67
CBH														
Mean		31.45	32.64	33.10	28.44	29.80	30.55	31.56	31.28	30.33	30.80	30.02	—	30.34
S.D.		0.21	1.18	—	1.00	—	1.45	1.55	1.50	0.86	1.06	0.90	—	1.28

(continues)

TABLE II (*continued*)

Variable	"Species" N:	A 2	B 31	C 1	G 21	I 1	L 41	M 8	N 6	Q 33	R 4	T 19	U —	V 26
TRL														
Mean		16.65	18.11	17.20	16.81	17.90	17.42	17.25	17.50	17.35	17.38	17.73	—	16.89
S.D.		0.21	0.56	—	0.63	—	0.49	0.55	0.88	0.61	0.36	0.56	—	0.66
MLL														
Mean		8.70	9.59	9.00	8.55	9.00	8.94	8.85	8.90	8.84	9.05	9.40	—	8.64
S.D.		0.14	0.39	—	0.32	—	0.26	0.28	0.53	0.35	0.40	0.29	—	0.35
BCW														
Mean		14.80	16.18	15.70	15.31	16.40	15.13	15.21	15.25	16.04	15.85	16.14	—	15.17
S.D.		0.57	0.83	—	0.49	—	0.90	0.87	0.55	0.55	0.52	0.65	—	0.68
BMW														
Mean		19.35	20.72	20.30	19.02	21.00	19.46	19.66	19.78	19.94	19.83	19.56	—	19.49
S.D.		0.21	0.74	—	0.49	—	0.64	0.84	1.04	0.52	0.29	0.46	—	0.65
TMJ														
Mean		24.65	26.24	23.70	23.28	23.30	23.93	22.86	22.43	24.20	22.83	23.77	—	22.99
S.D.		0.64	1.56	—	1.02	—	1.15	1.50	1.13	1.21	1.07	1.15	—	1.65

[a] "Species" categories are given in Appendix C, and follow those recognized by Hershkovitz (1983).
Variable abbreviations and descriptions are given in Appendix A.

TABLE III Group Means on First Three Canonical Variates

	Cranial data[a]			Pelage data[b]		
Taxon	CAN1	CAN2	CAN3	CAN1	CAN2	CAN3
A. azarae azarae	1.56	−1.71	1.55	—	—	—
A. azarae boliviensis	3.80	0.13	0.25	2.69	0.55	−4.37
A. brumbacki	−0.61	0.29	−1.33	−3.04	−0.61	0.48
A. infulatus	2.15	−0.68	2.58	1.21	−1.08	−0.49
A. lemurinus griseimembra	−1.84	0.49	1.39	−2.72	0.45	−1.38
A. lemurinus lemurinus	−0.57	−0.47	−0.76	−1.40	−1.03	0.99
A. miconax	0.37	−1.31	−0.45	3.81	−0.11	−0.22
A. nancymae	2.27	−0.76	−0.74	0.14	1.64	1.05
A. nigriceps	0.16	−0.98	−0.57	3.63	−0.71	0.71
A. trivirgatus	−0.49	2.58	−0.66	0.50	5.00	0.48
A. vociferans	−1.23	−1.04	−0.02	−2.00	0.32	1.78
Barranquilla origin	−0.72	0.10	0.44	−3.12	−0.49	0.25
Unsure of taxon	—	—	—	−1.94	0.19	−2.12

[a] Cranial data based on sample given in left column of Table I, including all variables listed in Appendix A except ANH (see text).
[b] Pelage data based on sample given in far right column of Table I, including all variables described in Appendix B except SIDENECK, TUFT, WHORL, HEADCHK, and RUMP (see text).

B. Pelage Data

The pelage data appear to be even better at discriminating some distinct groups, although as with the cranial data, no geographic populations of *Aotus* represent totally distinctive morphological populations (Table IV). Factor one of the principal components analysis reflects a basic division of the northern, gray-necked populations from the southern, red-necked populations, particularly in the western populations (Fig. 6). It is interesting to note that the two eastern populations, *A. trivirgatus* and *A. infulatus,* which are widely separated geographically by the eastern Rio Amazonas, fill intermediate positions between the otherwise discrete red-necked and gray-necked forms. The main variables that appear to account for this separation are ARM (weighted most heavily) and SIDENECK, and also LEG, NECK, and TAILRED. All of these measure the extension of orange-red hairs onto the various ventral body surfaces. *Aotus infulatus* appears to be less phaeomelanized on the arms and legs than other southern populations, and *A. trivirgatus* to be more so than other northern populations (see Table IV).

Factor two clearly separates out all but one specimen of *A. trivirgatus* (Fig. 7). This is heavily weighted by high values for TAILGRAY, and also low values for BELLY.

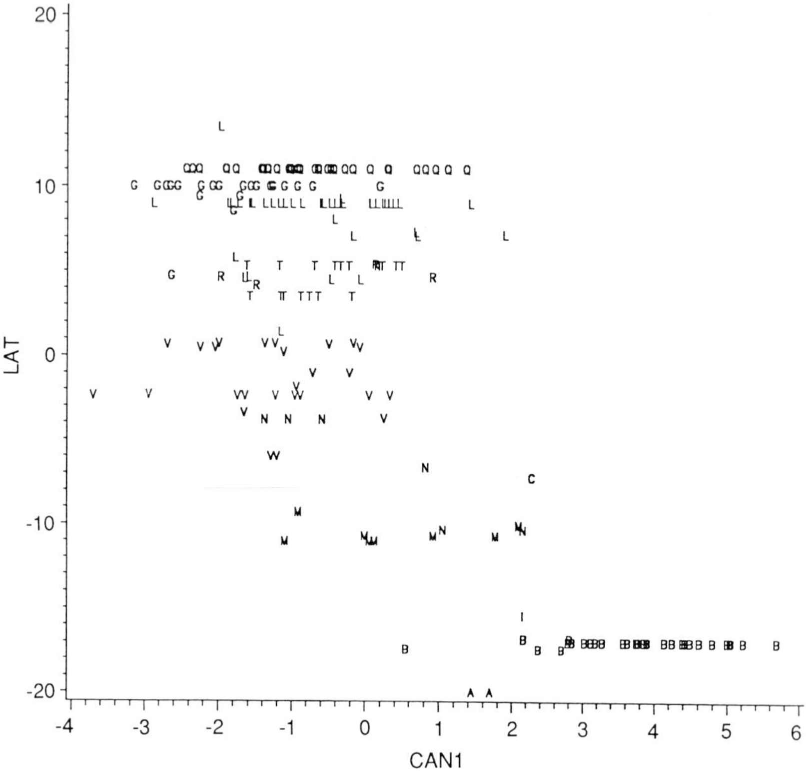

FIGURE 3 Canonical variates analysis of craniodental morphometric data, showing Canonical Variate 1 scores plotted against latitude (on Y axis). Letters as in Fig. 2 (see Appendix C). Note that as one proceeds south, population B (*A. azarae boliviensis*) demonstrates a marked nongradual (or nonclinal) break in morphology.

The canonical variates analysis very clearly indicates that several geographic populations of *Aotus* are quite distinctive, and that this difference appears abruptly and does not represent a geographic cline (see, e.g., Figs. 8 and 9). As in the principal components analysis, canonical variate one basically separates northern from southern populations of *Aotus*. However, with the exclusion of SIDENECK from the canonical variates analysis, *A. infulatus* and *A. trivirgatus* both fill a strongly intermediate location, along with the single specimen of *A. nancymae*. Although there is only one specimen of *A. nancymae* included in this analysis, the magnitude of its separation from the neighboring *A. miconax* and *A. nigriceps* suggests that increasing the sample size might not alter this basic division. The same could be suggested for the neighboring *A. infulatus* and *A. a. boliviensis*, although

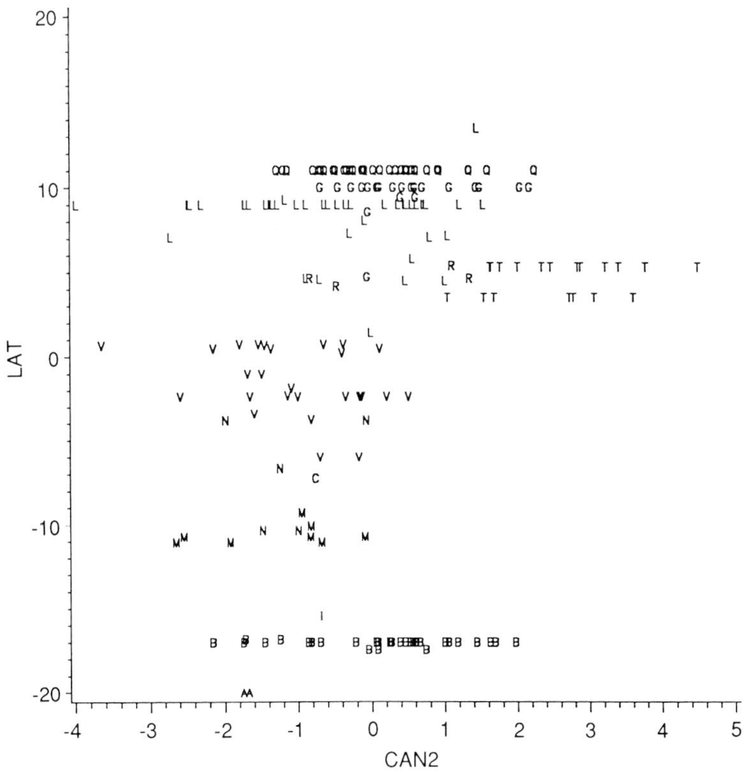

FIGURE 4 Canonical variates analysis of craniodental morphometric data, showing Canonical Variate 2 scores plotted against latitude (on Y axis). Letters as in Fig. 2 (see Appendix C). Note that as one proceeds south, population T (*A. trivirgatus*) demonstrates a marked break in morphology.

their separation is less marked. This canonical variate is due primarily to high loadings for differences in ARM and TAILRED, and much less so to differences in TAILBLK. In particular, the placement of *A. a. boliviensis* reflects its low values for TAILBLK, and the distinction of *A. nancymae* reflects the marked low value for TAILRED for this specimen (see also Table IV).

The second canonical variate separates *A. trivirgatus* from the remaining species included in the data set. This separation is due primarily to its high scores on TAILGRAY and also to low values for HEADCONV (see Fig. 10 and Table IV).

None of these multivariate analyses appears to clearly discriminate between the rest of the geographic populations of *Aotus* recognized by Hersh-

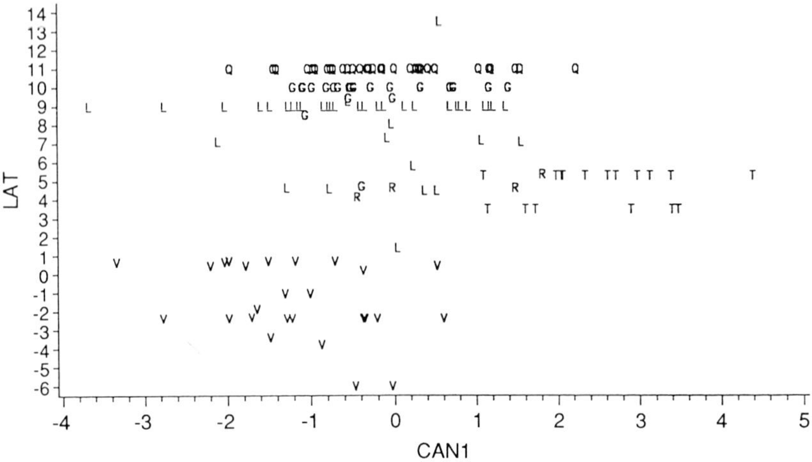

FIGURE 5 Canonical variates analysis of craniodental morphometric data for the populations north of the Rios Marañon/Amazonas only, showing Canonical Variate 1 scores plotted against latitude (on Y axis). Letters as in Fig. 2 (see Appendix C). Note that as one proceeds south, population T (*A. trivirgatus*) demonstrates a marked nongradual (nonclinal) break in morphology, against an otherwise clinal gradient to lower values as one proceeds south.

kovitz (1983). In particular, the northern populations of *A. lemurinus* (both subspecies), *A. brumbacki*, and *A. vociferans* appear as a broadly mixed single population. However, when only the northern groups are considered (with a reduced number of variables—see Methods discussion), some discrimination becomes apparent (Figs. 11 and 12). The first canonical variate segregates out *A. trivirgatus*, consistent with prior analyses, and again based on high values for TAILGRAY and low values for HEADCONV and also DORSSTRP. But the second canonical variate separates specimens classed in *A. l. lemurinus* from those in *A. l. griseimembra*, with no overlap. This is particularly clear when the second canonical variate is plotted on a north–south transect (see Fig. 12). This separation is due primarily to their differences in HEADDARK and much less so to differences in TAILBLK. *Aotus vociferans* clusters with *A. l. lemurinus*, as does *A. trivirgatus* (which is slightly more intermediate). It is interesting to note that the three specimens from Barranquilla included in this portion of the study all cluster with *A. l. lemurinus*, not *A. l. griseimembra*. However, it is important to notice that the six *A. brumbacki* specimens, from a confined and somewhat intermediate geographic location, range broadly morphologically and overlap the range of both subspecies of *A. lemurinus*.

TABLE IV Univariate Frequencies (as Percents) for Pelage Data[a]

Variable score	"Species": N:	B 1	C 1	G 11	I 23	L 22	M 9	N 8	Q 3	R 9	T 11	U 4	V 3
SIDENECK													
0		0	0	100	0	100	0	0	100	100	100	50	100
50		100	100	0	48	0	100	0	0	0	0	25	0
100		0	0	0	52	0	0	100	0	0	0	25	0
TUFT													
0		100	100	91	91	95	89	100	33	78	91	50	100
100		0	0	9	9	5	11	0	67	22	9	50	0
WHORL													
0		100	100	100	95	95	100	100	100	100	100	75	0
100		0	0	0	5	5	0	0	0	0	0	5	100
BELLY													
0		0	0	0	0	0	0	0	0	0	0	0	0
50		0	0	27	0	0	0	0	0	0	55	0	67
75		0	100	46	43	32	11	25	100	22	27	25	33
100		100	0	27	57	68	89	75	0	78	18	75	0
ARM													
0		0	0	50	0	14	0	0	0	67	9	25	0
25		0	0	20	41	81	0	0	100	11	0	25	67
50		0	0	10	18	0	0	0	0	0	64	0	33
75		0	100	20	27	0	11	38	0	11	27	25	0
100		100	0	0	14	5	89	63	0	11	0	25	0
LEG													
0		0	0	36	0	19	0	0	0	0	0	0	0
25		0	0	36	70	71	11	25	67	78	18	25	33
50		0	0	9	0	0	0	0	0	0	55	0	67
75		0	100	18	13	0	11	25	33	11	18	25	0
100		100	0	0	17	10	78	50	0	11	9	50	0

(continues)

TABLE IV (*continued*)

Variable score	"Species": N:	B 1	C 1	G 11	I 23	L 22	M 9	N 8	Q 3	R 9	T 11	U 4	V 3
NECK													
0		0	0	46	0	41	0	0	67	67	11	25	0
50		0	0	9	5	0	0	0	0	11	67	0	67
75		0	100	36	45	27	11	12	33	11	22	25	33
100		100	0	9	50	32	89	88	0	11	0	50	0
TAILRED													
0		0	100	91	4	67	0	0	100	89	40	75	67
50		0	0	0	18	0	0	0	0	11	50	0	33
100		100	0	9	78	33	100	100	0	0	10	25	0
TAILBLK													
0		0	0	18	0	0	0	0	0	0	0	25	0
10		100	0	9	0	0	0	0	0	0	0	0	0
20		0	0	0	0	0	0	0	0	0	0	0	0
30		0	0	27	4	6	0	0	33	0	0	25	33
40		0	0	9	0	6	0	0	33	0	0	0	0
50		0	0	0	0	6	0	0	0	0	0	0	0
60		0	0	0	48	11	12	12	33	50	0	50	0
70		0	0	36	9	0	0	25	0	17	0	0	0
80		0	0	0	35	50	38	50	0	17	10	0	33
90		0	0	0	0	22	25	0	0	0	60	0	0
100		0	100	0	4	0	25	13	0	16	30	0	33
TAILGRAY													
0		100	100	91	100	100	100	75	100	100	10	100	100
100		0	0	9	0	0	0	25	0	0	90	0	0

HEADCONV												
0	0	0	0	0	0	0	0	0	0	0	0	33
25	100	0	18	5	0	11	0	0	0	60	33	0
50	0	100	18	38	10	89	37	0	12	30	67	0
75	0	0	0	14	0	0	13	0	0	10	0	0
100	0	0	64	43	90	0	50	100	88	0	0	67
HEADDARK												
0	0	0	82	33	10	12	12	0	44	9	25	33
100	100	100	18	67	90	88	88	100	56	91	75	67
HEADWID												
0	100	100	100	70	75	33	37	67	56	60	75	33
100	0	0	0	30	25	67	63	33	44	40	25	67
HEADCHK												
0	100	0	73	17	33	22	12	67	75	60	50	33
50	0	100	18	35	17	11	50	0	13	0	0	33
100	0	0	9	48	50	67	38	33	12	40	50	33
RUMP												
0	0	0	0	9	0	33	50	0	0	0	25	0
50	0	0	30	35	9	11	13	33	0	18	0	0
100	100	100	70	56	91	56	37	67	100	82	75	100
DORSSTRP												
0	0	0	0	0	0	11	0	0	0	0	0	0
25	100	0	9	17	4	22	63	0	0	0	25	0
50	0	0	27	26	9	34	0	0	11	82	0	50
75	0	100	46	18	32	33	12	67	0	18	0	0
100	0	0	18	39	55	0	25	33	89	0	75	50

[a] Variables and score values are given in Appendix B. "Species" abbreviations are for species recognized by Hershkovitz (1983); abbreviations and associated species names are given in Appendix C. Sample sizes (*N*) are for the entire sample of skins examined; for many individual traits, the actual sample was decreased by one to nine skins, generally because of missing tails or portions of the head. Values in cells are percentages of sample.

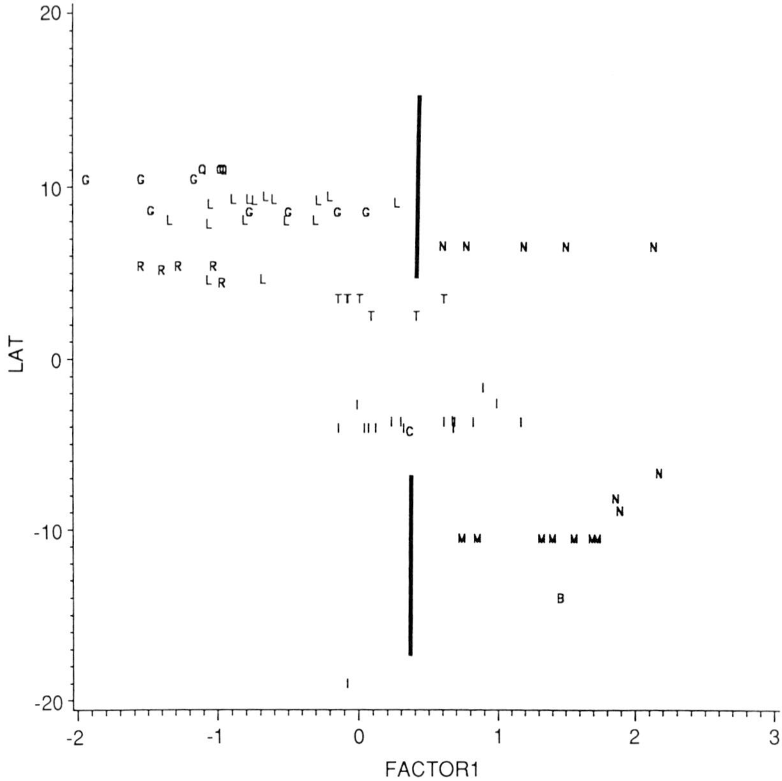

FIGURE 6 Principal components analysis of pelage data, showing Factor 1 scores plotted against latitude (on Y axis). Letters as in Fig. 2 (see Appendix C). Note the clear separation of the populations north of the Rios Marañon/Amazonas (G, L, R, V, plus Q) from those south of the rivers (M, C, N, A, B), indicated by solid line. Population T (northern) and I (southern) fall in intermediate positions (see text).

IV. DISCUSSION: TAXONOMY AND DISTRIBUTION

A. Data Analyzed Here

From these two data sets, the owl monkeys north of the Amazon appear to be clearly distinctive from those south of the Amazon, primarily on the basis of the presence in southern populations of red pelage on the side of the neck and increased red pigmentation throughout the ventrum, most notably extending onto the arms, legs, and proximal tail. There appear to be reverse clines (or morphological jumps, perhaps) operating as one proceeds east from the headwaters of the Amazon (where the north/south differences are most marked). As a result, the easternmost population

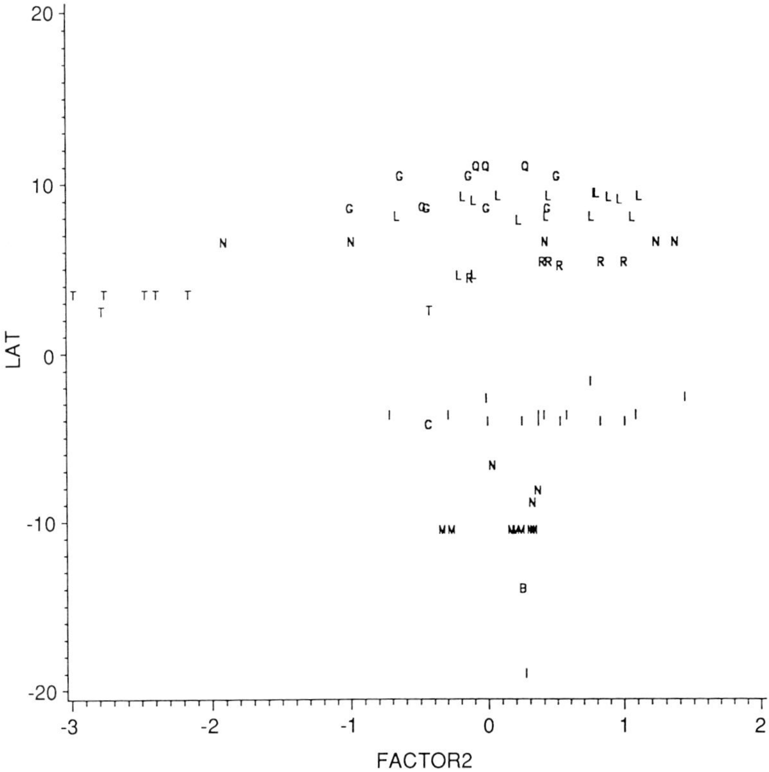

FIGURE 7 Principal components analysis of pelage data, showing Factor 2 scores plotted against latitude (on Y axis). Letters as in Fig. 2 (see Appendix C). Note that as one proceeds south, population T (*A. trivirgatus*) demonstrates a marked nongradual (or nonclinal) break in morphology. (Note: Population N, *A. nigriceps*, does not actually "bridge" the gap, as it is from south of a northward bend of the Amazon.)

south of the Amazon (*A. infulatus*) is the least red of that group, and the easternmost population north of the Amazon (*A. trivirgatus*) is the most red of the northern populations. Although these two eastern populations thus converge somewhat on each other morphologically, they are nonetheless highly distinctive in terms of degree of phaeomelanization as well as widely separated geographically by the wide lower Rio Amazonas.

Within each of these two areas, there appear to be at least two, and possibly more, geographically separate populations of *Aotus* that are also highly distinctive morphologically. The first of these is Hershkovitz's *A. trivirgatus*, east of the Rio Negro (to the Rio Trombetas) and north of the Rio Amazonas in Brazil, ranging roughly into the area of the Rio Caura/

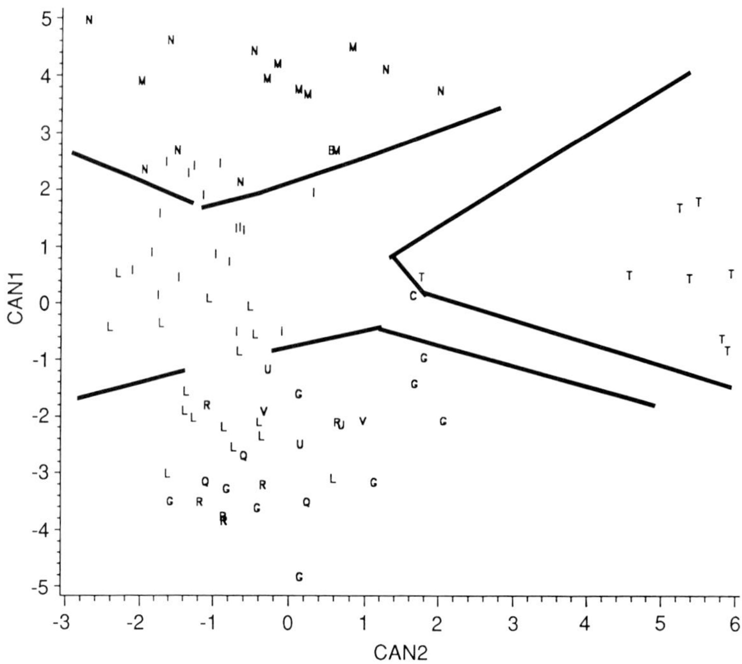

FIGURE 8 Canonical variates analysis of pelage data, showing Canonical Variate 1 scores plotted against Canonical Variate 2 scores. Letters as in Fig. 2 (see Appendix C). Several populations appear highly (although not totally) distinctive, indicated by solid lines: T at right; the southern M + N + B at top; and the northern G + R + V + Q (+L, less so) at bottom.

Rio Orinoco confluence in Venezuela, apparently restricted to the south bank of the Rio Orinoco. *Aotus trivirgatus* is highly distinctive in its large values for the length of the molar tooth row (MLL), bicaninal width (BCW), and length of the total tooth row (TRL). It shares these enlarged oral/dental dimensions with the population across the Rio Amazonas, *A. infulatus*, and the far distant Bolivian population, *A. a. boliviensis*, both of which differ from *A. trivirgatus* in many other ways. Additionally, *A. trivirgatus* differs from the other northern populations (and is similar to the southern populations in general) in large values for anion-lamba length (ANL), which may largely be the result again of its lengthened tooth row. *Aotus trivirgatus* specimens are also highly distinctive in the extension of grayish coloration onto the proximal tail (TAILGRAY), found otherwise here in only one specimen of *A. l. griseimembra* (out of many) and two specimens of the southern *A. nigriceps;* it is, however, missing from one specimen of *A. trivirgatus* from the center of the range (see Table IV and Fig. 13). *Aotus*

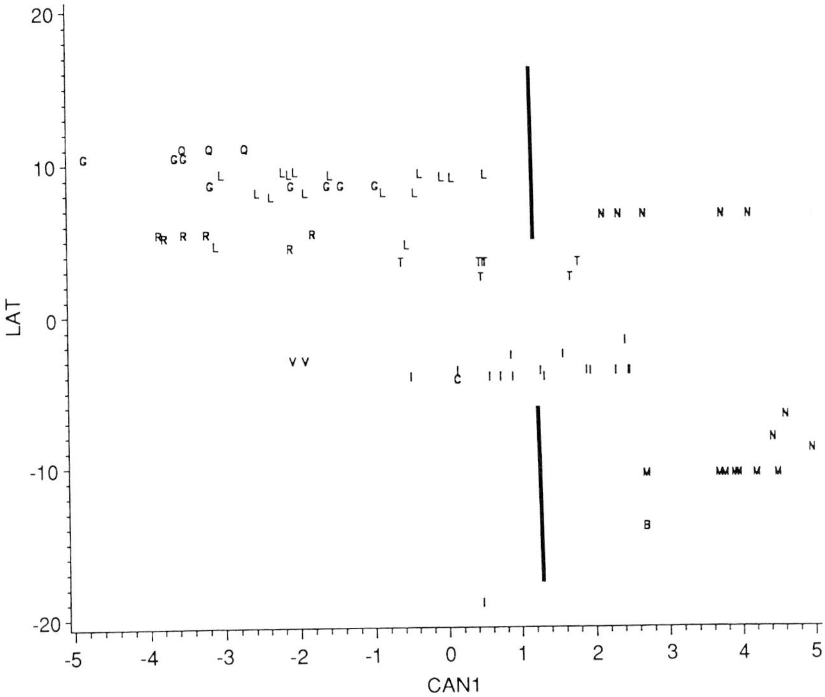

FIGURE 9 Canonical variates analysis of pelage data, showing Canonical Variate 1 scores plotted against latitude (on Y axis). Letters as in Fig. 2 (see Appendix C). Note the general separation, indicated by the solid line, of the populations north of the Rios Marañon/Amazonas (T, G, L, R, V, plus Q) from those south of the rivers (M, C, N, I, A, B), although there are intermediate populations (see text).

trivirgatus specimens also have generally parallel to only slightly converging stripes on the forehead (HEADCONV; Fig. 14), and a relatively narrow agouti stripe up the back, which is otherwise predominantly gray in color (DORSSTRP). As can be seen in Table IV, there is considerable variability across populations for DORSSTRP, although the stripe is narrower in a much larger proportion of *A. trivirgatus* than other populations. The belly is generally very light orange or buffy colored (BELLY) as opposed to deeper red/orange, this being shared with its western neighbor, *A. vociferans*. *Aotus trivirgatus* is otherwise quite distinctive from *A. vociferans*, particularly in its larger masticatory apparatus, grayer tail, and parallel head stripes, as opposed to head stripes that clearly converge posteriorly in the latter.

This description of the distinctive pelage features of *A. trivirgatus* agrees with that of Hershkovitz (1983). As he reported, no interscapular whorl was found on any of the specimens of *A. trivirgatus* examined, although

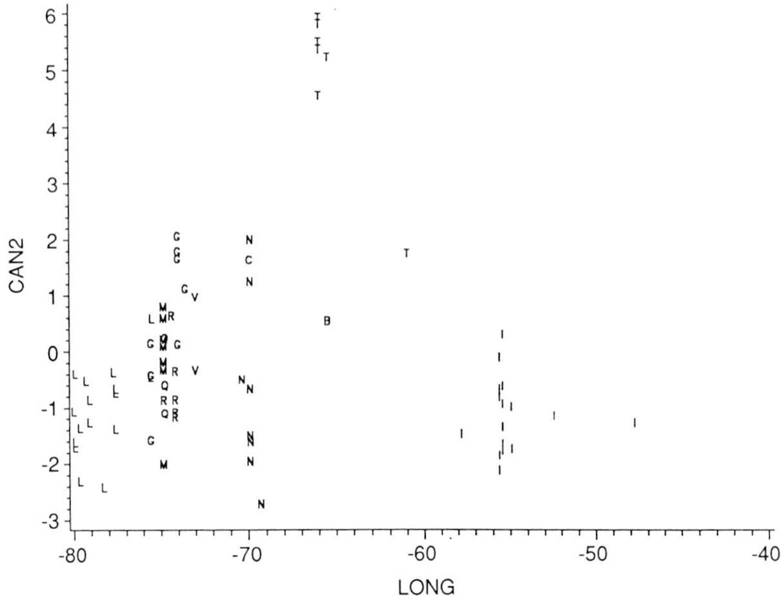

FIGURE 10 Canonical variates analysis of pelage data, showing Canonical Variate 2 scores plotted against longitude (on X axis). Letters as in Fig. 2 (see Appendix C). Note that as one proceeds east, population T (*A. trivirgatus*) demonstrates a marked nongradual (or nonclinal) break in morphology.

one did exhibit an interscapular tuft (see Table IV). However, tufts and whorls were also missing on specimens of other populations where he described them (the single specimen of *A. a. boliviensis* and most *A. brumbacki* examined) and present on a small number of specimens of populations where he indicated their absence (*A. l. lemurinus, A. l. griseimembra,* and *A. miconax*), making this feature of questionable taxonomic value. Hershkovitz did note "apparent aberrations" (p. 216) in the distribution of these traits as well.

The second distinctive population is *A. a. boliviensis,* extending south from the Rio Madre de Dios and Rio Guapore in Bolivia to an arc extending roughly from Lago Uberaba on the east to Sucre, Bolivia, on the west, and from the eastern slope of the Andes west to the Rio Paraguai. Unfortunately, the pelage of only a single specimen of *A. a. boliviensis,* and none of its suggested conspecific southern neighbor, *A. a. azarae,* was examined, making comparisons of pelage pattern highly speculative. However, the specimen is distinguished from the populations north of the Amazon by the extreme penetration of red pigmentation extending from its belly onto its ventral neck, arms, legs, and proximal tail. The redness of the ventrum is largely shared with most of the other southern populations, although least

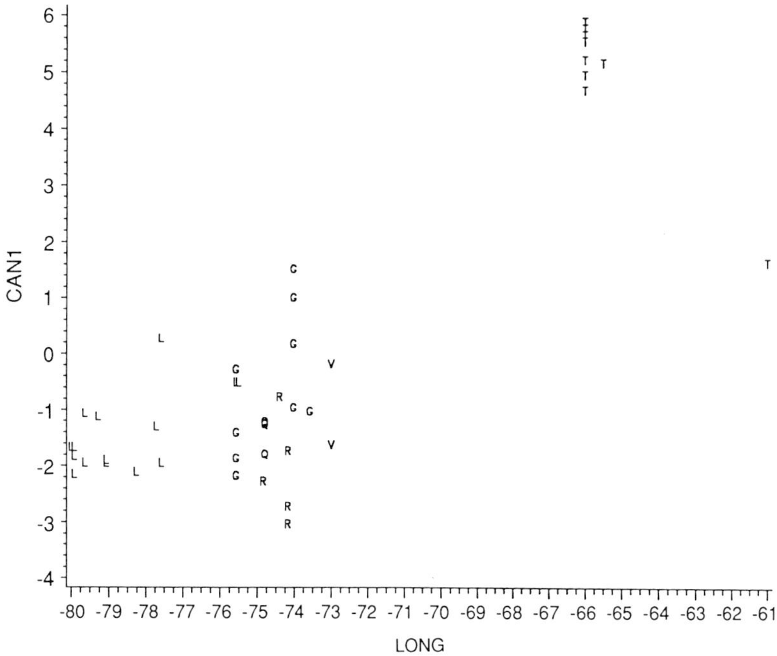

FIGURE 11 Canonical variates analysis of pelage data for the populations north of the Rios Marañon/Amazonas only, showing Canonical Variate 1 scores plotted against longitude (on X axis). Letters as in Fig. 2 (see Appendix C). Note that as one proceeds east, population T (*A. trivirgatus*) demonstrates a marked nongradual (or nonclinal) break in morphology.

so with its close neighbor to the east (and possible conspecific—see the following), *A. infulatus*. The examined skin has far less black coloration on its tail (being confined to just black-tipped hairs at the distal end of the tail) than the vast majority of other southern (red-necked) populations (and northern), although a specimen of *A. infulatus* and one of *A. miconax* show similarly lightly blackened tails (see Table IV and Fig. 15). In addition, its forehead stripes are more parallel than those of other southern populations, like those of *A. trivirgatus* in the north; however, many populations show some variability on this trait (see Table IV and Fig. 14).

The *A. a. boliviensis* specimen is, in general, very similar to *A. infulatus* in pelage features (see Table IV), differing primarily in the features of the broad agouti stripe to predominantly agouti dorsum (vs. gray—DORSSTRP), the convergent forehead stripes (HEADWID), blacker distal tail (TAILBLK), and considerably less extension of red-pigmented hairs down the length of the arms and legs (ARM, LEG) of most *A. infulatus*. However, the single specimen of *A. a. boliviensis* falls at the very end but within the distribution of *A. infulatus* on all of these pelage traits, and Pieczarka and Nagamachi (1988) describe the skins of *A. infulatus* and *A.*

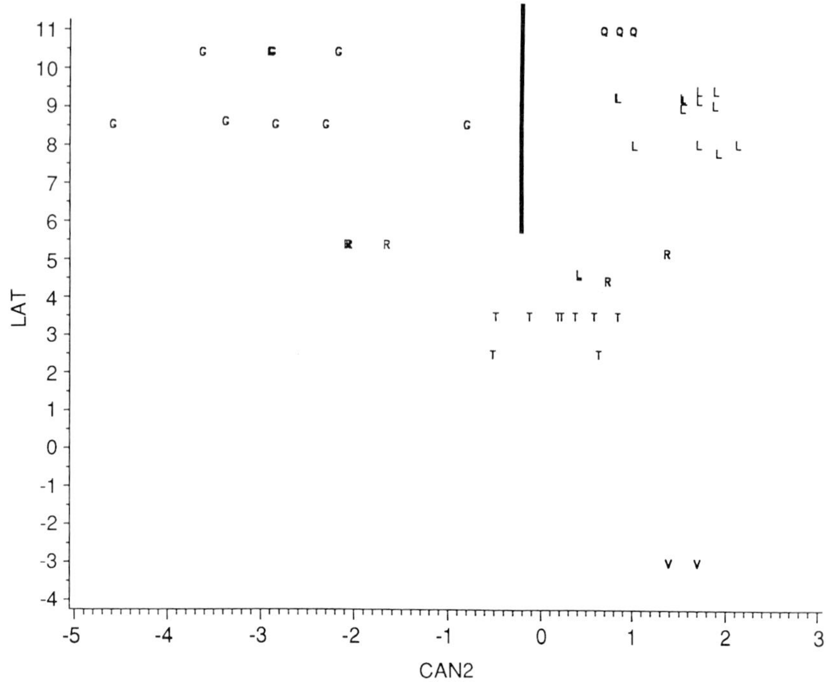

FIGURE 12 Canonical variates analysis of pelage data for the populations north of the Rios Marañon/Amazonas only, showing Canonical Variate 2 scores plotted against latitude (on Y axis). Letters as in Fig. 2 (see Appendix C). Note the clear separation of populations L (*A. lemurinus lemurinus*) and G (*A. l. griseimembra*) from each other (indicated by solid line), but their mutual overlap with population R(*A. brumbacki*) (see text).

a. boliviensis as identical. Although no skins of *A. a. azarae* were examined in this study, Mudry de Pargament *et al.* (1984) described the skins of these two neighboring subspecies—*A. a. azarae* and *A. a. boliviensis*—as almost identical. They specifically note the great extent of red pigmentation and the narrow, parallel head stripes in the *A. azarae* subspecies (both differences from *A. infulatus*). However, Mudry de Pargament *et al.* indicated a blacker tail for *A. a. azarae*, more similar to the *A. infulatus* condition; their observation suggests that the lighter distal tail in the specimen of *A. a. boliviensis* examined here is typical rather than unusual for the population.

Even more distinctive in *A. a. boliviensis* are its large values on nearly all the cranial and dental measures (for which there is a large sample size). In fact, the only measures in which it is not either the largest population or one of two or three largest are biorbital width (BOW) and bizygomatic width (BZW), and in these it is among the largest four groups (see Table IV). Although *A. trivirgatus* shares the large values for dental measures (both length of the tooth rows and breadth of the anterior palate), as

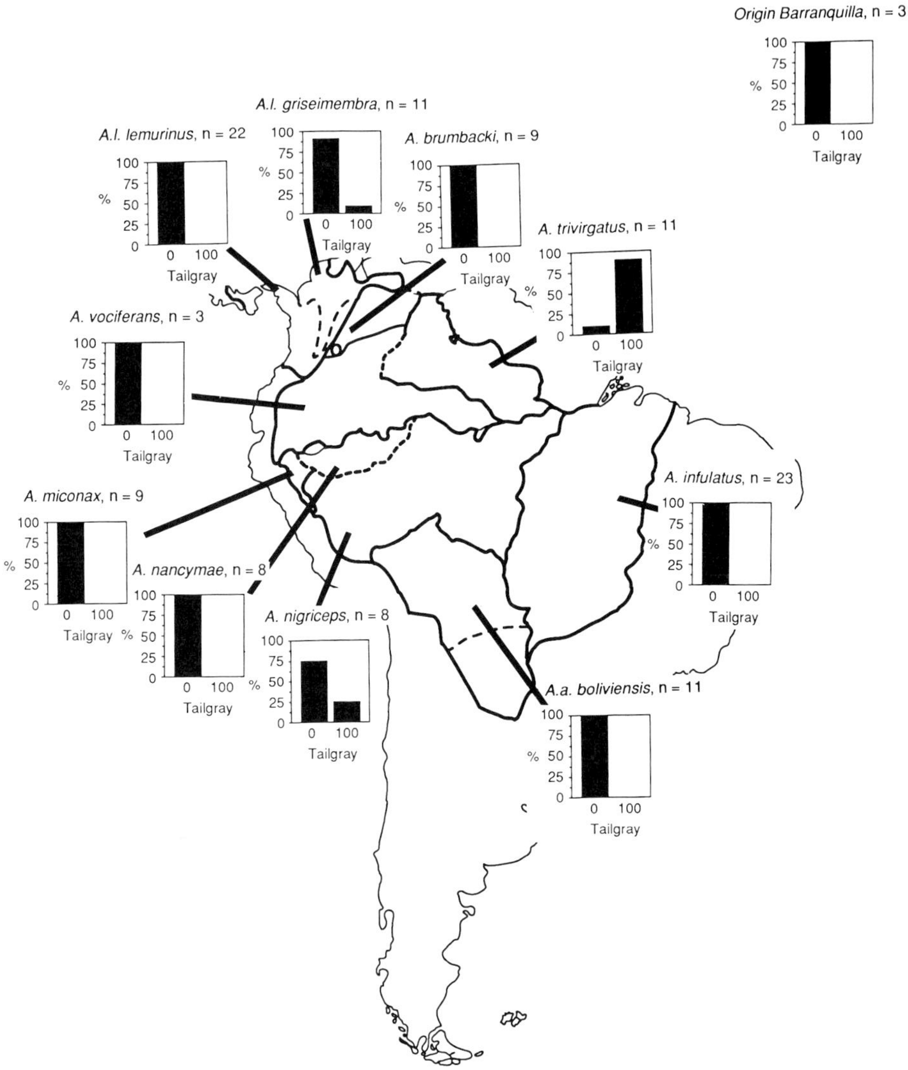

FIGURE 13 Univariate frequencies of TAILGRAY in geographically defined populations of *Aotus*, using taxonomy of Hershkovitz (1983). Note distinctiveness of *A. trivirgatus* (see Table IV, Appendix C, and text).

discussed earlier, it lacks the large values for other cranial measures. The single specimen of *A. nancymae*, on the other hand (but not *A. miconax*), shares many of the large values for cranial measures, including many height, width, and length measures of the posterior cranium and upper face (CBL, BOW, BZW, MSW, ANL, PSL, CBH, and BMW), but lacks large measures for the dentition, palate, and lower jaw.

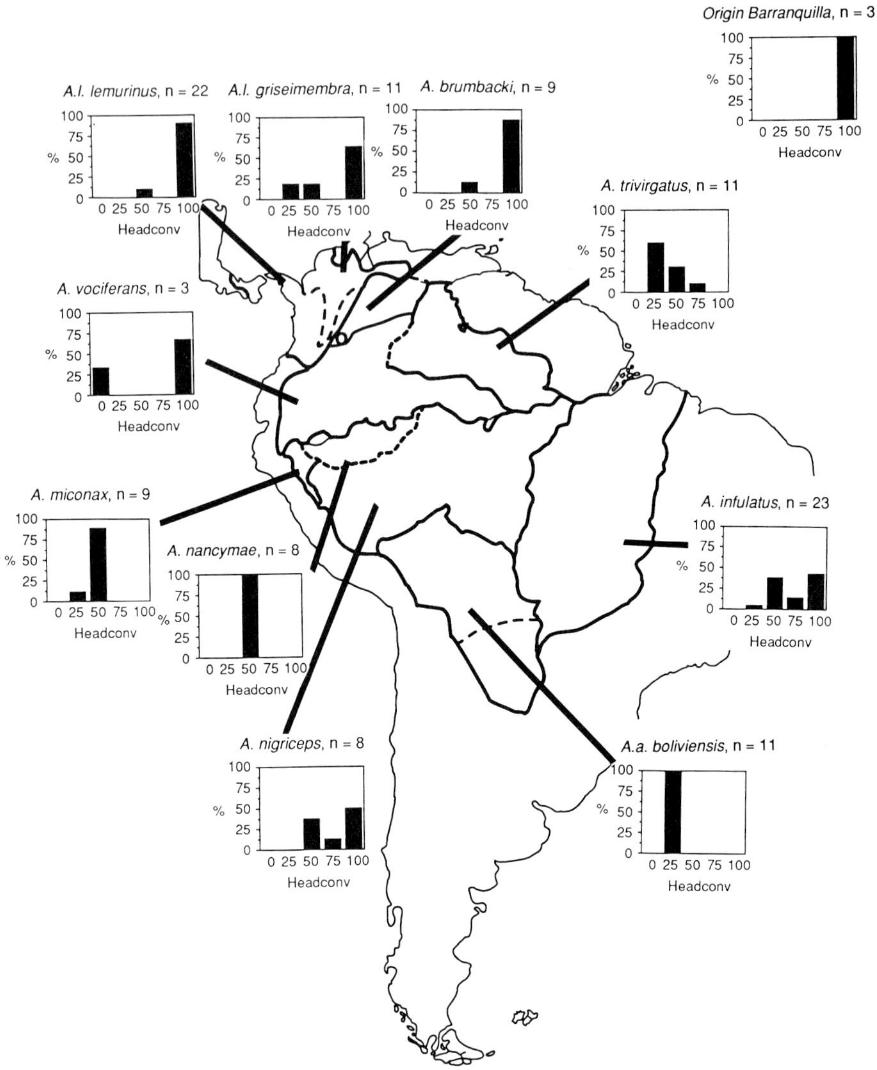

FIGURE 14 Univariate frequencies of HEADCONV in geographically defined populations of *Aotus*, using taxonomy of Hershkovitz (1983). Note the distinctiveness of *A. trivirgatus* in the north and of *A. azarae boliviensis* in the south (see Table IV, Appendix C, and text).

Although only two specimens of the more southern *A. a. azarae* are included in the cranial data set, it is intriguing that these specimens appear to be smaller and thus more similar to populations from north of the Rios Madre de Dios and Guapore than to their nearest neighbor, *A. a. boliviensis*. The two *A. azarae* populations are only similar to their hypothesized conspecific in their very large values for bizygomatic width (BZW), bregma-lambda (parietal suture) length (PSL), and temperomandibular joint height (TMJ). In fact, the large size of *A. a. boliviensis* is shared most closely, almost measure for measure, with the single specimen included of its neighbor to the east and north, *A. infulatus,* which is not distinguishable from *A. a. boliviensis* on the basis of the cranial and dental measures in this study.

A third potentially morphologically distinctive geographic population in this analysis is represented by the single specimens in each analyses of the northern Peruvian/western Brazilian *A. nancymae,* whose exact range is unclear (see Introduction). Of particular import in determining the distinctness of this population are comparisons to neighboring populations of *A. vociferans, A. miconax,* and *A. nigriceps.* As noted earlier, the specimen included in the craniodental analysis exhibits very large values for many cranial and upper facial dimensions, but not for the lower face and dentition (see Table IV). In most of these features, the neighboring populations show much smaller dimensions, although the differences are often not statistically significant given the small sample sizes.

Pelage features, particularly the red neck (SIDENECK, NECK) and reddish pigmentation on other aspects of the ventrum, clearly separate the specimen of *A. nancymae* examined here from *A. vociferans* (see Table IV), and unite it with its southern neighbors. In particular, the bright orange/red pigmentation of most *A. nancymae, A. miconax,* and *A. nigriceps* specimens extends well down the ventral surfaces of the arms (ARM) and legs (LEG) to the wrists and ankles (this being the sole difference from Hershkovitz's 1983 description of *A. miconax*). Thus the sample examined here does not support Hershkovitz's (1983) separation of these three species on the basis of differences he suggested in the extensiveness of red pigmentation on the arms, legs, and neck. A pelage difference between many of the southern groups seems to be the degree of black pigmentation on the distal end of the tail (see Fig. 15 and Table IV). The northwestern populations, *A. miconax* and especially *A. nancymae,* are characterized by the distal end of the tail being very black, with the distal one-fourth to one-half plus of the tail composed of totally black (rather than simply black-tipped agouti) hairs. The eastern (*A. infulatus*) and especially the southern (*A. a. boliviensis*) populations are characterized by less black, with most *A. infulatus* having half the tail black-tipped hairs or less than one-third all black hairs, and *A. a. boliviensis* only black-tipped hairs on the very tip of the tail. Though sample sizes here are in some cases small, these descriptions agree closely with those of Hershkovitz (1983). However, a determination of whether these primary differences represent a stepwise change or are clinal depends in large part on the pigmentation of the distal tail in the intermediate *A. nigriceps* population, for which Hershkovitz did not provide a conclusive description. As can be seen here (Fig. 15), specimens of this population fall intermediate in blackness of the tail between the *A. miconax* and *A. infulatus* populations. It is the extremely black tail of the single *A. nancymae*

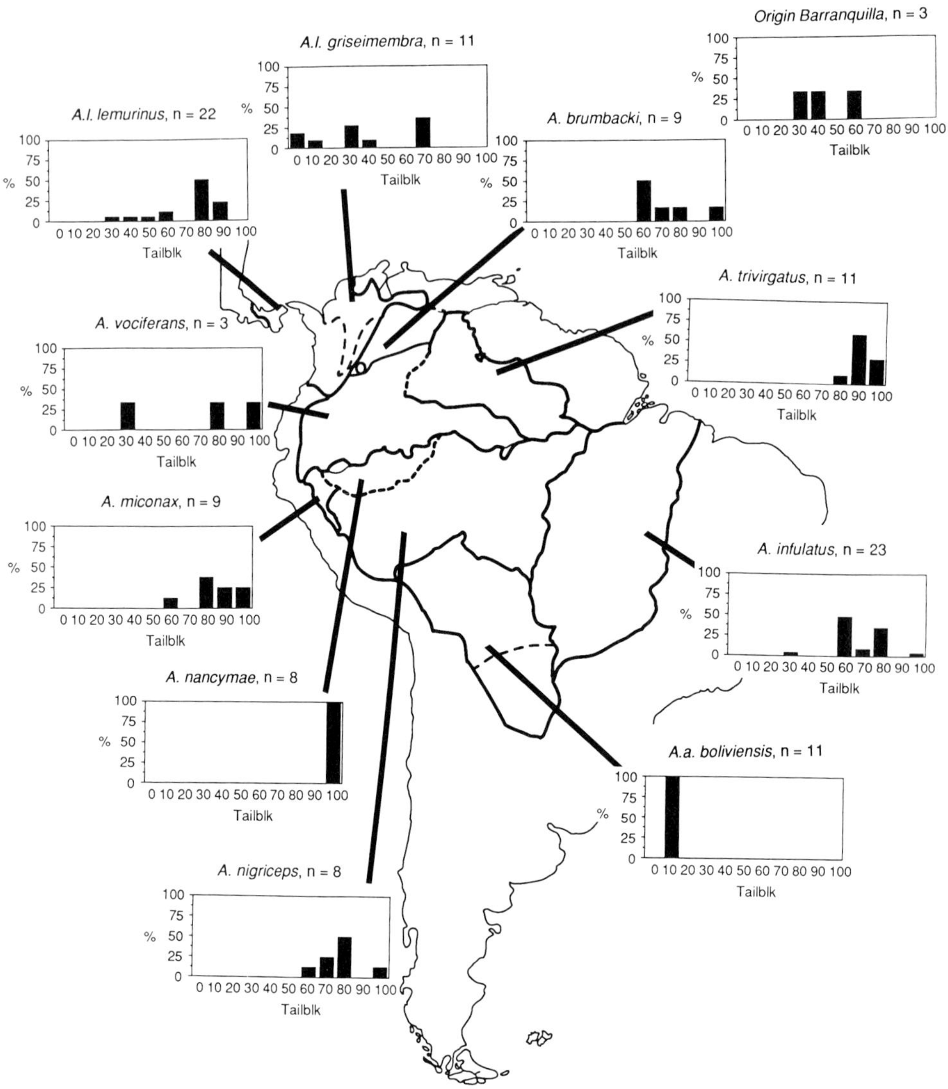

FIGURE 15 Univariate frequencies of TAILBLK in geographically defined populations of *Aotus*, using taxonomy of Hershkovitz (1983) (see Table IV, Appendix C, and text).

skin and the very gray/agouti tail of the single *A. a. boliviensis* skin that "fall out" as unusual; these may only represent one "tail" (end) of the range of variation of these two populations.

In nearly all the southern owl monkeys, the proximal fourth of the tail has red-pigmented hairs, the only exception being a more limited area of red pigmentation in less than one-fifth of the *A. infulatus* skins examined (see Table IV: TAILRED). However, the single *A. nancymae* skin examined has no extension of red pigmentation onto the tail. Although this is only a single specimen, the difference is striking and "drives" the separation of this specimen in many of the multivariate analyses. In contrast, Hershkovitz (1983) described the proximal third of the tail of *A. nancymae* specimens he has observed as distinctly reddish in coloration; however, he also noted the tail as lacking red hairs on the ventrum (p. 216). Without larger samples of these populations, any interpretation of the *A. nancymae* pelage as distinctive would seem unwarranted.

Finally, from the study of cranial and pelage features here, the northern populations of *A. vociferans*, *A. lemurinus* (both subspecies), and *A. brumbacki* are not clearly distinctive from one another. There is some indication of separation between the two subspecies of *A. lemurinus*, *A. l. lemurinus* and *A. l. griseimembra*, based on pelage features, particularly the much darker head stripes and more blackly pigmented distal tail in *A. l. lemurinus* (see Table IV). However, as can be seen in Fig. 15, there is considerable overlap in these populations for the blackness of the distal tail, and *A. brumbacki* specimens bridge the gap morphologically for both features and are close neighbors geographically. This suggests that though differences do occur across the northwestern portion of the range of *Aotus*, the differences are clinal and variable, and do not represent categorical shifts indicative of reproductive isolation or speciation.

B. Chromosomal Data

The importance of karyotype information in the study of geographic variation in *Aotus* can not be understated, as it formed the primary basis for Hershkovitz's (1983) division of the genus and of Brumback's (1973, 1974) before that. It is clear that *Aotus* is an extremely and unusually diverse genus karyotypically, with $2n$ ranging from 46 to 58, and pertinent data and sources are summarized in Table V. The tendency of several recent workers has been to construct separate species for each minor chromosomal variant. This is consistent with long-standing observations and assumptions in mammalian biology: most vertebrate populations are chromosomally monomorphic, and studies in some mammals have indicated that even single, simple chromosomal changes lead to greatly reduced fertility via meiotic malsegregation of homologous pairs (see discussion in Nachman and Myers, 1989). These models of chromosomal species all generally center on presumed negative heterosis or lowered fitness of the heterozygotes.

However, these assumptions of negative heterosis and/or sterility are rarely tested. Several studies have suggested that chromosomal hybrids in some taxa may be viable and fertile, in some cases with little direct evidence of lowered fitness (e.g., Paraguayan marsh rats—Nachman and Myers, 1989; trivalents but not multivalent hybrids in sportive lemurs—Ishak *et*

TABLE V Known Karyotypes of *Aotus* Populations[a]

Taxon	Ma's karyotype	2*N* M	2*N* F	FN	Acro.	Meta/Submeta.	Y
A. lemurinus lemurinus	VIII, IX	55, 56 (both sexes)		62	44, 46	9, 8	M
*A. lemurinus*1 *griseimembra*	IV, III, II	52, 53, 54 (both sexes)		62	38, 40, 42	12, 11, 10	M
A. brumbacki		50	50	58	10	38	A
A. trivirgatus	?	?	?	?	?	?	?
A. vociferans[b]	V	46, 47, 48 (both sexes)		60			M-50% A-50%
A. vociferans[c]		46	46	60	16	28	A-100%
A. nigriceps	VII	51	52	66	34	16	Y/auto[d]
A. nancymae[b]	I	54	54	72	32	20	M
A. nancymae[c]		54	54	72	32	20	A
A. miconax	?	?	?	?	?	?	?
A. infulatus	XII?	49	50	61/60?	35,36?	13,12	Y/auto, SM[d]
A. azarae azarae		49	50	61/60?	35,36?	13,12	Y/auto, A
A. azarae boliviensis	VI	49	50	61/60?	35,36?	13,12	Y/auto, SM[e]

[a] Taxonomy according to Hershkovitz (1983; see text). Abbreviations are as follows: Ma's karyotype = karyotype designations of Nancy Ma (1981), e.g., K-II; 2*N* = diploid chromosome number (males and females separately); FN = fundamental number; Acro. = number of acrocentric autosomes; Meta/Submeta = number of biarmed autosomes; Y = nature of Y chromosome, either metacentric (M), submetacentric (SM), acrocentric (A), or fused to one member of a homologous autosomal pair (Y/auto). Sources are: *A. azarae azarae,* Mudry de Pargament *et al.* (1984); *A. azarae boliviensis,* Brumback (1974), Ma *et al.* (1976), Pieczarka *et al.* (1993); *A. brumbacki,* Brumback *et al.* (1971), Yunis *et al.* (1977); *A. infulatus,* Pieczarka and Nagamachi (1988), Pieczarka *et al.* (1993); *A. lemurinus griseimembra,* Ma *et al.* (1976); *A. lemurinus lemurinus,* Ma *et al.* (1978). *A. nigriceps,* Ma *et al.* (1980).

[b, c] Alternate references: *A. nancymae,* [b]Brumback (1973), [b]Ma *et al.* (1976), [c]Pieczarka *et al.* (1992); *A. vociferans,* [b]Ma *et al.* (1985), [c]Pieczarka *et al.* (1992).

[d] Y fused on short arm of autosome (see text).

[e] Y inserted on long arm of autosome (see text).

al., 1992; gibbons—Stanyon *et al.*, 1987; and squirrel monkeys—Silva *et al.*, 1992). In at least one of these studies, the differences between the karyotypic morphs are more complex than a single change or a single class of changes, involving combinations of Robertsonian fusion/fissions, pericentric inversions, and additions of both terminal and centromeric euchromatic material (Nachman and Myers, 1989). Thus, one must approach the interpretation of the many chromosomal variants in *Aotus* with caution.

In fact, several proposed species of *Aotus* are themselves karyotypically diverse, such as *A. lemurinus* and *A. vociferans* (Aquino and Encarnación,

1988; Galbreath, 1983; Hershkovitz, 1983), and Ma and associates now have 15 hybrids of various karyotypic forms produced under laboratory settings, including hybrids of red-necked and gray-necked individuals (see Ma, 1981; Ma *et al.*, 1985). However, these laboratory hybrids are all F1 individuals, and it is not yet certain if any of them would be fertile. The evidence in the wild demonstrates that some *Aotus* populations can absorb some degree of chromosomal/karyotypic variation without speciation or cross-infertility whereas others cannot. Populations in Peru of *A. vociferans* and in Colombia and Panama of *A. lemurinus* (*A. l. griseimembra* and *A. l. lemurinus*, respectively) have all been shown to support chromosomal variability that does not pose a barrier to fertility or fitness for the hybrids in each population, the result of either a fusion or fission event in the past that resulted in a polymorphism retained across all three populations (Ma, 1981; Ma *et al.*, 1985; Ma and Harris, 1989). *Aotus infulatus* and *A. azarae boliviensis* have been shown to hybridize, and a marked zone of hybridization with a fairly well established population including its own distinct autapomorphy has been identified in Rondonia (Pieczarka *et al.*, 1993).

However, an enclave of *A. vociferans* exists south of the Rio Amazonas along the Rio Purus, within the distribution of *A. nigriceps*, with both groups retaining their distinct appearance (Hershkovitz, 1983). *Aotus vociferans* and *A. nancymae* have exhibited sympatry without any evidence of hybridization in appearance or karyotype in a region of river meanders (Pieczarka *et al.*, 1992). Therefore, chromosomal differences must be analyzed carefully, noting each distinct karyotypic morph and its distribution and watching for potential areas of hybridization. Where hybrids are found, the possibility of negative heterosis must be carefully assessed. Current work on chromosomal changes in other taxa, described earlier, suggests that no blanket assumptions can be made about loss of fitness or fertility on the basis of type or amount of chromosomal change.

Ma (1981) proposed a complex sequences of evolutionary events to explain current karyotypic variation in *Aotus*, including periods of isolation, development of new karyotypes through fusion, fission, and inversion events, subsequent recontact and free hybridization, and finally migration of portions of this newly polymorphic population into various ecosystems with subsequent selection coupled with further rearrangements. While she suggested that many of these final populations, although geographically isolated, may not be reproductively isolated, Galbreath (1983) and Hershkovitz (1983) suggested that in fact the different karyotypes would be reproductively isolated, with Galbreath presenting a more simplified phylogeny of nonreticulate branching based solely on karyotype. Since then, considerable new data have been presented on chromosomal structure for various populations.

Although differences among the populations are varied and complex, some generalities can be drawn. Considerable numbers of Robertsonian changes (either fusion or fission events) have occurred, along with less frequent inversions. Changes in the Y chromosome in particular have occurred repeatedly; both C- and G-banding studies have greatly improved our current knowledge of homologous chromosome segments across species, although no consistent cross-population numbering system is in current use (Ma, 1983; Ma and Gerard, 1988; Ma *et al.*, 1985, 1989a,b; Pieczarka *et al.*, 1993). Both Ma (1981) and Galbreath (1983) suggested that the ancestral *Aotus* karyotype consisted of 54 chromosomes, although the current evidence would not be inconsistent with an ancestral morph of 52 chromosomes either. A $2n = 54$ is close to the highest number in any *Aotus* population, and suggests that fusion of acrocentrics to form metacentrics has been a more common pattern of change than fissions; this is supported by findings of the conservation of gene linkages of acrocentrics and only single arms of metacentrics across other platyrrhines and primates (e.g., Ma, 1983; Ma and Gerard, 1988; Ma *et al.*, 1985, 1989a,b). In addition, the primitive morph of the Y chromosome would be acrocentric, preserved consistently in such divergent platyrrhines as *Alouatta, Cebus, Saimiri,* and *Callithrix* (Mudry *et al.*, 1990).

The two northernmost populations, *A. l. lemurinus* and *A. l. griseimembra*, are both polymorphic ($2n = 55$, 56, and $2n = 52$, 53, 54, respectively). According to Ma (1981), in both areas, there appear to be one morph with two acrocentric autosomes (B13 and B14) and one morph with one metacentric autosome (A1) corresponding to a fused version of the two acrocentrics. Karyotypes of uneven number (55, 53) appear to represent hybrids of the two morphs, suggesting that there should be individuals of $2n = 54$ karyotype within the *A. l. lemurinus* populations. The widespread presence of this polymorphism across two otherwise slightly varying populations suggests that it is under the control of strong stabilizing selection (Ma, 1981; Galbreath, 1983), perhaps the result of selection for the hybrid form or, alternatively, "nonselectively stabilized" (Ma and Harris, 1989, p. 34) (see the following). In all, the Y chromosome is metacentric (derived from the ancestral morph). The two populations differ by the presence of two acrocentrics (B28 and B29) in all *A. l. lemurinus*, which is matched by a single submetacentric chromosome (A2) in all *A. l. griseimembra*. Because no direct areas of sympatry of these morphs are known, it is not certain but suspected that this fusion/fission event also does not impact on heterozygote fitness or meiotic success.

A similar stable polymorphism is present in *A. vociferans*, $2n = 46$, 47, 48 (Ma *et al.*, 1985), again due to a single fusion (or fission) event that is maintained in both the ancestral and derived state, with hybrids present (acrocentrics X-23 and X-24 to metacentric V-1). As Ma *et al.* (1985; Ma and Harris, 1989) note, the widespread polymorphism in chromosome number found among the gray-necked forms appears to be the result of a single chromosomal rearrangement that has been selectively tolerated in the different populations. In addition, they find that there is no evidence for any selective advantage or disadvantage for the heterozygotes over either homozygous condition in each case. Other fission/fusion events led to the larger difference in chromosome numbers between the northern and Peruvian populations. In additon, whereas Ma *et al.* (1985) describe 50% of their *A. vociferans* sample as having a metacentric Y chromosome, Ma *et al.* (1985b) list it as acrocentric only, consistent with the findings of Pieczarka *et al.* (1992) of 100% acrocentric Y chromosomes.

Ma *et al.* (1985) suggested that *A. vociferans* is derived from the eastern Colombian population *A. brumbacki* ($2n = 50$) via a single fusion event. *Aotus brumbacki* is somewhat intermediate geographically between *A. vociferans* and *A. lemurinus,* and it shares the acrocentric Y chromosome found in the Peruvian *A. vociferans.* Ma *et al.* further predicted that if more specimens from eastern Colombia are sampled, they will exhibit the same polymorphism seen in *A. lemurinus* and *A. vociferans.* Thus, Ma and her colleagues predict that the actual diploid number of *A. brumbacki* may be $2n = 48, 49, 50$. This situation would then directly parallel that found in the northern Colombian/Panamanian populations.

Since the various gray-necked populations considered species by Hershkovitz (1983) are all allopatric at present, there is no direct evidence that any of the additional fusion or fission events characterizing them would cause reproductive isolation. Thus, the chromosome evidence does not appear sufficient at present to dictate the separation of the various populations north of the Amazon into separate species. It is unfortunate, in light of the evidence presented in this study, that nothing is yet known of the karyotype of the highly distinctive eastern population, *A. trivirgatus.*

South of the Amazon, the situation is somewhat different. *Aotus nancymae* ($2n = 54$) would appear to have developed a number of alterations, including at least two inversions and a centric fission, from any current reconstruction of the common ancestor (Ma, 1981). Both Ma (1981) and Galbreath (1983) suggested this may have occurred as this population became isolated in Peru, south of the Rio Amazonas/Rio Marañon. Currently, as a result of river meanders, this population appears to have spread back north across the river into an enclave between the Rios Tigre and Pastaza, where it remains distinct from *A. vociferans* (Aquino and Encarnación, 1988; Pieczarka *et al.*, 1992). In one area, they do not occur in the same locations, with *A. vociferans* surrounding the area of *A. nancymae,* the latter either filling a previously vacant region or displacing *A. vociferans* (Aquino and Encarnación, 1988). In another area, they are more sympatric (Pieczarka *et al.*, 1992). In either case the two forms remain highly distinctive in pelage and karyotype, suggesting that they are reproductively isolated. Ma feels their karyotypic differences are sufficiently critical in nature so as to preclude any interfertility (Nancy Ma, personal communication).

All the other southern populations share with *A. nancymae* at least one reciprocal translocation and possibly one fission event from the proposed ancestor (Galbreath, 1983). Since nothing is known of the karyotype of *A. miconax,* unfortunately, karyotypes cannot bring evidence to bear on the distinctiveness of this population. Although chromosome number and geographic placement would appear to put *A. nancymae* in a good position for giving rise to the other southern populations, it appears the presumed derived centric fission and two inversions that this population exhibit are not shared with the other southern groups (Galbreath, 1983; Ma, 1981), removing it as a possible ancestral group in its present form.

The remaining groups to the east and south all share a very distinctive alteration, the fusion of the Y chromosome with one homologue of an autosomal pair. This almost certainly occurred only once and originally onto the short arm of the autosome (Galbreath, 1983; Ma *et al.*, 1980; Ma, 1981; Pieczarka and Nagamachi, 1988; Pieczarka *et al.*, 1993; as confirmed by genetic marker analysis—Ma *et al.*, 1989b), leading to one less chromosome in the karyotype of males than females. This appears to be fixed throughout all the populations of *A. nigriceps,* both subspecies of *A. azarae,* and *A. infulatus,* with no indication of polymorphism. In addition, these groups all share one additional autosomal fusion event that did not occur in *A. nancymae.* The complete penetrance of these karyotypic changes suggests a real break in gene flow within the southern group, separating the eastern plus southern population from the western Peruvian population(s).

An enclave of *A. vociferans* exists south of the Rio Amazonas along the Rio Purus, within the distribution of *A. nigriceps,* with both groups retaining their distinct appearance (Hershkovitz, 1983). This distribution attests to the probable reproducitve isolation of the more central north- versus south-of-the-Amazon groups, concurring with the north versus south distinction in the more western populations, *A. vociferans* and *A. nancymae.*

Aotus a. azarae, A. a. boliviensis, and *A. infulatus* further share a unique karyotype, including a reduction in chromosome number from that found in *A. nigriceps* ($2n = 51, 52$) to $2n = 49, 50$, owing to the fixation of three autosomal rearrangements and a fusion event (Ma *et al.*, 1976—karyotype VI; Mudry de Pargament *et al.*, 1984; Pieczarka and Nagamachi, 1988). The karyotypes of all three appear to be almost identical, with the single exception of the structure of the Y/autosome. This chromosome is acrocentric with Y fused to the distal short arm in *A. a. azarae* (as well as *A. nigriceps*), but submetacentric in the other two. However, these last differ slightly as well, with an apparent very small pericentric inversion, not affecting the placement of the Y material, having occurred in *A. infulatus,* and a large pericentric inversion moving the Y segment from the short to the long arm having occurred in *A. a. boliviensis* (Pieczarka and Nagamachi, 1988). Although Pieczarka and Nagamachi suggest that the karyotypes of *A. a. boliviensis* and *A. infulatus* are more like each other than either is like that of the most southern population, *A. a. azarae,* in fact, all three are almost identical, and there is nothing in the original description of the karyotype of *A. a. azarae* (see Mudry de Pargament *et al.*, 1984) to indicate the relationship of its acrocentric Y/autosome to the submetacentrics found in the other populations. It could perhaps as easily be derived from the inverted chromosome found in the Bolivian group as from the supposed common ancestor of all three (see note added in proof).

Pieczarka and Nagamachi (1988) do not feel the differences between *A. infulatus* and *A. azarae boliviensis* are sufficient by themselves to indicate that speciation has occurred, particularly in light of the marked karyotypic variability present in northern populations of *Aotus*. Though at present, there is no evidence of polymorphism relating to the structure of the Y/autosome in any of the three geographically defined populations of *A. infulatus* and *A. azarae* subspecies, as opposed to the situation affecting autosomal variation north of the Amazon, *A. infulatus* and *A. a. boliviensis* have been shown to hybridize, and a marked zone of hybridization with a fairly well established population including its own distinct autapomorphy (the addition of more interstitial heterochromatin on chromosome B12) has been identified in Rondonia (Pieczarka *et al.*, 1993). This population is heteromorphic for the number of dark G-bands on the long arm of B12: 62% with 3, as in *A. infulatus*, and 38% with 4, as in *A. a. boliviensis*. Although this population is within the geographic area that Hershkovitz (1983) predicted should have *A. nigriceps*, in fact none of the animals sampled from this area (right bank of the Rio Jamari, south of the Rio Madeira, 63°30′W, 9°00′S) shows pelage or karyotypic characters typical of *A. nigriceps*.

C. Blood Protein Variation

Significant variability within *Aotus* has been documented for carbonic anhydrase II (CAII; Sampaio *et al.*, 1991). By examining electrophoretic patterns, they discovered that *Aotus* was more polymorphic than any of the other 11 platyrrhine genera examined, exhibiting 8 different alleles, in varying combinations in different populations. The distribution of the alleles in the 5 *Aotus* populations examined, corresponding to 4 of the species recognized by Hershkovitz (1983), is given in Table VI. This distribution can be interpreted in varying ways. Several alleles are sufficiently rare as to suggest that their presence is derived in those populations. These include CAII*3 in one population of *A. infulatus* (from the right bank of the Rio Tocantins); CAII*4 in another population of *A. infulatus* (from the left bank of the Rio Tocantins); CAII*22 in *A. nancymae;* CAII*23 in *A. azarae azarae* (the subspecific allocation is clear from the reported origin of these specimens); and CAII*5 in *A. vociferans*. However, the autapomorphic distribution of these alleles also makes them uninformative about relationships between the tested populations.

TABLE VI Allelic Distribution for Carbonic Anhydrase II Electrophoretic Patterns[a]

Taxon	II-3	II-4	II-5	II-9	II-15	II-17	II-22	II-23
A. a. azarae	0	0	0	0	1	0	0	1
A. infulatus left bank	0	1	0	1	1	0	0	0
A. infulatus right bank	1	0	0	1	1	0	0	0
A. nancymae	0	0	0	0	1	1	1	0
A. vociferans	0	0	1	1	0	1	0	0

[a] Data taken from Sampaio *et al.* (1991). *Aotus infulatus* were sampled from right and left banks of the Rio Tocantins.

This leaves alleles CAII*9, CAII*15, and CAII*17 as potentially informative. It seems clear that CAII*15 is a shared feature of all the southern populations. The disjunct distributions of the other two alleles led Sampaio *et al.* to derive the cladogram shown on the right in Fig. 16, based presumably on the assumptions that *A. a. azarae* and *A. nancymae* share the loss of CAII*9, and *A. nancymae* and *A. vociferans* independently derive CAII*17. However, the arrangement shown on the left is equally parsimonious and is based on the assumption that *A. a. azarae* and *A. infulatus* share the loss

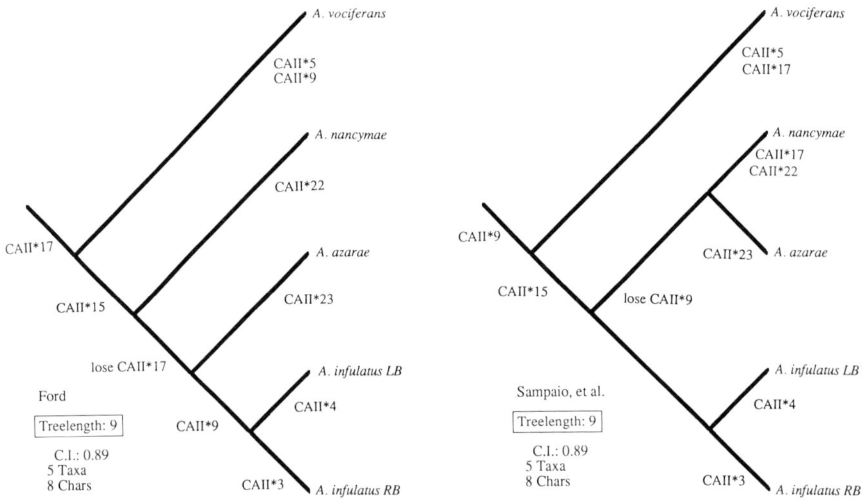

FIGURE 16 Alternate cladograms of relationships of five populations based on carbonic anhydrase II alleles. Right: offered by Sampaio *et al.* (1991); left: an equally parsimonious interpretation of the data (see Table VI and text).

of CAII*17, and *A. infulatus* and *A. vociferans* independently derive CAII*9. This alternate arrangement is also in closer agreement with the distribution of pelage, craniodental, and karyotypic data.

Schneider *et al.* (1989) also found significant differences between the right- and left-bank *A. infulatus* populations in heterozygosity levels of the phosphogluconate dehydrogenase (PGD) and glucose phosphate isomerase (GPI) polymorphic systems. In particular, the east-bank population is considerably more heterozygous.

D. Species Concepts as Applied to *Aotus*, and the Evolution of Variation in *Aotus*

There are currently a series of competing definitions of species, which vary significantly in their implications for how one operationally identifies a species and the nature of characters that one emphasizes. There are four primary competing definitions; these have been thoroughly discussed by Endler (1989) and Kimbel (1991; Kimbel and Martin, 1993), among others. The "traditional" Biological Species Concept (BSC) defines species on the basis of their reproductive isolation from other species (Mayr, 1970). Operationally, one seeks potential hybrid zones in the field; if the populations exist sympatrically without significant hybridization, then they can be considered separate species. The absence of areas of sympatry for taxa under study would render this definition almost (or entirely) impossible to apply. In addition, the literature is full of examples of highly distinctive populations (seemingly "good species") that nonetheless hybridize in areas of sympatry.

The Recognition Species Concept (RSC) defines species on the basis of their shared fertilization systems (Paterson, 1985, and elsewhere). Operationally, species are recognized as individuals that share a Specific Mate Recognition System (SMRS), that is, one or a series of traits that facilitate the identification of appropriate mates. Species should ideally be defined by the presence of significant primary or secondary sexual characteristics or of highly visible signaling traits. Areas of sympatry are not necessary for the application of this definition, and some hybridization in peripheral areas of sympatry would not necessarily violate the integrity of separate species identity. However, polymorphism across species in identifiable, potential SMRS traits is necessary. In addition, this concept argues that variation in non-SMRS traits is not useful in defining species; these traits may and often will vary independently of speciation events.

The Cohesive Species Concept (CSC) defines species as individuals that maintain a coherent, cohesive identity in space and time. It was developed by Templeton (1989), in an attempt to modify the RSC so that it would be applicable to fossils and to asexually reproducing organisms. Although it overcomes some of the constraints of the RSC, it provides a more difffuse and vague set of criteria for recognition of a species, particularly if one is in fact dealing with extant, sexually reproducing taxa.

Finally, paleontologists have continued to develop Simpson's (1961) Evolutionary Species Concept into its most recent form, the Phylogenetic Species Concept (PSC), which defines a species as a group of individuals that are diagnosably distinct, with a shared parentage (or lineage) through time (Cracraft, 1983, 1989). Operationally, this is very similar to the CSC. Though these concepts parallel the RSC in allowing hybridization, they add the flexibility of allowing species identification on the basis of any distinctive attributes, whether or not they appear to be a part of a Specific Mate Recognition System. This allows an individual researcher tremendous latitude in selecting key defining features or identifying "diagnosable" populations (also known as morphospecies). However, this latitude can introduce a dangerous amout of researcher bias and could easily lead to numerous subspecies or clinal variants being identified as valid species, particularly if they have disjunct distributions in either space or time.

In the case before us, populations of owl monkeys are largely disjunct geographically. In addition, many potential areas of sympatry are as yet unsampled (e.g., the eastern border of *A. trivirgatus,* or the southern and western, and perhaps eastern, boundaries of *A. nancymae;* see Fig. 1). It is therefore difficult to test for reproductive isolation under the biological species concept for most putative species. What is apparent from the data presented and analyzed here is that, for many characters and across large portions of its range (e.g., the northwestern Amazon Basin), owl monkey populations intergrade and exhibit largely clinal distributions, which do not appear under any definition of species to allow the differentiation of identifiable, valid, distinct population groups. However, where distinctive populations are apparent, the defining features are ones that may well be important signaling factors for the recognition of mates, such as the convergence of highly visible head stripes or overall size. Aside from the major separation of red-necked from gray-necked groups on either side of the Amazon River, the many variations seen in the degree of red coloration of body parts seem to be insignificant in distinguishing discrete populations. On the other hand, degree of black or gray pigmentation on the tail does appear partially to distinguish groups, and this might serve as a readily apparent signal even in a nocturnal animal.

Thus, many subtle pelage variations that are most likely not readily apparent to other owl monkeys (at least at some distance, at night) are clinal in distribution and not species specific. The dental and cranial variations, although perhaps significant adaptations for survival, may not be critical or even useful for mate recognition, and there is tremendous variability in most of these features across much of the range of *Aotus*. However, at least three populations do stand out with greatly increased dental dimensions, in conjunction with other distinguishing features (*A. trivirgatus* and *A. infulatus/azarae boliviensis*). Overall size differences, however, may indicate potentially species specific variation. The observed pattern of clinal or random variation in nonmate recognition traits, for the most part, but population-specific (species specific?) variation in likely or potential mate recognition traits, conforms to the predicted patterns of trait variation in Paterson's Recognition Species Concept. This increases the likelihood that the populations defined by these characters represent valid species regardless of the ultimate existence or nonexistence of total reproductive isolation of any of these populations.

The data reviewed here support the recognition of at least five species, possibly seven (Figs. 17 and 18). There is very sound evidence for Hershkovitz's separation of *Aotus* into two basic groups, the gray-necked group found primarily north of the Amazon, which is susceptible to malaria to varying degrees, and the red-necked group (many of which have highly red-to-orange pigmentation extending well onto the ventral surface of all extremities) found south of the Rios Amazonas and Marañon, members of which are resistant to malaria. At the very least, these two groups should be considered separate at the specific level; if only these two are recognized, they would be *A. trivirgatus* (north) and *A. azarae* (south). Karyotypic evidence would suggest that the ancestral form was most likely gray-necked, with a diploid number of 54 (possibly 52) and a separate, acrocentric Y chromosome, lacking the reciprocal translocation and likely fission event characterizing the red-necked populations. Therefore, the ancestral population would have been located somewhere in the Amazon Basin north of the Rios Marañon/Amazonas. Hershkovitz's (1977) theory of metachromism argues that evolution will tend to proceed from agouti, or multicolored, hairs to animals with more saturated, uniquely colored hairs (either darker or lighter). Application of this theory would suggest that the grayer pelage of the northeastern population (his *A. trivirgatus*) is derived from the more generally mixed agouti hairs of the western populations on the north bank. Location of a "center of origin" within the broad range of the northwestern *Aotus* populations is far more problematic.

From this putative ancestor, north of the river, there has been great diversification with the development of polymorphisms in many morphological, pelage, and karyotypic characters. These populations have spread north into Panama and along river systems draining into the Gulf of Mexico, but they have not crossed to the western slope of the Andes or into the central or lower Orinoco Basin. The variations have become fixed to greater or lesser degrees in populations separated by rivers or mountainous areas, but for the most part there is no compelling evidence that speciation has occurred in these northern populations.

The one exception would be the most eastern population, located east of the Rio Negro and south of the Rio Orinoco. This group appears to be distinctive in the larger size of its masticatory apparatus, the extensive

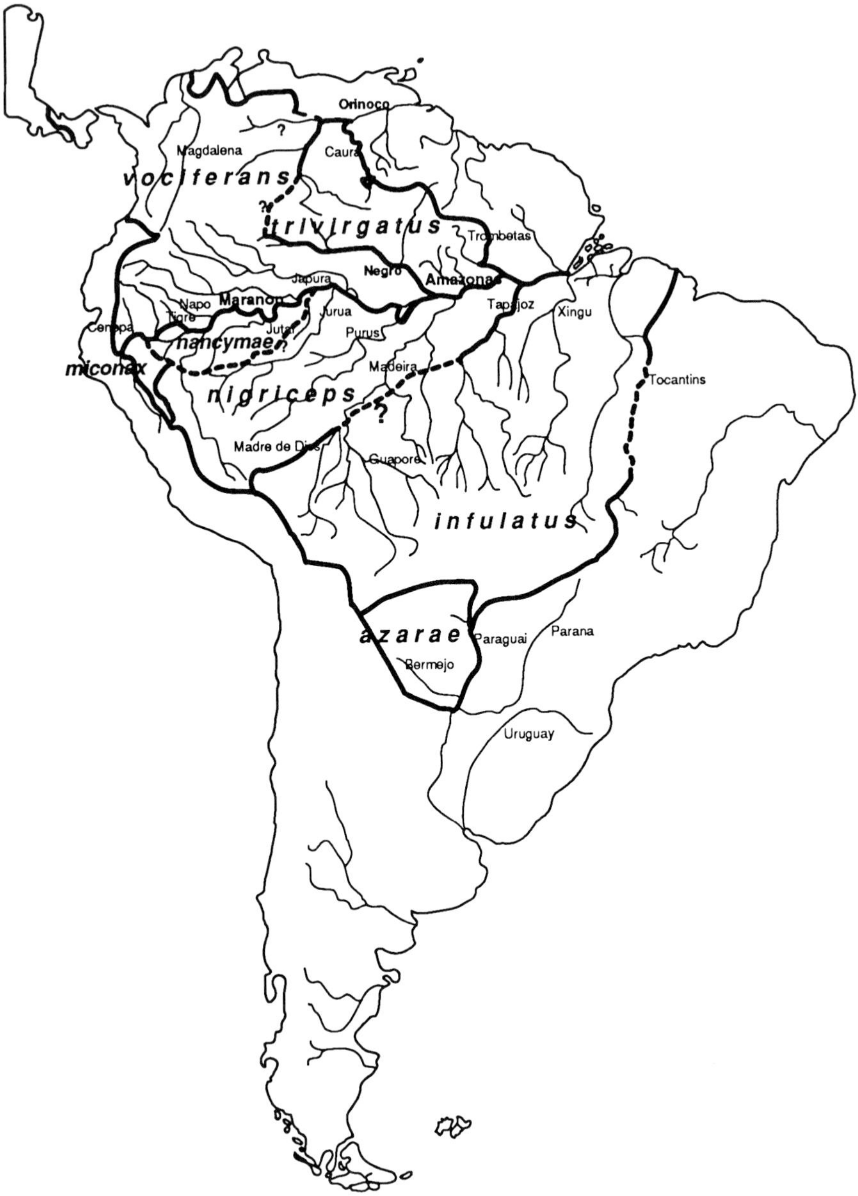

FIGURE 17 Distribution of seven species provisionally recommended here. As discussed in the text, the distinctiveness of *A. nancymae* from *A. miconax* and of *A. infulatus* from *A. azarae* are considered questionable at present, as are the likely precise boundaries between these populations. However, current data make the unique identities of these populations/species more defendable than possible divisions within *A. vociferans*. The boundary between *A. nigriceps* and *A. infulatus* is moved to incorporate new data from a population in Rondonia (Pieczarka *et al.*, 1993; J. Pieczarka, personal communication; see text).

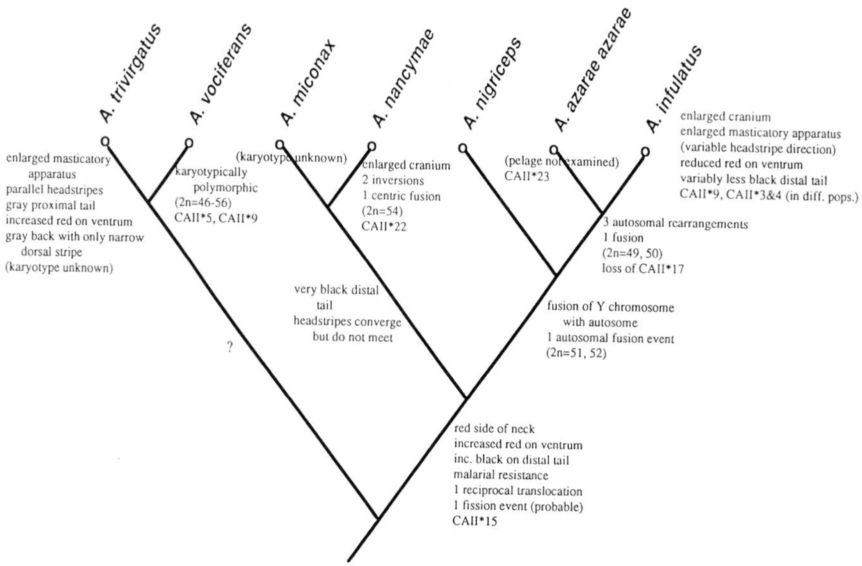

FIGURE 18 Cladogram of relationships between the five to seven species of *Aotus* recognized as a result of this study, indicating the polarity and evolution of traits (see text).

gray coloration, particularly the grayish hairs on the proximal tail and a predominantly gray dorsum with a narrow agouti stripe up the center, and roughly parallel stripes on the forehead. It has also increased the level of phaeomelanin in the hairs on the ventrum, particularly of the arms and legs. Unfortunately, the karyotype of specimens from this region is virtually unknown (Galbreath, 1983; Hershkovitz, 1983), as is the allelic component for the CAII system, and there are no known areas of sympatry with more western populations. It is most likely a derived population that migrated east (or south) from its ancestral region, becoming cut off by two large rivers (the Rios Negro and Orinoco) and developing its unique attributes in isolation. Although reproductive isolation cannot be proved at present, the magnitude of morphologic separation of this population from the others documented here, particularly when contrasted with the lack of such separation for the other northern populations, argues that this represents a distinctive species. The parallel head stripes, in particular, would serve as a highly visible and effective signal to a prospective mate (as part of a SMRS), as might the grayer back and, particularly, proximal tail.

This eastern species is then *A. trivirgatus* Humboldt 1811, and the remaining gray-necked forms are *A. vociferans* Spix 1823. Evidence reviewed here makes even subspecific differentiation within *A. vociferans* of doubtful utility except on karyotypic grounds, since the intergrading nature of both cranial and pelage characteristics across much of its range strongly suggests clinal gradients of shifting change in attributes, possibly associated with the movements of populations into wetter versus drier regions (as suggested by Thorington and Vorek, 1976) or areas of higher or lower altitude. Karyotypic variability, possibly reflecting chromosomal plasticity, is apparently a key attribute of this species, which may facilitate the adaptation of regional populations to altered environmental conditions. It will be interesting to learn more about the variability within this species of the CAII system (two derived alleles are clearly present, CAII*5 and CAII*9, in addition to the widespread CAII*17), as well as in *A. trivirgatus*.

Once populations of *Aotus* crossed south of the Amazon, probably somewhere in the headwaters of the Amazon, where the rivers pose slightly less imposing barriers (perhaps in Peru), more discrete differentiation appears to have occurred. All of these populations are marked by high penetration of phaeomelanin on the ventral surfaces of the throat, proximal limbs, and tail. In addition, they probably developed increased black pigmentation on the distal tail, although variability both north and south of the Amazon makes this less certain. All southern populations also have descended from an ancestor that was marked by a reciprocal translocation and a probable fission event in the karyotype, the evolution of the carbonic anhydrase II*15 allele, and the development of resistance to malaria as they spread into the southern Amazonian neotropics. Work on *A. nancymae* would suggest that the southern groups all share descent from a form with an acrocentric Y chromosome (see Table V; Pieczarka *et al.*, 1992).

The small samples considered here do not allow a concrete statement on the differentiation of *A. nancymae* and *A. miconax* based on morphological traits, and unfortunately there is no karyotypic information on *A. miconax*. However, this study is suggestive that *A. nancymae* is distinctive from *A. miconax* (and other neighboring populations) in its cranial dimensions. Other than this, no clear case for a separation can be made, and Hershkovitz (1983) suggested that further study might indicate that *A. miconax* and *A. nancymae* are conspecific. Though the single specimen of *A. nancymae* included in the two data sets analyzed here falls out as very distinctive from *A. miconax*, the differences may not be as marked as they appear. The *A. nancymae* specimen shares extension of red pigmentation onto the neck, arms, and legs, but it is lighter in color, more buffy/light orange, and it totally lacks extension of reddish hairs onto the tail. However, Hershkovitz (1983) described the proximal third of the tail of *A. nancymae* specimens he observed as distinctly reddish in coloration, and some variability in the darkness of the phaeomelanin on the arms and legs could be possible. A final statement on the distinctiveness of these two populations may need to await more complete analysis of skins of both and information on the karyotypes of *A. miconax*, as yet unknown.

A potential confounding problem is the fact that many specimens examined here had locality information that did not adequately indicate from which bank of the Rio Huallaga they were recovered, making their assignation to species group uncertain (here they were placed in *A. miconax*). The proposed range of *A. miconax* is quite small, and undoubtedly many museum specimens share this lack of specificity in locality information. In addition, the study by Aquino and Encarnación (1988) suggests a westward expansion of the range of *A. nancymae* from that indicated by Hershkovitz (1983). They would extend the range up into the foothills of the Andes, thereby including the range of *A. miconax*. The magnitude of the difference in cranial dimensions is sufficiently suggestive, and the region is sufficiently poorly known, that I here take the conservative step of retaining both species provisionally. However, it is recognized as highly probable that *A. nancymae* is not distinct from *A. miconax*, as Hershkovitz (1983) suggested may prove to be the case, particularly as the primary potential difference does not involve features that may be significant in mate recognition.

Aotus nancymae (and its very similar neighbor, *A. miconax*) is geographically in a prime position to be a candidate for the "ancestral" population of the other southern populations. However, it is clearly derived from both northern populations and other southern populations in numerous features, including the presence of totally black hairs extending well up the distal end of the tail, and converging (but not touching) head stripes (these are both shared with *A. miconax*); the enlarged cranium; two derived inversions and one centric fusion karyotypically; and the presence of the CAII*22 allele (karyotypes and blood proteins are unknown for *A. miconax*). This suggests a possible example of centrifugal evolution, as discussed most thoroughly by Groves (1989), in which a derived population develops in the center of origin, pushing the ancestral populations peripherally (both east/southeast and west). In this case, however, the displaced populations themselves appear to have undergone subseqent change. The pattern of pelage and craniodental variation suggests that the eastward expansion may have occurred earlier than the westward, as *A. miconax* shares at least some derived features with the "central" (in terms of area of origin) *A. nancymae* that other southern populations lack.

As owl monkeys migrated farther south and east, a major chromosomal change occurred, leading to an autosomal fusion event and the fusion of the Y chromosome to the short arm of an autosome. This change is shared in a nonvariant state by all populations that continued to move in these directions. The population directly bordering *A. nancymae*/*A. miconax* to the east and south, though exhibiting this distinctive karyotype, is otherwise not much changed morphologically from the proposed early southern population (again, fitting the centrifugal evolution model). Hershkovitz (1983) even reported potential hybrids along the Rio Ucayali. However, the karyotypic distinction appears profound, so that a small area of hybridization, if real, should not compromise the separate species status of this group, *A. nigriceps*.

As owl monkeys moved yet farther east and south, clearly some further major differentiation occurred. However, as the preceding discussion makes clear, the pattern of these changes is somewhat confusing and obscure. Some population, upon crossing either the Rio Tapajos, Guapore, or Madre de Dios, further reduced the chromosome number by an autosomal fusion event, as well as attaining three other autosomal rearrangements, and in addition losing the CAII*17 allele (if it ever had it). Even though this eastern and southern group has been separated into as many as three separate populations, the evidence for their separation is not strong. In particular, the populations bordering *A. nigriceps* (Hershkovitz's *A. infulatus* and *A. azarae boliviensis*) are highly distinct from their northwestern neighbor in an enlarged cranium and enlarged masticatory apparatus. The combination of fixed differences from *A. nigriceps* in the structure of the Y chromosome and other chromosomes, the loss of the CAII*17 allele, and significantly larger size all suggest the presence of both pre- and postmating differences in mate recognition. Although not all portions of this range have been sampled, the eastern area appears to demonstrate the greatest variability, and throughout the range there are individuals with more parallel head stripes, reduced red pelage on the ventrum (especially in the east), and reduced black pigmentation on the distal tail (especially in the south). There are slight differences in the structure of the Y/autosome, but these are most likely intraspecific variants (Pieczarka and Nagamachi, 1988; Pieczarka *et al.*, 1993). In addition, at least the far eastern populations (there are no data throughout the rest of the range) share the derived presence of the CAII*9 allele and, in different areas, either the CAII*3 or the CAII*4 allele. At least some researchers have suggested that *Aotus* in this area (i.e., the eastern populations, especially east of the Rio Tocantins) may be undergoing rapid evolutionary change (Schneider *et al.*, 1989). This would be consistent with the view here that these are among the most distant from the ancestral core area and among the most derived; they may be fairly recent immigrants (in evolutionary time) to this portion of the neotropics and in the process of divergence/adaptation to comparatively new ecozones.

The southernmost members of this group, although sharing the pelage and karyotypic traits (albeit with their own unique variant on the Y/autosome), appear to lack the enlarged size found in the rest of the group. They are much more similar cranially and dentally to *A. nigriceps* and other *Aotus* populations north of the Rio Madre de Dios, even though widely separated from them geographically. In addition, the structure of the Y/autosome is slightly more distinctive than the variants found north and east, this population lacks any of the three CAII alleles found in the eastern population (CAII*3, CAII*4, and CAII*9), and it has its own unique allele, CAII*23. Hershkovitz (1983) described the pelage as particularly long, thick, and shaggy (yet his description of the tail of *A. a. azarae* does not distinguish it from many of his *A. infulatus* specimens). However, the long, thick coat may simply be a physiological response to the cooler temperatures associated with the higher latitude (ranging from 20 to almost 30 degrees south). Mudry de Pargament *et al.* (1984) described the pelage of Argentinian specimens as almost indistinguishable from that of Bolivian specimens.

If an increased sample of this southermost population of *Aotus* confirms findings here that it lacks the enlarged size of its northern and eastern neighbors, as well as the minor karyotypic derivation and CAII allele differences, there are two possible explanations. (1) A population of *A. nigriceps*, moving either east across the Rio Tapajos or south across the Rio Madre de Dios, became isolated, reducing the number of chromosomes and developing some minor pelage alterations. Eventually, a division occurred in this group, with a second population developing an enlarged cranium and enlarged masticatory apparatus. This second population was increasingly successful, and as it spread it pushed the ancestral, smaller type farther and farther south until it remained only as a relict population on the southern fringe of the range of the species (and genus). This view would suggest the early development of differences in mate recognition systems, including overall size, as a result of centrifugal evolution, with two species: the relict *A. azarae* and the large *A. infulatus*. Somewhere along the way, minor alterations occurred in various populations in the Y/autosome.

(2) The second possible explanation is that when a southern/eastern migrating population of *A. nigriceps* first became isolated and started to change, it also became enlarged. This single population was highly successful and spread widely south and east, with minor variations arising in local populations. The southernmost members of this population, for some undetermined reason, decreased the size of the cranium and masticatory apparatus, secondarily appearing more like the ancestral *A. nigriceps* (and other northern *Aotus*). This shift may or may not have included the development of distinctive, separate mate recognition systems. If not, then all the ($2n$ = 49, 50) *Aotus* belong in *A. azarae*.

There appears to be little evidence, at present, to lend greater weight to one or the other of these alternatives. The suggested pattern of centrifugal evolution in other *Aotus* southern populations suggests the first as more likely, but this is slim evidence at best. Until further information is brought to bear on this issue, these two distinctive populations are recognized as separate species: the southernmost, smaller-dimensioned group as *A. azarae* (including only *A. a. azarae*); the remaining, larger-dimensioned, and somewhat variable population as *A. infulatus* (including *A. a. boliviensis* of Hershkovitz, 1983). However, paralleling the situation in the western populations, it is recognized that these three may all prove to be subspecific or clinal variants of *A. azarae*.

V. CONCLUSIONS

The wide degree of variability in *Aotus* is not best explained as clinal variation across a single, highly polymorphic species. On the other hand, there is compelling evidence that broad clinal change has occurred in several areas of the range of this genus, in particular in both the northwestern portions of its range and the southeastern portions. Although the null hypothesis of 9 species/11 subspecies, as proposed by Hershkovitz (1983), was rejected, there is substantial support for many of the species he recognized. There is indeed a basic division within the genus of northern, gray-necked populations from southern, red-necked populations, a division that is firmly maintained even in the two areas where populations have crossed the Rio Marañon/Amazonas, forming pockets in the "territory" of the alternate form (*A. nancymae*/*miconax* surging north, and *A. vociferans* expanding south).

In addition, applying the Recognition Species Concept, there is good support for two species north of the Amazon, *A. trivirgatus* in the east and *A. vociferans* in the west (see Fig. 17). Specimens in collections reported from Barranquilla, Colombia, most likely include specimens drawn broadly from the range of *A. vociferans* and possibly include some specimens of *A. trivirgatus*. South of the Amazon, five species are suggested, although further study may indicate that only three are valid. These are *A. miconax* in the far west, *A. nancymae* (which may not be distinct from *A. miconax*), *A. nigriceps*, and *A. azarae* in the southern extreme of the range, and *A. infulatus* in the far east and extending west between *A. nigriceps* and *A. azarae* (*A. infulatus* may not be distinct from *A. azarae*).

The pattern of change is best accounted for by an origin somewhere in the northwestern portion of the generic range. Major differentiation in this genus then followed isolation of populations on opposite banks of large rivers, a dispersal/isolation rather than a vicariant biogeographic pattern. Particularly important as isolating agents or barriers to further dispersal have been the Rios Amazonas/Marañon, Negro, Orinoco, Trombetas, Tapajos, Guapore, and Madre de Dios. These are all, for the most part, extremely large rivers. However, other rivers (with the possible exceptions of the Rios Huallaga and Paraguai) do not appear to have served as effective barriers to owl monkeys, including the Rio Tocantins. Species boundaries are particularly unclear in the far western portions of the range—in the headwaters of the Amazon, and it is unclear what, if any, role rivers in this area may have played as isolating agents. The fact that distribution of species of *Aotus* can be explained by current large river systems suggests that differentiation within this genus may be fairly recent, and the large degree of karyotypic polymorphism in several species suggests that differentiation is ongoing.

The seven (probably five) species are the result of both eastward and southward migrations, and broad dispersal of the polymorphic central species, *A. vociferans*. Upon initial differentiation of the southern, red-necked group, centrifugal evolution may have operated in part to push more generalized populations east, as the derived *A. nancymae/miconax* evolved. This mechanism may have acted again, resulting in the peripheral isolation of the craniometrically generalized *A. azarae* south of the more derived *A. infulatus* (see also Fig. 18).

Further support for these seven (or five) species can best be attained through several avenues. Clearly, larger samples are needed of the southern populations for the features examined here (and in the past). The small samples used here, particularly confounded by the problem that most localities were represented by only single specimens, did not readily allow independent formulation of groups for testing, via a clustering technique. Therefore, this study was restricted to examining the distinctiveness of putative species/subspecies proposed elsewhere. This restriction requires that the conclusions be used with caution until further testing can be done. In addition, karyotypic information is sorely needed for *A. trivirgatus* and western *A. miconax*. And finally, other sources of data need to be examined; work by several Brazilian workers on various allelic systems is exciting and promising. Though other morphological features should be considered, postcranial features are often highly conservative in a genus without major differentiation in modes of locomotion, such as *Aotus*. It could be predicted that other metric data may at best reflect again the increased size of *A. trivirgatus* and of *A. infulatus*, and possibly *A. nancymae*. More informative may be a search for features that would be likely to be meaningful reproductive signals to a nocturnal monkey (SMRS), and thus under relatively strong control. Pelage coloration may not be a significant means of communication except for highly striking differences, such as convergent versus parallel stripes on the forehead, red or gray neck, or lots of black versus little to no black on the end of the tail. However, sounds and olfactory cues are undoubtedly highly significant as a means of communication for owl monkeys (Wright, 1989), and comparative studies of vocalizations may provide one of the best clues to differentiation in this very complex genus.

ACKNOWLEDGMENTS

Many individuals have contributed to this paper. My greatest debt is to Dr. Richard W. Thorington, Jr., who first suggested I undertake this project, graciously allowed me to use his large volume of cranial and dental data and to study specimens in his care, and provided needed critical commentary. Dr. Paulo E. Vanzolini of the Museu de Zoologia da Universidade de São Paulo also allowed me to study specimens in his care and provided invaluable assistance in tracking down coordinates for numerous localities, as well as a great deal of other gracious assistance during my brief visit to São Paulo. The paper has benefitted from discussions with and/or critiques by Drs. J. Baer, J. Carr, S. Keffer, N. S. F. Ma, J. C. Pieczarka, R. Thorpe, and P. Woehlke. Dr. A. Singh Tiwana and B. Holt, L. Davis, and D. Hobbs researched localities; P. Reed and P. Spielmacher helped with SAS (which was used for all the graphs); K. Fiorino assisted in the production of the maps (produced with Adobe Illustrator 88). To all of these individuals I am indebted. An earlier version of this paper was presented at a symposium organized by Drs. Thorington and Avila-Peres at the Congress of the International Primatological Society, Brasilia, August 1988.

APPENDIX A: CRANIAL AND MANDIBULAR MEASUREMENTS

(From Thorington and Vorek, 1976, p. 1010)

1. Condylo-basal length (CBL): from behind the occipital condyles to the anteriormost part of the premaxilla.
2. Biorbital width (BOW): greatest width across the orbits.
3. Postorbital width (POW): least width behind the orbits.
4. Bizygomatic width (BZW): greatest width across the zygomatic arches.
5. Mastoid width (MSW): greatest width of the skull at the mastoid.
6. Anion-lambda length (ANL): from the midline anterior to the first incisors to the posterior end of the interparietal suture.
7. Anion-bregma length (ABL): from the anion to the anterior end of the interparietal suture.
8. Bregma-lambda length (PSL): length of the parietal suture.
9. Condylo-bregma height (CBH): from the posterior border of the occipital condyles to the anterior end of the parietal suture.
10. Length of the tooth row (TRL): anterior edge of the upper canine alveolus to the posterior edge of the alveolus of the third upper molar.
11. Length of molar tooth row (MLL): anterior buccal edge of the crown of the first upper molar to the most distant point on the crown of the third upper molar.
12. Bicaninal width (BCW): between the buccal borders of the alveoli of the upper canines.
13. Bimolar width (BMW): the least width between the buccal sides of the first upper molars—below the crowns, above the alveoli.
14. Angular height (ANH): the height of the angular process of the mandible above the ventral edge of the ramus. When the mandible is sitting on a plane, this is the height of the process above the plane.
15. Height of the temperomandibular joint (TMJ): the height of the joint above the ventral edge of the ramus.

APPENDIX B: PELAGE COLORATION VARIABLES

1. Color of side of neck (SIDENECK): 0 = gray, 50 = red/orange, 100 = orange.
2. Interscapular tuft[1] (TUFT): 0 = absent, 100 = present.
3. Interscapular whorl[1] (WHORL): 0 = absent, 100 = present.

4. Ventrum (belly) (BELLY): 0 = gray, 50 = lt. orange/buffy, 75 = lt. orange, 100 = red/orange.
5. Ventrum (arm) (ARM): 0 = gray, 25 = orange only to elbow, 50 = lt. orange/buffy, 75 = lt. orange, 100 = red/orange.
6. Ventrum (leg) (LEG): 0 = gray, 25 = orange only to knee, 50 = lt. orange/buffy, 75 = lt. orange, 100 = red/orange.
7. Ventrum (neck) (NECK): 0 = gray, 50 = lt. orange/buffy, 75 = lt. orange, 100 = red/orange.
8. Extension of red pigmentation onto tail (TAILRED): 0 = none, 50 = slight extension onto proximal end of tail, 100 = proximal 1/4+ of tail red.
9. Extent of black pigmentation on distal end of tail (TAILBLK) (BA is agouti hairs tipped in black, BL is all black hairs; cline approximates increasing appearance of black tail but does not necessarily represent an evolutionary cline): 0 = none; 10 = BA at tip of tail; 20 = BL at tip of tail; 30 = BA 1/4–1/3 distal end of tail; 40 = BA 1/4–1/3 distal end, BL at tip; 50 = BL 1/4–1/3 distal end; 60 = BA 1/2+ distal end; 70 = BA 1/2+ distal end, BL at tip; 80 = BA mid 1/4–1/3 of tail, BL distal 1/4–1/3 of tail; 90 = BL 1/2+ distal end; 100 = BA for second 1/4+ of tail, followed by BL distal 1/2+.
10. Extension of gray pigmentation onto tail (TAILGRAY): 0 = none, 100 = some gray.
11. Direction of stripes on forehead (HEADCONV): 0 = any other, 25 = parallel, 50 = stripes converge but do not meet, 75 = stripes converge close to front, 100 = stripes converge far back.
12. Darkness of stripes on forehead (HEADDARK): 0 = light, 100 = dark.
13. Width of stripes on forehead (HEADWID): 0 = narrow, 100 = wide.
14. Extension of stripes on forehead onto cheeks (HEADCHK): 0 = none, 50 = onto cheeks only slightly, 100 = extends onto cheeks.
15. Color of rump (RUMP): 0 = gray, 50 = lt. orangish agouti, 100 = reddish agouti.
16. Color of dorsum (DORSSTRP): 0 = any other, 25 = gray, 50 = gray with narrow agouti stripe up spine, 75 = gray with wide agouti stripe up spine, 100 = agouti.

APPENDIX C: VALUES FOR "HERSH"

(Following subgeneric categories in Hershkovitz, 1983)
A = *Aotus azarae azarae*
B = *Aotus azarae boliviensis*
C = *Aotus nancymae*
G = *Aotus lemurinus griseimembra*
I = *Aotus infulatus*
L = *Aotus lemurinus lemurinus*
M = *Aotus miconax*
N = *Aotus nigriceps*
Q = Origin given as Barranquilla, Colombia (see text)
R = *Aotus brumbacki*
T = *Aotus trivirgatus*
V = *Aotus vociferans*
U = unsure of location/group

[1] Interscapular tuft and whorl described in detail in Hershkovitz (1983).

REFERENCES

Aquino, R., and Encarnación, F. (1988). Population densities and geographic distribution of night monkeys (*Aotus nancymae* and *Aotus vociferans*) (Cebidae: Primates) in Northeastern Peru. *Am. J. Primatol.* **14,** 375–381.

Benton, M. J. (1980). Geographic variation in the garter snakes (*Thamnophis sirtalis*) of the north-central United States, a multivariate study. *Zool. J. Linn. Soc.* **68,** 307–323.

Brumback, R. A. (1973). Two distinctive types of owl monkeys (*Aotus*). *J. Med. Primatol.* **2,** 284–289.

Brumback, R. A. (1974). A third species of the owl monkey (*Aotus*). *J. Hered.* **65,** 321–323.

Brumback, R. A., Stanton, R. D., Benjamin, S. A., and Lang, C. M. (1971). The chromosomes of *Aotus trivirgatus* Humboldt 1812. *Folia Primatol.* **15,** 264–273.

Cabrera, A. (1957). Catalogo de los mamiferos de America del Sud. 1. *Rev. Mus. Argent. Cienc. Nat. "Bernardino Rivadavia" Inst. Nac. Invest. Cienc. Nat., Cienc. Zool.* **4,** 1–308.

Cracraft, J. (1983). Species concepts and speciation analysis. *Curr. Ornithol.* **1,** 159–187.

Cracraft, J. (1989). Speciation and its ontology: The empirical consequences of alternative species concepts for understanding patterns and processes of differentiation. *In* "Speciation and Its Consequences" (D. Otte and J. A. Endler, eds.), pp. 28–59. Sinauer Assoc., Sunderland, MA.

de Cruz Lima, E. (1945). "Mammals of Amazonia. 1. General Introduction and Primates." Contrib. Mus. Paraense Emilio Goeldi Hist. Nat., Etnogr., Belem do Para, Rio de Janeiro.

Emmons, L. H. (1990). "Neotropical Rainforest Mammals: A Field Guide." University of Chicago Press, Chicago.

Endler, J. A. (1989). Conceptual and other problems in speciation. *In* "Speciation and Its Consequences" (D. Otte and J. A. Endler, eds.), pp. 625–648. Sinauer Assoc., Sunderland, MA.

Fleagle, J. G., and Mittermeier, R. A. (1980). Locomotor behavior, body size, and comparative ecology of seven Surinam monkeys. *Am. J. Phys. Anthropol.* **52,** 301–314.

Fooden, J. (1963). A revision of the woolly monkeys (genus *Lagothrix*). *J. Mammal.* **44,** 213–247.

Ford, S. M., and Davis, L. C. (1992). Systematics and body size: Implications for feeding adaptations in New World monkeys. *Am. J. Phys. Anthropol.* **88,** 415–468.

Galbreath, G. J. (1983). Karyotypic evolution in *Aotus*. *Am. J. Primatol.* **4,** 245–251.

Groves, C. P. (1989). "A Theory of Human and Primate Evolution." Oxford (Clarendon) Univ. Press Oxford.

Hall, E. R. (1981). "The Mammals of North America." Wiley, New York.

Hershkovitz, P. (1949). Mammals of northern Colombia. Preliminary report no. 4: Monkeys (Primates) with taxonomic revisions of some forms. *Proc. U.S. Natl. Mus.* **98,** 323–427.

Hershkovitz, P. (1977). "Living New World Monkeys (Platyrrhini) with an Introduction to Primates." Univ. of Chicago Press, Chicago.

Hershkovitz, P. (1983). Two new species of night monkeys, genus *Aotus* (Cebidae, Platyrrhini): A preliminary report on *Aotus* taxonomy. *Am. J. Primatol.* **4,** 209–243.

Honacki, J. H., Kinman, K. E., and Koeppl, J. W. (1982). "Mammals Species of the World: A Taxonomic and Geographic Reference." Allen Press, Lawrence, KS.

Ishak, B., Warter, S., Dutrillaux, B., and Rumpler, Y. (1992) Chromosomal rearrangements and speciation of sportive lemurs (*Lepilemur* species). *Folia Primatol.* **58**(3), 121–130.

Kimbel, W. H. (1991). Species, species concepts and hominid evolution. *J. Hum. Evol.* **20,** 355–371.

Kimbel, W. H., and Martin, L. B., eds. (1993). "Species, Species Concepts, and Primate Evolution." Plenum, New York.

Ma, N. S. F. (1981). Chromosome evolution in the owl monkey, *Aotus*. *Am. J. Phys. Anthropol.* **54,** 293–303.

Ma, N. S. F. (1983). Comparative gene mapping of the owl monkey, *Aotus*, with karyotypes V ($2n = 46$) and VI ($2n = 49,50$). *Cytogenet. Cell Genet.* **35,** 117–123.
Ma, N. S. F., and Gerard, D. S. (1988). Chromosome assignment of the gene loci ETS1 and THY1 in the owl monkey. *Cytogenet. Cell Genet.* **48,** 170–173.
Ma, N. S. F., and Harris, T. S. (1989). A putative homoecolog of human chromosome 12 in the owl monkey. *Cytogenet. Cell Genet.* **50,** 34–39.
Ma, N. S. F., Jones, T. C., Miller, A. C., Morgan, L. M., and Adams, E. A. (1976). Chromosome polymorphism and banding pattern in the owl monkey (*Aotus*). *Lab. Anim. Sci.* **26,** 1022–1036.
Ma, N. S. F., Rossan, R. N., Kelley, S. T., Harper, J. S., Bédard, M. T., and Jones, T. C. (1978). Banding patterns of chromosomes of two new karyotypes of the owl monkey, *Aotus*, captured in Panama. *J. Med. Primatol.* **7,** 146–155.
Ma, N. S. F., Renquist, D. M., Hall, R., Sehgal, P. K., Simeone, T., and Jones, T. C. (1980). XX/"XO" sex determination system in a population of Peruvian owl monkeys, *Aotus*. *J. Hered.* **71,** 336–342.
Ma, N. S. F., Aquino, R., and Collins, W. E. (1985). Two new karyotypes in the Peruvian owl monkey (*Aotus trivirgatus*). *Am. J. Primatol.* **9,** 333–341.
Ma, N. S. F., Harris, T. S., Watkins, P., and Tsui, L. C. (1989a). Owl monkey gene map: Evidence for a homologous human chromosome 7q region near the cystic fibrosis locus. *Genomics* **5,** 389–396.
Ma, N. S. F., Page, D. C., and Harris, T. S. (1989b). Molecular evidence of Y-autosomal translocations in owl monkeys. *J. Hered.* **80,** 259–263.
Mayr, E. (1970). "Populations, Species, and Evolution." Belknap Press, Cambridge, MA.
Mittermeier, R. A., and Coimbra-Filho, A. F. (1981). Systematics: Species and subspecies. *In* "Ecology and Behavior of Neotropical Primates" (A. F. Coimbra-Filho and R. A. Mittermeier, eds.), Vol. 1, pp. 29–110. Acad. Bras. Cienc., Rio de Janeiro.
Mittermeier, R. A., Rylands, A. B., and Coimbra-Filho, A. F. (1988). Systematics: Species and subspecies—An update. *In* "Ecology and Behavior of Neotropical Primates" (R. A. Mittermeier, A. F. Coimbra-Filho, and G. A. B. da Fonseca, eds.), Vol. 2, pp. 13–75. World Wildlife Fund, Washington, DC.
Moynihan, M. (1976). "The New World Primates". Princeton Univ. Press, Princeton, NJ.
Mudry, M. D., Slavutsky, I., and de Vinuesa, M. L. (1990). Chromosome comparison among five species of Platyrrhini (*Alouatta caraya, Aotus azarae, Callithrix jacchus, Cebus apella*, and *Saimiri sciureus*). *Primates* **31**(3), 415–420.
Mudry de Pargament, M. D., Colillas, O. J., and Brieux de Salum, S. (1984). The *Aotus* from northern Argentina. *Primates* **25,** 530–537.
Nachman, M. W., and Myers, P. (1989). Exceptional chromosomal mutations in a rodent population are not strongly underdominant. *Proc. Natl. Acad. Sci. U.S.A.* **86,** 6666–6670.
Napier, P. (1976). "Catalogue of Primates in the British Museum (Natural History). Part I: Families Callitrichidae and Cebidae." British Museum (Natural History), London.
Paterson, H. E. (1985). The recognition concept of species. *In* "Species and Speciation" (E. S. Vrba, ed.), Transvaal Mus. Monogr. No. 4, pp. 21–29. Transvaal Museum, Pretoria.
Paynter, R. A., Jr. (1982). "Ornithological Gazeteer of Venezuela." Bird Department, Museum of Comparative Zoology, Harvard University, Cambridge, MA.
Paynter, R. A., Jr., and Traylor, M. A., Jr. (1977). "Ornithological Gazeteer of Ecuador." Bird Department, Museum of Comparative Zoology, Harvard University, Cambridge, MA.
Paynter, R. A., Jr., and Traylor, M. A., Jr. (1981). "Ornithological Gazeteer of Colombia." Bird Department, Museum of Comparative Zoology, Harvard University, Cambridge, MA.
Paynter, R. A., Jr., Traylor, M. A., Jr., and Winter, B. (1975). "Ornithological Gazeteer of Bolivia." Bird Department, Museum of Comparative Zoology, Harvard University, Cambridge, MA.

Pieczarka, J. C., and Nagamachi, C. Y. (1988). Cytogenetic studies of *Aotus* from eastern Amazonia: Y/autosome rearrangement. *Am. J. Primatol.* **14,** 255–263.

Pieczarka, J. C., de Souza Barros, R. M., Nagamachi, C. Y., Rodrigues, R., and Espinel, A. (1992). *Aotus vociferans* × *Aotus nancymae:* Sympatry without chromosomal hybridization. *Primates* **33**(2), 239–245.

Pieczarka, J. C., de Souza Barros, R. M., de Farria, F. M., Jr., and Nagamachi, C. Y. (1993). *Aotus* from the southwestern Amazon region is geographically and chromosomally intermediate between *A. azarae boliviensis* and *A. infulatus. Primates* **34**(2), 197–204.

Rathbun, G. B., and Gache, M. (1980). Ecological survey of the night monkey, *Aotus trivirgatus,* Formosa Province, Argentina. *Primates* **21,** 211–219.

Sampaio, M. I. C., Schneider, M. P. C., Barroso, C. M. L., Silva, B. T. F., Schneider, H., Encarnación, F., Montoya, E., and Salzano, F. M. (1991). Carbonic anhydrase II in New World monkeys. *Int. J. Primatol.* **12,** 389–402.

SAS Institute, Inc. (1990). "SAS Language Version 6.0," 1st ed. SAS Institute, Inc., Cary, NC.

Schneider, M. P. C., Sampaio, M. I. C., Schneider, H., and Salzano, F. M. (1989). Genetic variability in natural populations of the Brazilian night monkey (*Aotus infulatus*). *Int. J. Primatol.* **10,** 363–374.

Silva, B. T. F., Sampaio, M. I. C., Schneider, H., Schneider, M. P. C., Montoya, E., Encarnación, F., and Salzano, F. M. (1992). Natural hybridization between *Saimiri* taxa in the Peruvian Amazonia. *Primates* **33,** 107–113.

Simpson, G. G. (1961). "Principles of Animal Taxonomy." Columbia Univ. Press, New York.

Stanyon, R., Sineo, L., Chiarelli, B., Camperio-Ciani, A., Haimoff, A. R., Mootnick, E. H., and Sutarman, Drh. (1987). Banded karyotypes of the 44-chromosome gibbons. *Folia Primatol.* **48,** 56–64.

Stephens, L., and Traylor, M. A., Jr. (1983). "Ornithological Gazeteer of Peru." Bird Department, Museum of Comparative Zoology, Harvard University, Cambridge, MA.

Tate, G. H. H. (1939). The mammals of the Guiana region. *Bull. Am. Mus. Nat. Hist.* **76,** 151–229.

Templeton, A. R. (1989). The meaning of species and speciation: A genetic perspective. *In* "Speciation and Its Consequences" (D. Otte and J. A. Endler, eds.), pp. 3–27. Sinauer Assoc., Sunderland, MA.

Thorington, R. W., Jr., and Vorek, R. E. (1976). Observations on the geographic variation and skeletal development of *Aotus. Lab. Anim. Sci.* **26**(6), 1006–1021.

Thorpe, R. S. (1976). Biometric analysis of geographic variation and racial affinities. *Biol. Rev. Cambridge Philos. Soc.* **51,** 407–452.

Thorpe, R. S. (1979). Multivariate analysis of the population systematics of the ringed snake, *Natrix natrix* (L.). *Proc. Royal Soc. Edinburgh, Sect. B: Biol. Sci.* **78,** 1–62.

Thorpe, R. S. (1980a). A comparative study of ordination techniques in numerical taxonomy in relation to racial variation in the ringed snake *Natrix natrix* (L.). *Biol. J. Linn. Soc.* **13,** 7–40.

Thorpe, R. S. (1980b). Microevolution and taxonomy of European reptiles with particular reference to the grass snake *Natrix natrix* and the wall lizards *Podarcis sicula* and *P. melisellensis. Biol. J. Linn. Soc.* **14,** 215–233.

Thorpe, R. S. (1983). A review of the numerical methods for recognising and analysing racial differentiation. *In* "Numerical Taxonomy: Proceedings of a NATO Advanced Studies Institute" (J. Felsenstein, ed.), Vol. G1, pp. 404–423. Springer-Verlag, Berlin.

Thorpe, R. S. (1984). Multivariate patterns of geographic variation between the island and mainland populations of the Eastern grass snake (*Natrix natrix natrix*). *J. Zool. Soc. London* **204,** 551–561.

Thorpe, R. S. (1988). Multiple group principal component analysis and population differentiation. *J. Zool. Soc. London* **216,** 37–40.

Thorpe, R. S., and McCarthy, C. J. (1978). A preliminary study, using multivariate analysis, of a species complex of African house snakes (*Boaedon fuliginosus*). *J. Zool.* **184,** 489–506.
Wright, P. (1981). The night monkeys, genus *Aotus*. *In* "Ecology and Behavior of Neotropical Primates" (A. F. Coimbra-Filho and R. A. Mittermeier, eds.), Vol. 1, pp. 211–240. Acad. Bras. Cienc., Rio de Janeiro.
Wright, P. C. (1989). The nocturnal primate niche in the New World. *J. Hum. Evol.* **18,** 635–658.
Yunis, E., Torres de Caballero, O. M., and Ramirez, C. (1977). Genus *Aotus* Q- and G-band karyotypes and natural hybrids. *Folia Primatol.* **27,** 165–177.

Note Added in Proof: Recent clarification of new information on the karyotypes of the southeastern populations from Julio Pieczarka (personal communication; also Pieczarka *et al.*, 1993) indicates that the Y chromosome fused to the short arm of an ancestral B12 autosome containing three bands, as is found in the easternmost *A. infulatus* populations. A later paracentric inversion in B12 broke one band, giving rise to the four banded variant found in the Bolivian populations (*A. a. boliviensis*, now *A. infulatus*), with both variants occurring in the Rondonian population (now included in *A. infulatus*). Thus, throughout *A. infulatus*, there is polymorphism in the B12 autosome, but the Y/autosomes are identical. Contrary to my text here, the Brazilian workers feel that the structure of this Y/autosome of *A. infulatus* is most like that of *A. nigriceps*, and it is the Y/autosome of *A. azarae azarae* that is alone distinctive. This may support the possible separate status of this southernmost population from *A. infulatus* [including *A. azarae boliviensis*]. However, it must be noted that Y is relatively free from selection to maintain consistency of structure, since it pairs with only a small portion of the X chromosome during meiosis. Therefore, variations in the Y/autosome alone are unlikely to result in reproductive isolation. I am grateful to Sr. Pieczarka for this clarification.

2

Owl Monkey Populations in Latin America: Field Work and Conservation

Rolando Aquino and Filomeno Encarnación
Centro de Reproducción y Conservación de Primates del Instituto Veterinario de Investigaciones Tropicales y de Altura Universidad Nacional Mayor de San Marcos, Iquitos, Perú

I. INTRODUCTION

Members of the neotropical taxon *Aotus* are nocturnal. The species live in humid and subhumid forests of primary, secondary, and remnant types from sea level to elevations above 3200 m in cloud forests.

The geographical range of the *Aotus* species, with its differences in habitats, provides both advantages and limitations for scientific field studies. Among these are conditions such as physiographical relief, accessibility of the forest, the ability of trappers to work in the field at night, availability of equipment, and human pressure on forest habitats. Yet the demand for *Aotus* as animal models for biomedical research is an incentive for performing field studies. The purposes of such studies include pelt collection, by means of trapping, for taxonomic purposes; genetic sampling; enumerating populations; and evaluating natural populations in their natural habitat.

Currently, there is extensive information on the geographical distribution of most owl monkeys. Exceptions are *Aotus a. azarae* in Argentina and *A. nigriceps* in Puerto Bermudez and Manu National Park in Perú. For these species only limited data on home range, minimum ecological area, population density, and behavior are available.

This chapter discusses the species *A. nancymae* and *A. vociferans* with regard to habitat evaluation, census information, capture techniques, population structure, population recovery, and conservation status. Our studies took place in the lowland forests of northeastern Perú, between the Tigre, Marañón, Amazonas, and Napo river basins. Information on populations of three other species of *Aotus* located in Perú are also included in this report.

II. HABITAT AND FOREST PREFERENCES

The *Aotus* species inhabit the primary, secondary, and remnant tropical forests from as far north as Panama in Central America to as far south as Argentina and Paraguay in South America, at elevations ranging from sea level to 3200 m above sea level in Colombia (Eisenberg, 1989). Owl monkeys demonstrate preferences for varying plant communities and formations, which differ according to the altitude and latitude of their geographical range (Rathbun and Gache, 1980; Wright, 1978; Zunino *et al.*, 1986; Hernandez-Camacho and Defler, 1989).

The harvesting and cropping of owl monkeys in the lowland Amazonian forests of northeastern Perú since 1979 have allowed us to determine their preference for two types of habitats. Following the nomenclature given by Encarnación (1985), owl monkeys most commonly live in (a) "bajial" forest–lowland forest that is flooded seasonally or (b) "altura" forest–highland forest that is never flooded.

The plant formations in bajial forest grow on the Amazonian lowland plain, which is subject to seasonal inundation by water that sometimes reaches 7 m in height from whitewater and blackwater rivers, lakes, and streams (Encarnación, 1985). These types of forests appear to be the optimal habitat for owl monkeys based on the observation of high population densities (Aquino and Encarnación, 1988). This habitat preference may be linked to ecological factors such as (a) high productivity of food resources, (b) predominance of trees with thickened trunks and branches that provide hollows that can be used as sleeping sites, and (c) exuberant growth of liana tangles and tufts of epiphytes and hemiepiphytes on trunks and branches, which may also serve as sleeping sites or refuges. The altura forest or Brazilian "terra firme" forest grows slowly on highland terrain transected by small valleys with a diversity of slopes from which whitewater rivers originate. The trees in these habitats have hardwood trunks with few hollows. There are also few lianas, epiphytes, or hemiepiphytes. Thus, the population density of owl monkeys in the altura forest is lower than that in the bajial forest (Aquino and Encarnación, 1988).

In the course of their daily activities, owl monkeys occupy all forest strata from near ground level to the crowns of the tallest trees (Moynihan, 1964; Durham, 1975; Wright, 1978, 1985). They use the superior strata, including emergent tall trees, to forage for those plant species that comprise the largest part of their diet, whereas the lower and middle strata are used for displacement activities. Occasionally, the monkeys descend to the ground in open forest to utilize their complete home range when it is divided by small roads and cultivated areas, to reach farms with orchards of fruiting trees, or to escape predators.

III. POPULATION CENSUSES

Among the different censusing methods developed and applied to primates (Terborgh, 1985), the transect method provides the best results for *Aotus* living in the Perúvian lowland Amazonia. Other methods, such as direct counts of groups in the area and detection by vocalizations, are of limited use. The transect method, developed by Southwick and Cadigan

(1972) and modified by Freese *et al.* (1982) for neotropical primates, consists of using trials previously cut for this purpose or paths made by woodcutters to attempt to see and hear the monkeys. This method for enumerating *Aotus* monkeys requires walking at night in the forest or canoeing along the shores of rivers and lakes. The transect method permits coverage of extensive areas in a short amount of time while estimating the relative population densities of owl monkeys inhabiting the area. This information can then be compared to data derived from other areas.

The censuses are usually carried out on calm nights, without threat of storms or rain, between the hours of 3:00 and 6:00 am, or between 6:00 and 9:00 pm, local time. During these hours, owl monkeys travel a great deal while foraging for food (Garcia y Braza, 1987; Aquino and Encarnación, 1988). Usually, two observers walk slowly and quietly, stopping frequently to listen for sounds produced by these nocturnal primates. Typically, owl monkeys vocalize when they leap through the trees, drop fruits, fruit particles, or seeds, or become alarmed. The noises and vocalizations of owl monkeys are unique and easily differentiated from those of other nocturnal mammals or arboreal animals; however, verification by flashlight and/or binoculars is desirable. Population densities are then estimated from the data variables obtained during the census. Data variables obtained during the census include (a) the distance of the observer from the animal; (b) the perpendicular distance of the transect; and (c) group size.

A. Population Densities in Perú

Evaluation of owl monkey populations in Perú has been directed toward three of the five recorded species to the country. Through direct detection by vocalization of *A. nigriceps* in the southernmost part of Perú, Wright (1978, 1985) estimated 40 individuals/km^2 in altura forest and 36 to 40/km^2 in bajial forest. Aquino and Encarnación (1988), using the transect method, recorded 46.3 individuals/km^2 in bajial forest and 24.2/km^2 in altura forest for *A. nancymae* and 33 individuals/km^2 in bajial forest and 7.9/km^2 in altura forest for *A. vociferans* (Table I). The low densities estimated for both species in the altura forest suggest that they are better adapted to bajial forest. These differences in densities may relate to the availability of diurnal sleeping sites. The trees in the altura forest have hardwood hearts, resulting in few resting hollows available for use by the monkeys. The bajial forest, on the other hand, has trees with predominantly softwood hearts, which provide hollows that are used by both *A. nancymae* and *A. vociferans*. In addition, the bajial forest contains many concavities formed by branches, liana tangles, or epiphyte tufts that are used by *A. nancymae* as sleeping sites (Aquino and Encarnación, 1986b).

Recent censuses of *A. vociferans* were carried out by Aquino *et al.* (1992) in the Nanay and Santa María river basins near the border with Ecuador.

TABLE I Population Densities of *Aotus* in South America

Species	Forest type	Country/methods	Groups/km^2	Individuals/km^2	References
Red-neck group					
A. azarae	Gallery	Argentina/transect	6	18	Rathbun and Gache (1980)
A. azarae	Gallery	Argentina/transect	5.4	12.7	Zunino *et al.* (1986)
A. azarae	Lowland	Bolivia/plot	68.9	242.5	Garcia and Braza (1989)
A. azarae	Gallery	Paraguay/transect	3.3–4.6	8.8–14.4	Stallings (1989)
A. nancymae	Lowland	Perú/transect	13.5	46.3	Aquino and Encarnacion (1986a)
A. nancymae	Highland	Perú/transect	5.9	24.2	Aquino and Encarnacion (1988)
A. nigriceps	Hillside	Perú/transect		40	Wright (1978)
A. nigriceps	Lowland	Perú/transect		36–40	Wright (1985)
Gray-neck group					
A. lemurinus	Gallery	Colombia/transect		18–150	Heltne (1977)
A. lemurinus	Gallery	Colombia/transect	2.5–9.5	15–33	Green (1978)
A. vociferans	Lowland	Perú/transect	10	33	Aquino and Encarnacion (1988)
A. vociferans	Highland	Perú/transect	2.4	7.9	Aquino and Encarnacion (1988)
A. vociferans	Lowland	Perú/count	9–14	30–52	Aquino *et al.* (1992)

Using vocalization and direct group counts in plots of 1.9 and 1.0 km^2, respectively, they counted 30 individuals/km^2 in the Nanay Basin and 52/km^2 in the Santa María Basin. These differences are the direct result of major forest alterations produced by human settlements and hunting pressure in the Nanay Basin, which is located near Iquitos, Perú. These influences were absent in the Napo River Basin.

B. Population Densities in Other Countries

According to the distribution map published by Hershkovitz (1983), owl monkeys inhabit nine countries; however, population densities recorded are known for some species only in Argentina, Bolivia, Colombia, and Perú (Table I). The estimated population density varies according to the census method used and the forest type. Rathbun and Gache (1980) and Zunino *et al.* (1986), using the transect method in gallery and high tropophyllous forests in Argentina, estimated the population density of *A. azarae* at between 12.7 and 18 individuals/km^2. Garcia and Braza (1989), using the plotted method, found 242.5 individuals/km^2 for *A. azarae* in seasonally flooded forest in Bolivia. Stallings (1989) reported 8.8 to 14.4 individuals/km^2 for *A. azarae* in the gallery forests from Teniente Enciso and Defensores del Chaco National Parks in Paraguay, respectively. The differences in population densities of *A. azarae* are the result of hunting pressure, intensive deforestation that decreased food sources, and unusually cold weather in the high tropophyllous forests (Zunino *et al.*, 1986).

In Colombia, Heltne (1977), using the transect method, estimated 18 to 150 individuals/km^2 for *A. lemurinus* in the northernmost gallery forests. Green (1978) estimated 15 to 33 animals/km^2 between the Cauca and Magdalena river basins, also in Colombia.

IV. HARVESTING WILD POPULATIONS

Wild populations of *Aotus* are occasionally harvested to provide controlled exports of nonhuman primates for use in biomedical research, captive breeding programs, and zoological exhibits.

A. Selection of Field Personnel

Trapping owl monkeys, unlike the trapping of diurnal primate species, requires strenuous physical exertion. During the day, owl monkeys occupy a diverse array of sleeping sites that are difficult to locate in daylight. The sleeping sites consist of hollows, holes, lianas, bamboo tangles, or tufts of epiphytes, hemiepiphytes, and hemiparasitic plants (Aquino and Encarnación, 1986b). Capturing owl monkeys requires locating the animals at night while they are most active, then following them to their sleeping site. As a consequence, personnel responsible for trapping owl monkeys must have specialized forest skills. Among these skills are (a) the ability to orient themselves at night, (b) optimal auditory and visual acumen, (c) being adept in walking and canoeing at night, (d) being skillful at swimming and climbing trees, and (e) excellent balance and skill in using tree-cutting tools in precarious perches on the ground or in a canoe. These skills are important enough to influence the outcome of a trapping expedition. Personnel selected for trapping must already possess these specialized forest skills, which can be augmented through training sessions.

B. Locating Sleeping Sites

Although sleeping sites may be difficult to find during the day, some can be sighted by trappers by walking trails and paths in lines parallel to the selected area, or by traversing rivers, streams, or lake shores adjacent to the forest. They use cutlasses or axes to knock on tree trunks that appear hollow or to push on branches or liana tangles and tufts of epiphytes or hemiepiphytes. The noise or the unexpected movement of the foliage causes the animals to appear in the openings to their holes and hollows ready to escape. This technique is effective in locating the sleeping sites of members of the red-necked phenotype such as *A. nancymae,* whose hollows are usually visible. However, it is not useful for members of the gray-necked phenotype such as *A. vociferans,* whose hollows are generally hidden among tufts of epiphytes and hemiepiphytes (Hershkovitz, 1983; Aquino and Encarnación, 1986b).

Finding their sleeping sites at night has proven to be the best method for trapping owl monkeys. Animals are lcoated by field personnel traveling quietly at approximately 1.5 km/hr. Vocalizations and other noise made by animals moving through the canopy and dropping fruit and seeds aid in locating them. The best time is from 3:30 to 5:45 am, when they are most active. Individual groups are followed until they are observed entering their sleeping site. One of the most important factors in locating animals is the ability of field personnel to track them at night over sometimes difficult terrain.

C. Usual Trapping Techniques

1. Physically Tiring Animals The technique of tiring the monkeys is limited to the low tree forest with open stratum consisting of *Erythrina* sp. and *Cecropia* sp. successional plant associations. The animals are slowly pursued until they become tired. Signs of tiring include loss of equilibrium, falling while attempting to leap over obstacles, and frequent stops to rest, during which they lie flaccidly in ventral recumbency. Their fatigue facilitates direct and rapid capture. Utilizing this technique may take four trappers up to 2 hr to capture a single animal.

2. Covering Holes of Sleeping Sites When tree hollows have two holes that are lower than 10 m above ground level, monkeys can be captured by covering the holes of sleeping sites. A strong, hard pole with a piece of cloth attached to one end is used as a stopper. Two such poles are needed. A trapper standing on the ground pushes the pole with the cloth into the lower hole to cover it. Another trapper immediately climbs up the tree and secures the sleeping site exit so the animals cannot escape. Catching the animals in this manner requires experience and dexterity to avoid being bitten or allowing an animal to escape.

3. Isolation of a Tree Sleeping Site The technique of isolating a sleeping site is most successful when 10 or more people are available to assist in the capture. Following location of a sleeping site, a clearing approximately 10 to 25 m in diameter is made around the sleeping site tree. One or two trappers then climb the tree and make noises or poke around until the monkeys exit the sleeping site by leaping over branches or running to the ground. Other trappers in the cleared area then catch the animals by hand, restraining the monkey as quickly as possible to avoid being bitten.

4. Use of Curare Darts impregnated with curare and shot from cerbatana (blow guns) are also used to capture owl monkeys. Darts coated with psychotropic substances that are shot from pistols or rifles do not provide effective results. The poisonous substance, locally named "ampi," is used as a sedative. It consists of a mixed extract of *Chondodendron tomentosum* (Menispermaceae) and *Psychotria* sp. (Rubiacae) with bactrachotoxin from the collared frog (Dendrobatidae). In the body, this substance effectively blocks neurotransmitters, resulting in muscular relaxation. A large dose may result in respiratory paralysis, leading to death by anoxia. Traps baited with fruit soaked in strong liqueurs have failed to produce good results.

D. Temporary Holding Station and Transportation to Quarantine Center

1. Transport from the Capture Site to the Temporary Holding Station Individual cloth sacks or cages have worked well for moving captured monkeys from the trapping site to a temporary holding station in the field. The Centro de Reproducción y Conservación in Iquitos uses boxes measuring 150 × 35 × 35 cm. Each box consists of a wood frame covered by 0.5-in. galvanized wire mesh that is used to divide the box into smaller compartments measuring 30 cm on each side. These boxes are strong and spacious, even though they may be awkward to transport into the forest when the capture site is some distance from the river shore. In some cases, cloth sacks are better for transporting animals long distances; they are also useful for placing the animals in cages at the holding site. To ensure that the monkey has sufficient air, a small piece of bamboo cane is placed inside the sack.

2. Feeding and Cleaning at the Temporary Holding Site The animals must be fed at least twice each day, typically between 6:00 and 7:00 am, and again between 6:00 and 7:00 pm. The diet utilized by Centro de Reproducción y Conservación de Primates has wheat as the basic constituent. In addition, the animals receive wild fruit collected in the forest. Water is provided *ad libitum* from bottles attached to the cages with wire or placed in bracket holders. Each compartment in the holding box is provided with a water bottle.

Cleaning the holding cages is important to prevent accumulation of uneaten food and to decrease floor moisture resulting from fermentation of uneaten food and excreta. Hair loss and abrasions may result from allowing the cages to remain damp or soiled. Proper sanitation measures should also aid in preventing infestations of coprophagous insects.

3. Health Concerns during Temporary Holding Owl monkeys adjust well to captivity in relatively small holding cages. Health problems encountered during the holding period, which can last up to 2 weeks following capture, are usually limited to slight injuries and bruises, which frequently do not require attention. However, direct exposure to sunlight for prolonged periods may result in ocular problems, such as conjunctivitis. Also, rapid changes in ambient temperature from hot to cold, including the cooling associated with rain, frequently results in pneumonia. For these reasons, animals must be protected from direct sun and sheltered from other weather or temperature extremes.

4. Transportation from the Temporary Holding Site to the Quarantine Facility Transportation to the quarantine facility must be accomplished as rapidly as possible. High-speed motor boats are recommended if holding sites are situated near a river. Animals are transported in cages containing sufficient food and water for the duration of the trip.

V. EFFECTS OF CROPPING

A. Data on Population Recovery

Between 1981 and 1987, in an area limited by the flooded forest of the Tahuayo River Basin, three croppings of *A. nancymae* were carried out at 36-month intervals. The croppings took place during the flood season in March and April, and all work was done from canoes. All activities related to cutting and clearing of vegetation around sleeping sites took place at a height of about 7 m above ground level. Circular clearings around the sleeping site did not exceed 15 m in radius. Experienced workmen and field guides with knowledge of the area were employed to facilitate the capture of groups of monkeys. Only groups found within 200 m from the river bank along 5 km of the Caño Huaisi, a small tributary of the Tahuayo River, were captured.

Estimates of population densities before and after each cropping are detailed in Table II. During the initial cropping in 1981, 59% of the population was removed. Twenty-nine individuals remained, most of which were organized into family groups. Five of those groups included reproductive pairs of adult animals. In 1984, 3 years after the first cropping, the population had increased by 169% of the remaining population and had reached 110% of the 1981 precropping level. The second cropping removed 84% of the population, leaving only 12 individuals. Although, in subsequent years through 1987, the population increased by 433% of the 1984 postcropping level, it had not yet reached the precropping level when the third cropping was carried out (Table II).

B. Mechanisms of Population Recovery

After the first cropping in 1981, both immigration and reproduction among remaining animals played important roles in population recovery. Among groups of owl monkeys, the relative contribution to population recovery of immigration versus natural reproduction is still uncertain. The presence of immature individuals in the five groups that remained following the initial cropping in 1981 provided evidence of natality among the remnant animals. Natality was also observed in the immigrant groups.

Those remnant groups containing four or five individuals were primarily

TABLE II Periodic Densities and Population Recovery of *Aotus nancymae* between 1981 and 1987 in Lowland Flooded Forests of Perúvian Amazonia

Period of cropping	Total population size	Individuals/km^2	Population affected	% Removed	% Increase
1981 before cropping	71	35.5	—	—	—
1981 after cropping	29	14.5	42	59.0	—
1984 before cropping	78	39.0	—	—	169.0
1984 after cropping	12	6.0	66	84.0	—
1987 before cropping	64	32.0	—	—	433.3

composed of subadults, juveniles, and infants belonging to the birthing seasons of 1982, 1983, and 1984, respectively (Table III). Births occurred in reproductive pairs that were most likely native to the area prior to cropping. The rest of the population increase appeared to result from solitary immigration of animals that were expelled from their original groups that lived adjacent to the study area. Immigration occurred immediately after cropping in 1981 and continued until 1984. The presence of juveniles and infants in groups with three or four individuals suggested that the majority of reproductive pairs began producing offspring after 1983, indicating that the contribution of immigrating animals to population recovery was greater after 1982.

During population recovery after the second cropping in 1984, immigration played a greater role in the recovery than births. Groups were composed of regrouped individuals that, with the exception of those groups listed as 3, 8, 16, and 19 in Table IV, were the product of active and intensive immigration occurring shortly after 1986. This conclusion is based on the observation that infants belonging to the 1987 birth season were present in each complete group that was trapped at that time. The high percentage of infants in these groups also suggests that until 1987 population recovery was slow. The dynamics of population recovery are the subject of current and planned field studies.

C. Forest Recovery

Owl monkeys were trapped in seasonal flooded forests called "tahuampa" and restinga," and also in permanent swamp forests. Cropping in some swamp forests, also referred to as "tahuampa," was carried out during the wet season, from March to May, because the area was difficult to traverse during the dry season. Clearings were made 6 to 8 m above ground level (Fig. 1). In 1984, in order to isolate 10 sleeping site trees, 636 plants were cut, including both trees and shrubs. Of the 636 plants, 92% survived the clearing procedure (Table V). The species that died were principally *Cecropia* sp. (11.5%), *Hura crepitans* (9.6%), and *Campsiandra laurifolia* (7.7%) (Table VI). Regrowth of the surviving plants was quite rapid, and by 1987 the forest had resumed its precropping appearance. After 3 years, in some cases the forest had recovered sufficiently to permit monkeys to use tree sleeping sites that they had used just prior to the first cropping (Fig. 2).

The restinga and tahuampa seasonally flooded forests can be accessed from the ground, permitting trapping to take place from July to November. Alterations to the forest were made at ground level. Three years after areas were cleared in 1984, the secondary forest, consisting mostly of budding trees and seedlings, had taken over the areas. These were markedly different from the natural vegetation in both morphology and height (Fig. 3). In this case, original sleeping site trees were no longer available for diurnal resting.

TABLE III Group Size and Composition of *Aotus nancymae* Population in 1984 in Lowland Flooded Forests of Perúvian Amazonia

	Adult		Subadult		Juvenile		Infant			
Group number	M	F	M	F	M	F	M	F	Trapped	Group size
1	1			1	1				3	4
2	1					1			2	3
3	1	2			1		1		5	5
4	2	1				1	1		5	5
5	1	1		1	1		1		5	5
6	1	1	1					1	4	4
7	1	1		1				1	4	4
8	1	2			1		1		5	5
9	1							1	2	3
10	1								1	2
11	1						1		2	3
12	1	1	1						3	5
13	1	1				1			3	3
14	1	1							2	2
15	1	1		1		1			4	4
16	1	1	1		1			1	5	5
17		1			1				2	3
18	1	1				1			3	3
19		1							1	3
20	1	1				1			3	3
21	1	1							2	4
Total 21	10	18	3	4	6	6	5	4	66	78

TABLE IV Group Size and Composition of *Aotus nancymae* Population in 1987 in Lowland Flooded Forest of Perúvian Amazonia

	Adult		Subadult		Juvenile		Infant			
Group number	M	F	M	F	M	F	M	F	Trapped	Group size
1		1						1	2	3
2		1					1		2	3
3	1	1		1				1	4	4
4	1	1						1	3	3
5	1				1			1	3	4
6	1				1				2	3
7	1	1					1		3	3
8	2	1				1		1	5	5
9	1							1	2	3
10	1	1						1	3	3
11	1	1						1	3	3
12					1		1		2	4
13	1	1							2	3
14	1							1	2	3
15	1	1						1	3	3
16	2	1						1	4	4
17	1	1							2	3
18	1	1					1		3	3
19			1					1	2	4
Total										
19	17	13		2			3	1	52	64

FIGURE 1 Clearing around the sleeping site to trap owl monkeys in floodable forest during high water season. (Photograph by R. Aquino.)

TABLE V Record of Trees and Shrubs Affected by Clearing in *Aotus nancymae* Sleeping Site Isolation in 1984, in Lowland Flooded Forest of Perúvian Amazonia

Sleeping site tree number	Trees and shrubs affected by cutting	Trunks with reiterations after 3 years	% Recuperation
1	77	71	92.0
2	46	44	95.6
3	46	40	87.0
4	43	42	97.6
5	70	61	87.0
6	99	91	92.0
7	78	76	97.4
8	86	74	86.0
9	50	48	96.0
10	41	37	90.0
Total	Mean ± 1S.D. = 63.6 ± 20.8 (636)	Mean ± 1S.D. = 58.4 ± 18.7 (584)	Mean: 92

TABLE VI Cut and Dead Trees after Trapping *Aotus nancymae* by Isolation in Lowland Flooded Forests of Perúvian Amazonia

Family	Species	Common name	Frequency	%
Bombacaceae	*Pachira aquatica*	Punga	2	3.8
Euphorbiaceae	*Hura crepitans*	Catahua	5	9.6
Flacourtiaceae	*Laetia* sp.	Timareo	2	3.8
Lauraceae	*Ocotea* sp.	Mohena	1	1.9
Lecythidaceae	*Couroupita amazonica*	Ayahuma	3	5.8
	Eschweilera sp.	Machimango	3	5.8
Leguminosae	*Campsiandra laurifolia*	Huacapurana	4	7.7
	Inga sp.	Shimbillo	3	5.8
	Macrolobium sp.	Pashaco	2	3.8
Moraceae	*Brosimun* sp.	Murure	2	3.8
	Cecropia sp.	Cetico	6	11.5
	Ficus sp.	Renaco	1	1.9
Rubiaceae	*Genipa americana*	Huito	2	3.8
Sapotaceae	*Pouteria* sp.	Caimitillo	3	5.8
Sterculiaceae	*Guazuma crinita*	Bolaina	2	3.8
Unknown			11	21.2
Total			52	100.0

FIGURE 2 Sleeping site and surrounding forest with evidence of new growth 3 years after clearing. (Photograph by R. Aquino.)

FIGURE 3 A sleeping site 3 years after clearing to trap owl monkeys during the dry season. Bud trees and seedlings are present in the surrounding vegetation. (Photograph by R. Aquino.)

VI. POPULATION STRUCTURE

A. Age Categories

Age categories were established by Aquino and Encarnación (1986a) for *A. nancymae* according to the following criteria: (a) length and wear of canine teeth, (b) presence or absence of permanent molars, (c) stage of overall dental wear, (d) development of the subcaudal and pectoral glands, and (e) genital pigmentation. The subcaudal gland, which first manifests itself during the postjuvenile stage of both sexes, is situated along the ventral region of the base of the tail. It is a triangular area with the apex directed caudally. This gland produces an odoriferous, greasy, gray secretion that sticks to the pelage. The pectoral gland, located on the ventral aspect of the thorax, is a concave, triangular excrescence with rounded corners, which points caudally. This gland occurs in both sexes but is well developed only in males.

Light-brown pigmentation of the external genitalia occurs in adults of both sexes. In males, pigmentation begins from the caudal part of the testicles and in females from the caudolateral part of the vulva. Pigmentation of the external genitalia is most intense in adults, although it is also observed in younger age groups.

Adults (A): Teeth are generally worn and stained with tartar. In some individuals, incisors are missing, and canines are blunt and flattened. The subcaudal gland is well developed in both sexes, with abundant secretion that sticks to the pelage. The external genitalia are intensely pigmented. The pectoral gland is prominent and, in males, secretes a greasy substance. The females' mammary glands may be distended, with protuberant nipples exuding milk in some cases.

Subadults (SA): Dentition is usually normal and complete. Canines are not worn and are sharp. The subcaudal gland is incompletely developed and secretes a substance that lightly adheres to the pelage. The external genitalia are only partially pigmented and, in males, the pectoral gland is prominently developed and secretes a greasy substance. The female mammary glands are less prominent, covered with hair, and have small nipples.

Juveniles (J): Dentition is incomplete in most cases, comprising only permanent or deciduous canines, and the third molar is absent. The external genitalia are only lightly pigmented. The subcaudal gland has not fully developed and secretes only a light-colored substance.

Infants 2 (I2): Dentition is incomplete and deciduous, with only the first molar present in some cases. External genitalia are generally unpigmented. The subcaudal gland is undeveloped and secretes only a light-colored substance.

Infants 1 (I1): Dentition is deciduous, with no molars. External genitalia are unpigmented, and the testicles are undescended in male animals. The subcaudal gland does not secrete any substance.

B. Groups Size and Composition

According to the literature and field experience, owl monkeys form family groups consisting of two to six individuals. Published data report groups of up to four individuals for *A. lemurinus* (Heltne, 1977; Green, 1978), two to five individuals for *A. nigriceps* (Wright, 1978), and one to four individuals for *A. azarae* (Rathbun and Gache, 1980; Zunino *et al.*, 1986).

Aquino and Encarnación (1990) determined mean groups sizes to be 3.9 and 3.3 individuals for *A. nancymae* and *A. vociferans*, respectively. These data are based on 142 groups of *A. nancymae* trapped in the Amazonas and Maranon river basins between 1981 and 1987, and 88 groups of *A. vociferans* captured in the Napo and Nanay river basins between 1983 and 1987. Family groups were composed of two to six individuals in *A. nancymae* and from two to five individuals in *A. vociferans* (Fig. 4). Groups of four (35.2%) or five (27.5%) were most common in *A. vociferans*, whereas groups of three (38%) or four (39%) were most common in *A. nancymae*.

Groups of *A. nancymae* usually consisted of a reproductive pair and as many as three, or in rare cases four, offspring ranging in age from subadult

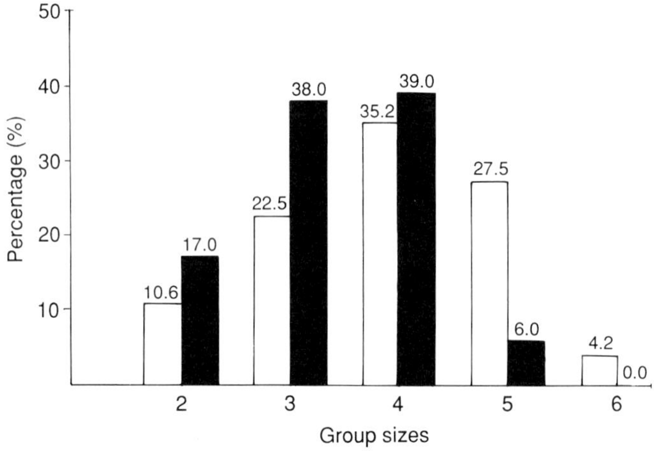

FIGURE 4 Group size of *A. vociferans* and *A. nancymae* populations in the northeastern lowland forest of Perúvian Amazonia. □, *A. nancymae*, n = 142; ■, *A. vociferans*, n = 82.

to infant. A few groups also included one or two young adults of opposite sexes but never of the same sex. Adult female offspring were nulliparous, and nonreproductive adult offspring were tolerated within family groups. Between March and June the groups ranged in size from four to five individuals, with 60% of these groups having ambulant and nonambulant infants. Between July and November the groups averaged two individuals, and during these same months solitary individuals were occasionally sighted (Aquino and Encarnación, 1990).

Groups of *A. vociferans* usually consisted of a male : female pair with two, or rarely three, offspring ranging in age from subadults to infants. Groups sometimes contained young-adult male offspring that appeared to be tolerated by the breeding male within the family group. Between January and March groups of *A. vociferans* were composed of two to three individuals and often contained juvenile animals. No observations were made between April and September. Groups of four individuals were common between October and December, with the majority of nonparental individuals being nonambulant infants. In a few groups, offspring ranged in age from subadults to juveniles but did not include ambulatory infants.

In *A. nancymae*, during the height of the birth season, adults and infants comprised 60 and 11% of the population, respectively. In *A. vociferans*, adults and infants comprised 63 and 9.4% of the population, respectively. The sex ratio in all ages of 39 groups of *A. nancymae* was 1 : 1.3 (M : F), and in 38 groups of *A. vociferans* the ratio was 1 : 1.1 (Table VII).

TABLE VII Age Composition of 39 Groups of *A. nancymae* Compared with 38 Groups of *A. Vociferans* Trapped in Lowland Forests of Peruvian Amazonia

	Age	Males	Females	Subtotals	%	Sex ratio
Aotus nancymae						
	Adult	41	43	84	60.0	1:1
	Subadult	11	12	23	17.0	1:1
	Juvenile	4	13	17	12.0	1:3
	Infant II	2	6	8	6.0	1:3
	Infant I	2	5	7	5.0	1:2.5
	Total	60	79	139	100.0	1:1.3
Aotus vociferans						
	Adult	42	38	80	63.0	1:0.9
	Subadult	7	12	19	15.0	1:1.7
	Juvenile	6	10	16	12.6	1:1.7
	Infant II	1	0	1	0.8	
	Infant I	4	7	11	8.6	1:1.7
	Total	60	67	127	100.0	1:1.1

C. Breeding Season and Litter Size

Nonambulant infants of *A. nancymae* were between 5 and 30 days old, whereas ambulant infants ranged between 1.5 and 3 months of age. Based on 75 groups of *A. nancymae* trapped during the years 1981 to 1987 between March and April, 47% had nonambulant infants weighing between 75 and 200 g, and 17% had ambulant infants weighing between 225 and 400 g. Thus *A. nancymae* births were concentrated between December and March in our study areas (Table VIII).

Aotus vociferans were not captured between March and June. Data gathered during the remaining 8 months showed that the greatest number of births occurred between November and January, suggesting a breeding period similar to that of *A. nancymae* (Table VIII).

In both owl monkey species only one infant is usually produced per gestation. We have a single record of twin *A. vociferans* captured in the Napo River Basin during January, 1987. The infants were removed from the male's back and found to weigh 125 and 150 g, respectively. The group included one lactating adult female, which was presumably the mother.

TABLE VIII Bimonthly Comparisons of Infant Composition in Groups of *A. nancymae* and *A. vociferans* Trapped and Observed from 1981 to 1987 in Lowland Forests of Peruvian Amazonia

	Months	Number of groups	Groups with 1 infant	%	Groups with 2 infants	%	Total
Aotus nancymae							
	Jan–Feb	—	—	—	—	—	—
	Mar–Apr	75	32	47	13	17	64
	May–Jun	33	1	3	21	64	67
	Jul–Aug	32	0	0	13	41	41
	Sep–Oct	78	0	0	0	0	0
	Nov–Dec	76	6	8	1	0	8
Aotus vociferans							
	Jan–Feb	39	14	36	4	10	46
	Mar–Apr	—	—	—	—	—	—
	May–Jun	—	—	—	—	—	—
	Jul–Aug	22	2	9	5	23	32
	Sep–Oct	7	2	29	1	14	43
	Nov–Dec	35	15	43	0	0	43

VII. SELECTION OF SLEEPING SITES

A. General Description

According to Aquino and Encarnación (1986b), tree holes are the most common sleeping sites of owl monkeys. Tree hole sleeping sites are natural refuges and were not created or modified by the animals in any manner. Nor were other sleeping sites, which lacked walls, modified by the animals. Tree holes were characterized by one to three orifices or access routes into the internal cavity. Holes with four or five orifices were rarely observed and then only in trees of *Calycophyllum spruceanum*, *Campsiandra laurifolia*, and *Eschweilera* sp. However, in the latter, only one or two of these routes were used by the monkeys as entries and exits; the others are used only if necessary. Access holes were generally circular, with diameters varying between 14 and 30 cm.

The internal cavity was generally cylindrical in form, having the same or somewhat larger diameter than the orifice. The depth of the cavity varied between 0.6 and 13.4 m. Some cavities contained roots of hemiepiphytes and hemiparasites that formed steps or ladders for the animals. The cavity was usually clean and more or less dry. Organic remains, if present, were found decomposing at the bottom of the cavity. These remains often served as substrates for fly larvae and other insects. The walls of the internal cavity and orifice were often worn and polished as a result of the constant trafficking of animals in and out of the cavity. The walls were also impregnated with odoriferous substances secreted by the subcaudal and pectoral glands of the animals (Aquino and Encarnación, 1986b).

The sleeping sites were grouped into five main types with 13 variant forms (Fig. 5). The following types coincide with those described by other authors: tree holes (Thorington *et al.*, 1976; Rathbun and Gache, 1980; Wright, 1981; Hershkovitz, 1983); dense tangles of climbers and vines (Thorington *et al.*, 1976; Rathbun and Gache, 1980; Wright, 1978, 1981; Hershkovitz, 1983); bamboo and dense thickets (Wright, 1981); and concavities between the branches of trees, many of which were protected by dense entanglements of climbers, vines, root mats, and the foliage of hemiepiphytes (Aquino and Encarnación, 1986b). A fifth type, described in this chapter, was in holes of termite nests, many of which were located in tree trunks and branches.

Aotus nancymae were found in all types and variants of sleeping sites, whereas *A. vociferans* were found only in tree holes, which was consistent with the observations of Hershkovitz (1983) that the "grey neck" owl monkey of northern Colombia was only found in tree holes. This peculiarity of *A. vociferans* suggests a certain selectivity of habitat and also might explain the absence or low population densities of the species in highland forests, where hardwood tree species predominate, and trees with suitable nesting holes are rare (Aquino and Encarnación, 1988).

B. Principal Types of Sleeping Sites

1. Type A—occurs in holes of tree trunks. Access is by the thin trunks and branches of neighboring trees, climbers, or vines. It affords optimal conditions for shelter and security from predators.

a. Form Aa: found in the apical position in the dry trunk of a palm tree. Entrance is gained by the trunk or a proximal frond. The superposition of branches and leaves of adjacent trees prevents the direct entry of water into the hole (Fig. 5Aa).

b. Form Ab: located in the apical position in the trunk of dry or senescent trees that have shoots or "reiterations" (Halle *et al.*, 1978). The entry of rainwater into the hole is prevented by the superposition of branches of taller neighboring trees (Fig. 5Ab).

c. Form Ac: the subaxial position on senescent or dry trunks where the entrance corresponds to the knot of a fallen branch (Fig. 5Ac).

d. Form Ad: the subapical position with one or more entrances in dry or senescent trunks. Woody epiphytes whose roots reach into the cavity and form steps or shelves are used for resting or daytime sleep (Fig. 5Ad).

e. Form Ae: the lateral position over thick, dry, or senescent branches of trees of *Macrolobium*, *Eschweilera*, or *Calycophyllum spruceanum* (Fig. 5Ae).

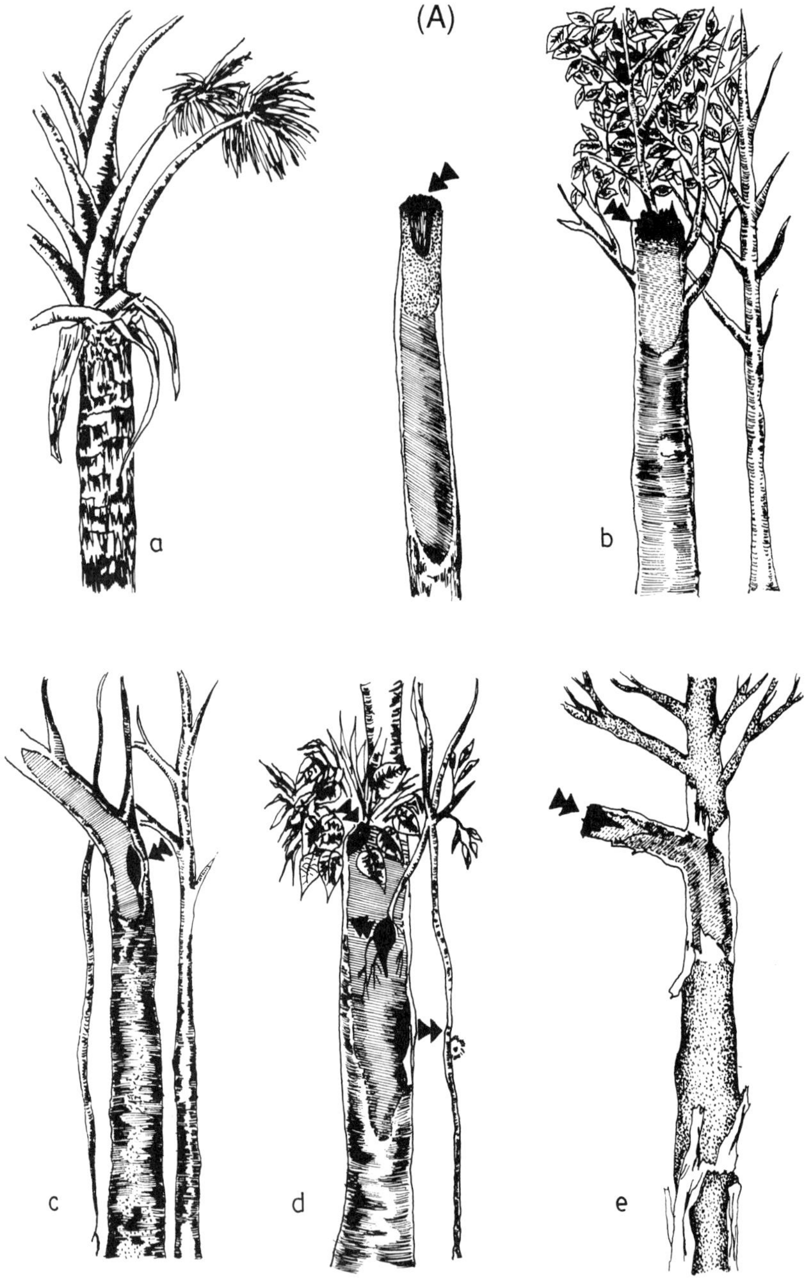

FIGURE 5 (A) Types and forms of sleeping site holes. Aa: apical position in palm trunk; Ab: apical position in dicotylendonous tree trunk; Ac: subaxial position; Ad: subapical position; Ae: lateral position. (B) Complex sleeping sites in concavities with climbers and vines. Ba: axial concavity; Bb: subaxial concavity. (C) Complex sleeping sites with climbers and epiphytes. Ca: subaxial position; Cb: subapical position.

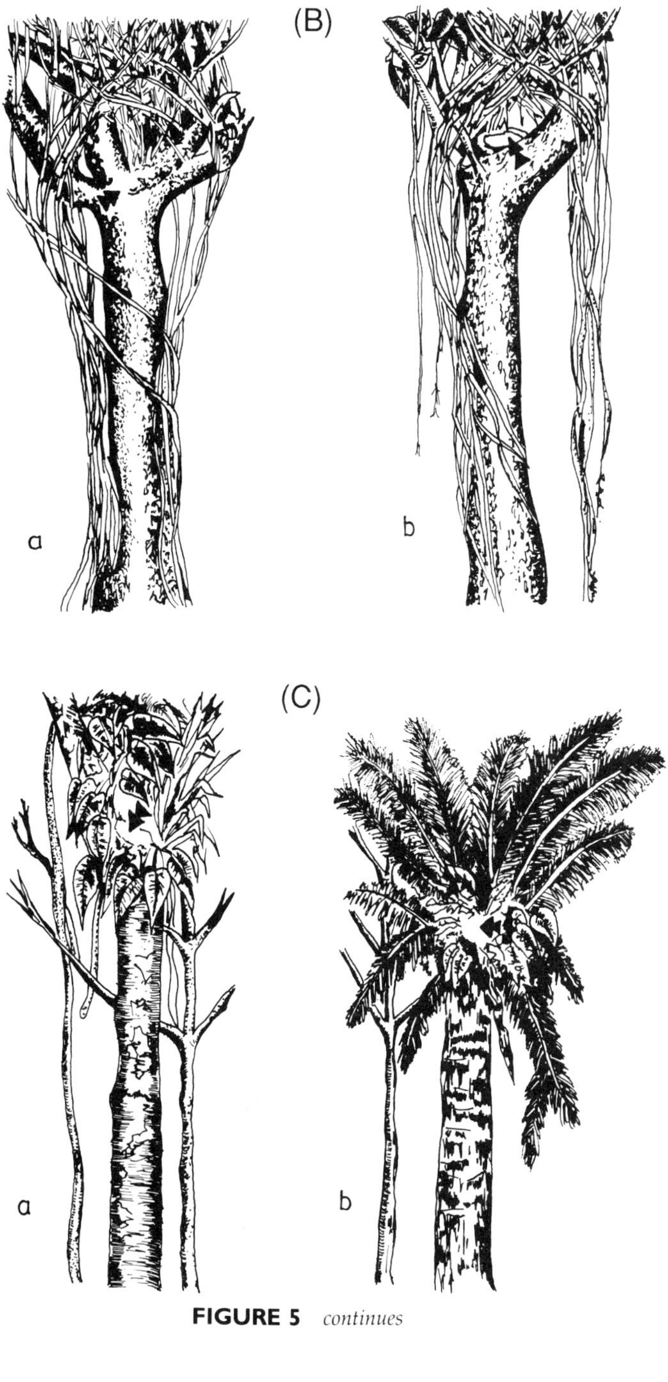

FIGURE 5 *continues*

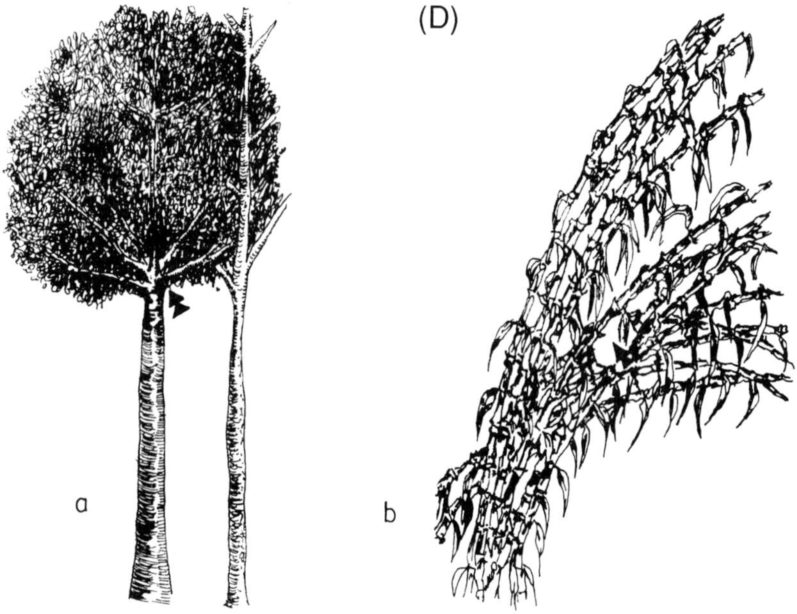

FIGURE 5 (continued) (D) Simple sleeping sites in thickets and dense foliage. Da: axial position; Db: irregular position. (E) Holes into termite nests: Ea: termite nest in subaxial position; Eb: termite nest held on or between branches.

2. Type B—these complex sleeping sites consist of concavities formed by branches supporting entanglements of climbers and vines that serve as access routes and provide protection and shelter. Branching patterns are principally of the Koriba, Prevost, or Nozeran models (Halle *et al.*, 1978).

a. Form Ba: an axial concavity formed by more than four branches (Fig. 5Ba).

b. Form Bb: a subaxial concavity formed by the loss or fall of a branch (Fig. 5Bb).

3. Type C—complex sleeping site composed principally of epiphytes, hemiepiphytes, and hemiparasites, which occur almost always in parabiosis with ants. Access is by vines or the branches of neighboring trees.

a. Form Ca: the subaxial position occurring over a slight concavity protected by diverse epiphytes and hemiepiphytes. The concavity is produced by the fall of a branch or a branch in the Nozeran model (Halle *et al.*, 1978) (Fig. 5Ca).

b. Form Cb: the subapical position without concavity. The sleeping site is at the base of the diverse epiphytes and hemiepiphytes that grow on a palm tree of the intermediate model of the Holttum and the Coner model (Halle *et al.*, 1978) (Fig. 5Cb).

4. Type D—simple sleeping sites positioned among thickets and dense foliage; they have defined access sites.

(E)

a. Form Da: the axial position with protection provided by the foliage or branches of trees principally in the models of Aubreville, Mangenot, or Massart (Halle *et al.*, 1978) (Fig. 5Da).

b. Form Db: an irregular position with protection from foliage and thickets of dense shrubs or trees with poliaxial branching as occurs in bamboos of the McClure model or *Myristicaceae* of the Aubreville model (Halle *et al.*, 1978) (Fig. 5Db).

5. Type E—located in termite nests. Access is usually by the branches of neighboring trees, climbers, or vines. This type of sleeping site likewise affords optimal conditions for shelter and relative security from predators.

a. Form Ea: the subaxial position of the termite nest, usually between the trunk or lower part of a branch. The orifice is in the lower part of the nest to prevent the direct entry of rainwater into the hole (Fig. 5Ea).

b. Form Eb: corresponds to an irregular position. The termite nest is located on a branch or at the conjunction of two branches of different trees. The orifice is always in the lower part of the termite nest (Fig. 5Eb).

Form Ac, characterized by the presence of epiphytes and woody hemiepiphytes (*Ficus, Coussapoa, Clusia, Souroubea,* and other species) with root systems in the holes, can be confused with form Bb when they are covered by herbaceous or semiwoody climbers and some succulent or herbaceous epiphytes.

In dry or senescent trees, the orientation of the cavity is generally descending with respect to the entrance or opening. However, in forms Aa and Ad the cavity can occur with an ascending orientation relative to the entrance. In forms Ba and Bb, protection from rainwater is complemented by the foliage, the decomposition and detachment of cork, and the decomposition of arthropod and other remains, which sometimes serve as the substrates for certain epiphytes.

In form Ca, the foliar scars and spathes of palms such as *Scheelea cephalotes, S. bassleriana,* and others called "shapaja" constitute the base of the sleeping site. The site's form is often complemented with *Philodendron, Ficus, Polypodium,* and *Nephrolepis* species among the epiphytes.

Form Da sleeping sites frequently occur among small trees of *Rheedia, Ocotea, Iryanthera, Cordia, Sapium,* and other species. Form Db is found in thickets of *Guadua superba* in Loreto, and over *Guadua* sp.("paca") between the Manuripe and Acre rivers in Madre de Dios. Forms Ea and Eb are frequently found only in vegetational communities of *Erythrina* sp. growing in permanent swamp areas.

C. Trees Containing Holes

Nest holes used by *A. nancymae* were found in 134 old and senescent trees, consisting of 23 species representing 14 plant families. Among these the authors found 21.6% of the holes in trunks and branches of *Eschweilera* sp. 14% in *Macrolobium acaciaefolium*, and 13.4% in *Calycophyllum spruceanum* (Table IX). Nest holes used by *A. vociferans* were found in 71 trees, consisting of 18 species representing 10 plant families. Among these, 18.3% of the holes were found in *Eschweilera* and *Diplotropis* sp. 15.5% in *Calycophyllum spruceanum*, and 11.2% in *Buchenavia* sp. (Table IX).

D. Distribution of Sleeping Sites in Relation to Forest Strata

The ages of individual trees and their densities and crown shapes were factors determining the locations of sleeping sites. The following forest strata were identified by Aquino and Encarnación (1986b): (a) understory (7.1–10 m above ground level), (b) lower story (10.1–19 m), (c) middle

TABLE IX Trees with *Aotus* Nest Holes Observed in the Northeastern Lowland of Peruvian Amazonia

		Aotus nancymae		*Aotus vociferans*	
Family	Species	Frequency	%	Frequency	%
Arecaceae	*Astrocaryum* sp.	3	2.2	1	1.4
	Guilielma sp.	0	0.0	2	2.8
	Iriartea sp.	3	2.2	0	0.0
	Mauritia flexuosa	6	4.4	1	1.4
Bombacaceae	*Bombax* sp.	1	0.7	0	0.0
Combretaceae	*Buchenavia* sp.	5	3.7	8	11.2
Chrysobalanaceae	*Parinari* sp.	0	0.0	3	4.2
Sterculiaceae	*Guazuma* sp.	1	0.7	0	0.0
Euphorbiaceae	*Hura crepitans*	2	1.5	1	1.4
	Mabea speciosa	0	0.0	1	1.4
Lecythidaceae	*Eschweilera* sp.	29	21.6	13	18.3
	Couroupita sp.	2	1.5	0	0.0
Leguminosae	*Campsiandra* sp.	7	5.2	2	2.8
	Clitoria sp.	4	3.0	1	1.4
	Diplotropis sp.	0	0.0	13	18.3
	Erythrina sp.	3	2.2	1	1.4
	Macrolobium sp.	19	14.0	6	8.4
Meliaceae	*Guarea* sp.	2	1.5	0	0.0
Moraceae	*Brosimum* sp.	1	0.7	0	0.0
	Coussapoa sp.	14	10.0	0	0.0
	Maquira biflora	3	2.2	0	0.0
	Ficus sp.	6	4.4	2	2.8
Olacaceae	*Minquartia* sp.	2	1.5	0	0.0
Polygonaceae	*Coccoloba* sp.	2	0.7	0	0.0
Rubiaceae	*Calycophyllum* sp.	18	13.4	11	15.5
	Coutarea sp.	0	0.0	1	1.4
Sapotaceae	*Manilkara* sp.	1	1.7	1	1.4
Tiliaceae	*Sloanea* sp.	1	0.7	0	0.0
Total		134	100.0	71	100.00

story (19.1–25 m), (d) upper story (25.1–31 m), and (e) emergent trees (31.1–37 m).

In the flooded forest of the Tahuayo River, 21.4% of the sleeping sites of *A. nancymae* were in the shrub stratum of the understory, 64.3% were in the lower story, and 14.3% were in the middle story. No sleeping sites were found in either the upper story or emergent trees. This contrasts with the forest of the Nanay and Napo rivers, where 11.4% of the sleeping sites of *A. vociferans* were found in the understory, 54.3% in the lower story, 20% in the middle story, 8.6% in the upper story, and 5.7% in emergent trees (Fig. 6).

Each forest stratum had different forms of sleeping sites. The Aa, Da, Db, and Ea forms were found in the understory, the Ab, Ac, Ba, Bb, and Eb forms in the lower story, and the Ad, Ae, Ca, and Cb forms in the higher strata. The variation in position of sleeping sites in relation to height within the forest canopy is summarized in Fig. 6. *Aotus nancymae* appeared to prefer sites lower than 19 m, whereas *A. vociferans* often also occupied the middle and upper stories including emergent trees. Forest physiognomy, degree of disturbance, topography, and water drainage are factors that may

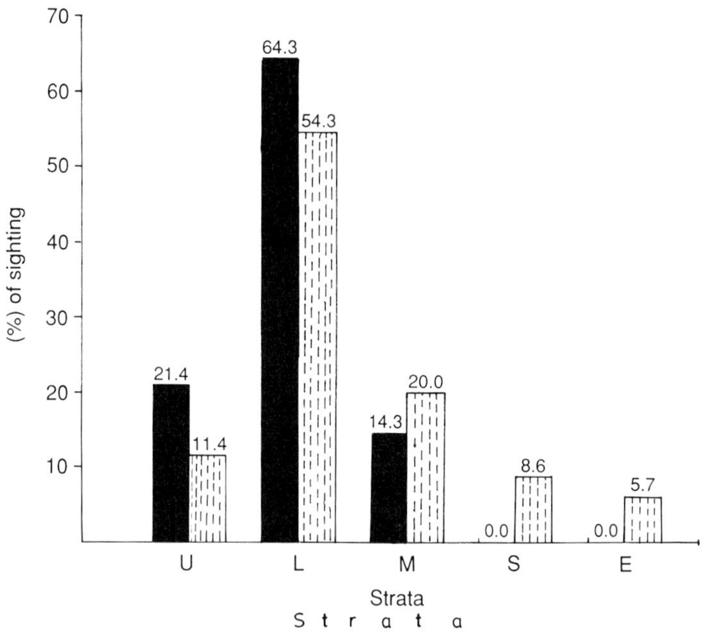

FIGURE 6 Frequency and distribution of owl monkey sleeping sites by forest strata in the northeastern lowland forest of Peruvian Amazonia.

influence the selection of sleeping sites by owl monkeys. These observed differences in habitat preference by different species of *Aotus* in the natural environment require further study.

E. Factors Influencing Selection

Principal factors affecting the selection of sleeping sites include (a) protection and refuge from predators, (b) accessibility, (c) shelter from the weather, and (d) sufficient space.

1. Protection and Refuge from Predators The intrinsic security of sleeping sites is assured by the position, location, and provision for escape if threatened by a predator. This requires isolation of the lodging tree, optimum visibility in all directions, and easy access and exist routes that permit escape when necessary. Isolation is achieved by surrounding bodies of water, such as lakes, rivers, streams, or swamps, which serve as barriers, for example, in the Bombonaje area on the Amazonas River in a *Maurita* palm swamp area or "aguajal," the sleeping site of a family group with six members. Included in the group was an infant that was found in the dry trunk of an isolated *Mauritia flexuosa,* surrounded by open swamp and a few scattered trees. The only access route to the lodging tree was via a large leaf of a *Euterpe precatoria* palm tree.

2. Accessibility Lodging trees are generally mature senescent, or dead decaying, trees. When sleeping sites are located on isolated trunks, the diameter usually does not exceed 30 cm, thereby facilitating vertical movement by short hops. When the trunk diameter is greater than 30 cm, access is facilitated by climbers and vines hanging from the tree or by other trees in close proximity that have branches leading directly or indirectly to the sleeping site.

3. Shelter from the Weather Shelter is afforded by leaves and branches that provide nearly total protection from direct rainfall. Sleeping sites in holes of senescent or dry tree trunks offer optimal conditions by virtue of the subapical, subaxial, or lateral position of the openings (see Fig. 5). In sleeping sites positioned at the bifurcation of principal branches, among vines, climbers, and epiphytes, or in thickets, protection from falling water is afforded by the accumulation of decaying organic matter. This material fills the interstices and forms an impermeable cover at contact points or nodes in the vegetation. Protection may also be provided by laminar superposition of foliage, which directs runoff laterally. This protection from rainwater has the additional benefit of providing protection from wind, direct sunlight, and temperature extremes.

4. Sufficient Space to Accommodate the Family Group When sleeping sites are located in holes, the entrance is of variable dimensions, ranging in width from 14 to 35 cm, which permits the simultaneous entry of two or more individuals. The internal cavity can accommodate up to five individuals, or in exceptional circumstances six. If the cavity is spacious, other arboreal and nocturnal mammals may use it simultaneously with the monkeys.

VIII. HABITAT INTERACTIONS

A. Nocturnal Competition for Food

1. Intergroup Competition Some competitive interactions between groups of owl monkeys were observed by Moynihan (1964) in Panama and by Rathbun and Gache (1980) in Argentina. These occurred most frequently when home ranges overlapped and fruits were scarce. Since 1981, we have observed many cases of intergroup competition in *A. nancymae* and *A. vociferans;* however, in no case did we observe physical aggression. Feeding competition was observed in the Nanay River Basin between two groups of *A. vociferans* while they were separately eating fruit from a tree for a period of approximately 15 minutes. Suddenly, one adult from each group, possibly males, began to emit aggressive vocalizations in a clear oral dispute, which lasted 4 minutes. At the end of that time, one group left the area; the other remained for an additional 15 minutes before departing.

2. Competition with Other Nocturnal Mammals Other nocturnal mammals have behavioral patterns similar to *Aotus* and occupy the same forest strata. They use the same sleeping sites and eat the same foodstuffs. Competitors include *Potos flavus, Bassaricyon gabbi, Caluromys bistriatus, Didelphys* sp. *Philander opposum,* and some *Chiroptera* species. *Potos flavus, Bassaricyon gabbi,* and the *Chiroptera* are the main frugivorous competitors of owl monkeys.

B. Sleeping Site Competition

We have observed many cases of competition with other mammals that have similar nesting habits. The majority were among *A. vociferans* and *Potos flavus* in various circumstances and localities; occasionally competition occurred among *A. nancymae* and *Coendu bicolor*. In all cases, the owl monkey groups were prevented from entering the sleeping site. Competition occurred primarily when sleeping sites were small and shallow, with only one entrance.

C. Sleeping Site Cohabitation

Cohabitation is defined as the sharing, by two or more species, of a sleeping site hole with two or more access orifices and deep internal cavities. In addition to six cases of cohabitation previously reported by Aquino and Encarnación (1986b), we have observed nine other cases in various circumstances and localities. Of the 15 cases of cohabitation encountered to date, three were between owl monkeys and *Potos flavus*, two with *Bassaricyon gabbi*, three with *Coendu bicolor*, one with *Tamandua tridactyla*, one with *Isothrix bistriatus*, and the remainder with chiropterans. Owl monkeys always occupied that part of the cavity nearest the entrance, whereas the interior part was occupied by the other species. In cases where the coinhabitants were chiropterans, the interior of the cavity had shelves, or gradients, where the chiropterans could rest while suspended or hanging. With the exception of the chiropterans, all other coinhabitants were observed to arrive at the sleeping site at least 15 to 20 minutes before, and to leave between 20 and 25 minutes after, the owl monkeys. Cohabitation may also occur where root mats of hemiepiphytes and hemiparasites form lattices used as substrata by the occupants.

D. Sleeping Tree Co-occupation

Co-occupation is defined as the use of one or more sleeping sites that may be located in the same tree but are independent of one another. The joint utilization of a sleeping tree between owl monkeys and other nocturnal animals has not been previously described in the literature. On two occasions, we have observed *A. vociferans* sharing a sleeping tree with *Coendu bicolor* and *Caluromys lanatus*. The first instance was in a *Macrolobium acaciaefolium* tree, with the owl monkeys occupying a sleeping site hole with two access orifices that were 15 and 23 m, respectively, above ground level, and the *C. bicolor* occupying a sleeping site situated in a concavity on the same branch. The second occurrence was in a *Clitoria arborea* tree. The owl monkeys occupied a subaxial-position hole located 14 m from the ground. In the same tree, the sleeping site of *C. lanatus* was composed of an entanglement of climbers with dry, woody branches, located 8 m above the owl monkey sleeping site.

IX. CONSERVATION

A. Threatening Factors

As with all arboreal fauna, owl monkey populations rely on the integrity of the forest they inhabit. The need for agricultural land has prompted intensive clearing and deforestation, particularly in the extra-Amazonian regions of southern Panama, northern Colombia, and northern Argentina. Populations of *A. lemurinus, A. brumbacki,* and *A. azarae* are seriously threatened following drastic habitat reduction in those regions. Other factors that adversely affect owl monkey populations include subsistence hunting and trapping for use in biomedical research.

1. Deforestation Accelerated and intensive clearing of large areas, prompted by agriculture and cattle grazing, has resulted in the removal of sleeping site trees and trees that directly produce foodstuffs, such as flowers, fruits, and nectars, or act as indirect food sources by serving as substrates for insects, birds, and small mammals.

Selective deforestation for wood products also decreases the amount of usable habitat. These traditional practices must be discouraged in forests where owl monkeys and other arboreal mammals live. Previous inventories describing the diversity of the flora and fauna that inhabit forested areas will permit the use of appropriate resource management practices and assist in promoting alternative uses of existing forests (Vasquez and Gentry, 1989).

2. Hunting In Latin American countries, especially in forested regions such as the Amazonia, technical assistance and support for increasing the productivity of land already under cultivation or being using for grazing do not exist. The lack of support for agricultural activities is a major contributor to the chronic economic instability experienced by these countries. Subsistence hunting is regulated by the government, which designates which species can be harvested. However, because of the economic crises being experienced in many countries, commercial hunting for meat, skins, skulls, teeth, and other ornaments also occurs. Thus, primate species, despite their small muscle mass, are taken in large numbers by subsistence hunters as substitutes for the larger animals taken by commercial hunters for their higher market value. In the human settlements of the Perúvian Amazonia with populations greater than 1000 inhabitants, virtually all mammals, birds, and other game animals historically utilized for meat have been depleted by hunting.

3. Trade Between 1968 and 1972 the United States imported 20,869 owl monkeys, primarily from Colombia and Bolivia. These included *A. lemurinus, A. vociferans, A. nancymae,* and *A. azarae* (Mack and Eudey, 1984). The number exported between 1976 and 1980 was drastically reduced to 3300 individuals, owing to government bans on exportation of nonhuman primates. From 1981 to 1990, Perú, under an agreement with the Pan American Health Organization, exported 1788 owl monkeys. The total number of 24,169 animals, cited by Mack and Eudey (1984), represented only one-third of the total captured during that time period according to Castro

(1976). The other two-thirds died prior to exportation, so that the total number of animals removed from the wild was actually closer to 75,000. Improvements in trapping and transportation techniques by the staff of the Centro de Reproducción y Conservación de Primates de Iquitos have substantially reduced capture-associated mortality, which partially facilitates recovery of feral owl monkey populations.

B. Protected Areas

Conserving biodiversity should be the top priority in neotropical regions. According to Myers (1987), the owl monkey species that inhabit Panama, Colombia, and the southern and western portions of Brazilian Amazonia are likely to be threatened or endangered by the year 2000, the causes including increased cattle grazing, land cultivation, and timber harvesting. Habitat alteration has also been accelerated by road construction in the forests of Brazil, Bolivia, and Perú. In the Pacific forest of Colombia and in the Ecuadorian Amazonia, habitat alteration has been hastened by the increasing numbers of banana and coffee plantations.

Aotus lemurinus, with two taxa (Hershkovitz, 1983; Hernandez-Camacho and Defler, 1989), inhabits part of Central America and northern South America from sea level to 3000 m above sea level in primary, secondary, and remnant forests among coffee plantations. The *A. l. lemurinus* inhabits the Pacific coast of southern Panama and most of Colombia (Hernández-Camacho and Defler, 1989; Eisenberg, 1989). It is protected in Los Katios, Purace, Las Orquideas, and Cueva de los Guacharos National Parks in Colombia (IUCN, 1982; Hernández-Camacho and Defler, 1989) and, in Panamá, in the Darien National Park (IUCN, 1982). The *A. l. griseimembra* habitat extends from the central and northern portions of Colombia to the Andean part of Venezuela (Ferreira, 1989-1990 and the Maracaibo Basin (Eisenberg, 1989). It is protected in the Tayrona, Paramillo, and Sierra Nevada de Santa Marta National Parks in Colombia (IUCN, 1982; Hernández-Camacho and Defler, 1989), and in El Tama National Park (IUCN, 1982; Bodini, 1989; Bodini and Pérez-Hernández, 1989) and possibly Sierra del Perija National Park (IUCN, 1982) in Venezuela. In Colombia, from 1960 to 1970, the spread of coffee plantations and capture for trade purposes seriously depleted the natural populations of the two taxa (Mack and Eudey, 1984).

Aotus brumbacki, which is also endemic to Colombia, inhabits humid and subhumid forests as far north as midnortheastern Venezuela (Hernandez-Camacho and Defler, 1989; Eisenberg, 1989). This species is seriously threatened by progressive clearing on the Andean foothills and the slopes of the La Meta River Basin. It is possibly protected in Chingaza National Park (Hernández-Camacho and Defler, 1989).

Aotus trivirgatus inhabits the right bank of the Negro and Amazonas river basins from Brazil to Venezuelan Amazonia and the Orinoco River (Hershkovitz, 1983; Bodini and Pérez-Hernández, 1986). Its habitat has been critically depleted, particularly in the lower and occidental parts of the Orinoco River Basin (Bodini and Pérez-Hernández, 1986). The species is possibly protected in Yacapana National Park (Cerda, 1987-1988) in Venezuela, and is protected in Pico de Neblina National Park in Brazil.

Aotus vociferans lives from the Sierra Nevada de La Macarena and Tomo River in Colombia to the eastern Andean slopes, from the Santiago River in the territories of Ecuador and Perú to the Amazonian lowland, and following the left bank of the Amazonas River to west of the Negro River in Brazil (Hershkovitz, 1983; Hernandez-Camacho and Defler, 1989; Aquino and Encarnación, 1990). Habitat loss has been accelerated in Colombia and Ecuador by human colonization and cultivation (F. Encarnación, personal observations). The species is protected in the Amacayacu and Tuparro National Parks and the La Macarena Natural Reserve in Colombia (Hernandez-Camacho and Delfer, 1989); in the Cuyabeno Faunal Reserve, Yasuni National Park, and possibly the Cayambe-Coca Ecological Reserve in Ecuador; and in Brazil probably in the Jau National Park. In Perú it is not yet protected (Aquino and Encarnacioń, 1994).

Aotus nancymae described by Hershkovitz (1983) inhabits the central part of the Peruvian lowland Amazonia, from upper and middle courses of Marañón, Huallaga and Ucayali rivers, following south of the Amazonas river, the Yavari river to Jaudiatuba river to the South of Amazonas (or Solimoes) river in Brazil. The habitat is not seriously affected in Perú and Brazil. In Perú it is protected in Pacaya Samiria national reserve and since 1991 in Tamshiyacu Tahuayo communal reserve (Aquino and Encarnación, 1994).

Aotus miconax, which is endemic to Perú, is known only through museum skin and skull collections (Aquino and Encarnación, 1994). It was found in the Chinchao cloud forest, located in the central part of the region of Chinchao, Huanuco, Perú. Habitat destruction began there in the eighteenth century with human colonization of the Amazonian forest. *Aotus miconax* could be protected in the Río Abiseo National Park in Perú.

Aotus nigriceps lives in the Amazonian and Madeiran forest in Perú, Brazil, and Bolivia (Hershkovitz, 1983; Brown and Rumiz, 1986). It also inhabits forest between the Huallaga, Ucayali, Yavari, Purus, and Madre de Dios river basins, and areas south of the Amazonas River to the Madeira River Basin and the northwestern part of the Tapajoz River Basin. The habitat of *A. nigriceps* is seriously affected in the western and northwestern portions of its geographical range. In Perú, this range corresponds to the upper Huallaga and Ucayali rivers, where human encroachment began in 1990 and involves timber extraction, cattle grazing, and crop cultivation. The most serious threat in Brazil is the Trans-Amazonian Highway, which has facilitated cattle ranching and hydroelectric development in Brazilian

Amazonia. The species is protected in the Yanachaga Chemillen Reserve and Manu National Park in Perú (Aquino and Encarnación, 1994); in the Manuripi-Heath Biological Reserve in Bolivia (Brown and Rumiz, 1986; T. Tarifa, personal communication); and possibly in the Amazonia National Park, Jaru Federal Biological Reserve, Ique Biological Station, and Río Acre Ecological Station in Brazil (IUCN, 1982).

Aotus infulatus is reported endemic to Brazil. Its habitat, east of the Tapajoz, Xingu, and Tocantins river basins (Hershkovitz, 1983), is threatended by clear-cutting. However, it appears that populations are not currently in danger. We have no information regarding protective measures for the species.

Aotus azarae has two taxa that inhabit the southernmost regions of the neotropics. Until 1982, this was the species most frequently trapped in Bolivia for exportation (Brown and Rumiz, 1986). In Bolivia, Argentina, and Paraguay the worst threat to the species has been trapping for the pet trade and the encroachment of human settlements (Brown and Rumiz, 1986; Stallings, 1989).

Aotus a. boliviensis lives in the southeastern Amazonian region of Perú and the Madeiran and Yungas regions of Bolivia. It is possibly protected in the Tambopata-Candamo Reserve and the Pampas del Health Natural Sanctuary in Perú (Aquino and Encarnación, 1994), and in the Pilon-Lajas Biosphere Reserve, Isiboro-Secure National Park, and Amboro Natural Reserve in Bolivia (Brown and Rumiz, 1986; T. Tarifa, personal communication).

Aotus a. azarae is found in the Gran Chaco forest between Bolivia, Argentina, and Paraguay. Until 1985, more than 93% of its habitat in Argentina showed serious deterioration due to cattle ranching (Zunino *et al.*, 1986). It is now protected in the Pilcomayo National Park in Argentina, where Zunino *et al.* (1986) estimated an ecological range of 39,000 km^2, of which 15% is currently under the protection of the government. In Paraguay, it inhabits the northern Chaco, where it appears to be restricted to small patches of forest (Stallings, 1989). There the species is protected in the 11,000 km^2 between Teniente Enciso, Defensores del Chaco, and Tinfunque National Parks. In Bolivia, *A. a. azarae* is found in the Chaco plain of Santa Cruz and, according to Brown and Rumiz (1986), in Tarija. It is protected in the 17,000 ha of the Chiquitos National Park of Santa Cruz La Vieja (T. Tarifa, personal communication).

REFERENCES

Aquino, R., and Encarnación, F. (1986a). Population structure of *Aotus nancymae* (Cebidae, Primates) in Peruvian Amazon lowland forest. *Am. J. Primatol.* **11,** 1–7.

Aquino, R., and Encarnación, F. (1986b). Characteristics and use of sleeping sites in *Aotus* (Cebidae, Primates) in the Amazonian lowland of Perú. *Am. J. Primatol.* **11,** 319–331.

Aquino, R., and Encarnación, F. (1988). Population densities and geographic distribution of night monkeys (*Aotus nancymae* and *Aotus vociferans*) (Cebidae, Primates) in northeastern Perú. *Am. J. Primatol.* **14,** 375–381.

Aquino, R., and Encarnación, F. (1990). Supplemental notes on population parameters of northeastern Perúvian night monkeys, genus *Aotus* (Cebidae). *Am. J. Primatol.* **21,** 205–221.

Aquino, R., and Encarnación, F. (1994). Primates of Perú. *Primate Rep.* **38** (in press).

Aquino, R., Puertas, P., and Encarnación, F. (1992). Las poblaciones de *Aotus vociferans* y *Aotus nancymae* en la Amazonia Perúana. *Alma Mater, Rev. Invest. UNSM* **3,** 77–82.

Bodini, R. (1989). Distribución geográfica y conservación de primates sub-humanos en Venezuela. *In* "La Primatología en Latinoamerica, Anales del Simposio de Primatología del IX Congreso Latinoamericano de Zoología, Arequipa, Perú, 9–15 Octubre, 1983" (C. J. Saavedra *et al.*, eds.), pp. 101–113. UICN/CSE-WWF, Washington, DC.

Bodini, R., and Pérez-Hernández, R. (1986). Proposición de regiones biogeográficas para Venezuela en base a la distribución de los géneros. *In* "A Primatología no Brasil 2. Anais do 2° Congreso Brasileiro de Primatología, Campinas, SP. 27/01/85 a 01/02/85" (M. Thiago de Mello, ed.), pp. 323–333. Soc. Bras. Primatol./Univ. Fed. Minas Gerais.

Bodini, R., and Perez-Hernandez, R. (1989). Distribution of the species and subspecies of cebids in Venezuela. *Fieldiana, Zool.* **39,** 231–244.

Brown, A. D., and Rumiz, D. I. (1986). Distribución y conservación de los primates en Bolivia. Estado actual de su conocimiento. *In* "A Primatología no Brasil 2. Anais do 2° Congreso Brasileiro de Primatología, Campinas, SP. 27/01/85 a 01/02/85" (M. Thiago de Mello, ed.), pp. 335–363. Soc. Bras. Primatol./Univ. Fed. Minas Gerais.

Castro, N. (1976). Guidelines for the conservation of primates in Perú. *Sci. Publ.—Pan. Am. Health Organ.* **317,** 216–234.

Cerda, J. (1987–1988). Contribución al conocimiento de los primates del parque nacional Yacapana, territorio federal amazonas, Venezuela. *Bol. Inf. Soc. Venez. Primatol.* **1,** 6.

Durham, N. (1975). Some ecological distributional and group behavioral features of Atelinae in southern Perú; With comments on interspecific relations. *In* "Socioecology and Psychology of Primates" (R. H. Tuttle, ed.), pp. 87–101. Mouton, The Hague, The Netherlands.

Eisenberg, J. F. (1989). "Mammals of the Neotropics: The Northern Neotropics, Panama, Colombia, Venezuela, Guyana, Suriname, French Guiana," Vol. 1, pp. 240–243. Univ. of Chicago Press, Chicago.

Encarnación, F. (1985). Introducción a la flora y vegetación de la Amazonia Perúana: Estado actual de los estúdios, medio natural y ensayo de una clave de determinacion de las formaciones vegetales en la llanura Amazónica. *Candollea* **40,** 237–252.

Ferreira, C. (1989–1990). Taxonomía de los Cebidos en Venezuela. *Bol. Inf. Soc. Venez. Primatol.* **1,** 7–8.

Freese, C., Heltne, P. G., and Whitesides, G. (1982). Patterns and determinants of monkey densities in Perú and Bolivia, with notes on distributions. *Int. J. Primatol.* **3,** 53–90.

Garcia, J. E., and Braza, F. (1987). Activity rhythms and use of space of a group of *Aotus azarae* in Bolivia during the rainy season. *Primates* **28,** 337–342.

Garcia, J. E., and Braza, F. (1989). Density comparisons using different analytic methods in *Aotus azarae*. *Primate Rep.* **25,** 45–50.

Green, K. M. (1978). Primate censusing in northern Colombia. A comparison of two techniques. *Primates* **19,** 537–550.

Halle, F., Oldeman, R. A., and Tomlinson, P. B. (1978). "Tropical Trees and Forest: An architectural Analysis." Springer-Verlag, Berlin.

Heltne, P. G. (1977). "Census of *Aotus* in Northern Colombia," Rep. PAHO Proj. AMRO-3171, pp. 1–11. Pan Am. Health Organ., Washington, DC.

Hernandez-Camacho, J., and Defler, T. R. (1989). Algunos aspectos de la conservación de primates en Colombia. *In La Primatología en Latinoamerica, Anales del Simposio de Primatología del IX Congreso Latinoamericano de Zoología,* Arequipa, Perú, 9–15 Octubre, 1983" (C. J. Saavedra *et al.*, eds.), pp. 67–100. UICN/CSE-WWF, Washington, DC.

Hershkovitz, P. (1983). Two new species of night monkeys, genus *Aotus* (Cebidae, Platyrrhini): A preliminary report on *Aotus* taxonomy. *Am. J. Primatol.* **4,** 209–243.

IUCN (1982). "IUCN Directory of Neotropical Protected Areas." Tycooly International Publishing Limited, Dublin.

Mack, D., and Eudey, A. (1984). A review of the primate trade. *In* "The International Primate Trade: Legislation, Trade, and Captive Breeding" (D. Mack and R. A. Mittermeier, eds.), Vol. 1. WWF-Traffic USA-IUCN, Washington, DC.

Moynihan, M. (1964). Some behavior patterns of platyrrhini monkeys. I. The night monkey (*Aotus trivirgatus*). *Smithson. Misc. Collect.* **146,** 1–84.

Myers, N. (1987). Trends in the destruction of rain forests. *In* "Primate Conservation in the Tropical Rain Forest" (C. W. Marsh and R. A. Mittermeier, eds.), Vol. 9, pp. 3–22. Liss, New York.

Rathbun, G. H., and Gache, M. J. (1980). Ecological survey of the night monkey *Aotus trivirgatus,* in Formosa Province, Argentina. *Primates* **21,** 211–219.

Southwick, C. H., and Cadigan, F. C., Jr. (1972). Population densities of Malaysian primates. *Primates* **13,** 1–18.

Stallings, J. R. (1989). Status y conservación de primates en el Paraguay. *In* "La Primatología en Latinoamerica, Anales del Simposio de Primatología del IX Congreso Latinoamericano de Zoologia, Arequipa, Perú, 9–15 Octubre, 1983" (C. J. Saavedra *et al.*, eds.), pp. 133–157. UICN/CSI-WWF, Washington, DC.

Terborgh, J. (1985). "Five New World primates: A Study in Comparative Ecology." Princeton Univ. Press, Princeton, NJ.

Thorington, R. W., Jr., Muckenhirn, N. A., and Montgomery, G. C. (1976). Movements of the wild night monkey (*Aotus trivirgatus*). *In* "Neotropical Primates: Field Studies and Conservation" (R. W. Thorington and P. G. Heltne, eds.), pp. 32–34. National Academy of Sciences, Washington, DC.

Vasquez, R., and Gentry, A. H. (1989). Use and misuse of forest-harvested fruits in the Iquitos area. *Conserv. Biol.* **3,** 1–8.

Wright, P. C. (1978). Home range activity pattern and agonistic encounters of a group of night monkeys (*Aotus trivirgatus*) in Perú. *Folia Primatol.* **29,** 43–55.

Wright, P. C. (1981). The night monkeys, genus *Aotus*. *In* "Ecology and Behavior of Neotropical Primates" (A. F. Coimbra-Filho and R. A. Mittermeier, eds.), Vol. 1, pp. 211–240. Acad. Bras. Cienc., Rio de Janeiro.

Wright, P. C. (1985). "The Cost and Benefits of Nocturnality for *Aotus trivirgatus* (the Night Monkey)." University Microfilm International, Ann Arbor, MI.

Zunino, G. E., Galliari, C. A., and Colillas, O. (1986). Distribución y conservación del Mirikina (*Aotus azarae*) en Argentina: Resultados preliminares. *In A Primatologia no Brasil 2. Anais do 2° Congreso Brasileiro de Primatología, Campinas, SP. 27/01/85 a 01/02/85*" (M. Thiago de Mello, ed.), pp. 305–316. Soc. Bras. Primatol./Univ. Fed. Minas Gerais.

3

The Behavior and Ecology of the Owl Monkey

Patricia C. Wright
Department of Anthropology,
State University of New York at Stony Brook,
Stony Brook, New York 11794

I. INTRODUCTION

Although Madagascar, continental Africa, and Asia support large numbers of nocturnal primates, tropical forests of the New World do not harbor any of these strepsirhine prosimians. The only nocturnal primate in the Neotropics is the owl monkey, *Aotus*. The nocturnal life-style of the genus *Aotus* is particularly interesting because morphological evidence indicates that the ancestral higher primate was diurnal. Why would a New World monkey, equipped with color vision and lacking the acute olfactory system of the Old World lemurs and lorises, evolve to become active primarily at night? The limitations of being night-active are many. Both insects and fruits are more difficult to locate in dim light. Rapid locomotion in an arboreal environment requires sharp eye–limb coordination during the day, and is even more challenging in the dark. And the nighttime is filled with a different set of environmental circumstances such as lower air temperatures, increased humidity, decreased air currents, moonlight phases, and a different set of competitors and predators.

Despite such apparent difficulties and disadvantages, in evolutionary terms the owl monkey is a success. What we know of its evolutionary history suggests a long period of adaptation to a nocturnal habit (Setoguchi, and Rosenberger, 1987), and today owl monkeys are widely distributed throughout South American forests from Panama south to northern Argentina. It should be noted that owl monkeys are absent from northeastern South America, the Guianian Shield.

In this chapter I will review the morphological adaptations of the owl monkey to a nocturnal life-style and discuss my research in the South American forest on the behavior and ecology of the only nocturnal monkey. I then summarize the behavior of owl monkeys in their natural forest environment.

A. Geographic Distribution of *Aotus*

Today the genus *Aotus* is widespread throughout most of the forested areas of South America from Panama to northern Argentina, excepting the northeastern Guianian Shield (Hershkovitz, 1983). Comparisons of morphological variation among populations (Thorington and Vorek, 1976) suggest that differences in pelage color are subtle and variation in tooth and skeletal morphology is slight. There is, however, extensive polymorphism in karyotypes, including natural hybrids (Yunis *et al.*, 1977). It has been suggested that these variations may have been caused by temporary reproductive isolation of populations in forest refuges during the Pleistocene (Ma, 1981) or by random genetic drift (Galbreath, 1983). Whether these 12 or more groups with different karyotypes should be distinguished as different species is still unresolved (Thorington and Vorek, 1976; Ma *et al.*, 1985; Hershkovitz, 1983; Ford, 1988).

B. Adaptations to a Nocturnal Life-style

1. Basal Metabolic Rate It has been suggested that nocturnal mammals are characterized by a low metabolic rate, allowing them to survive on much less energy (Crompton *et al.*, 1978). In contrast to all other New World monkeys studied, owl monkeys do have a low metabolic rate. Its basal metabolic rate (BMR) is 18 to 24% below the predicted value for a 1-kg mammal (Le Maho *et al.*, 1981). In comparison, the BMR value for *Saimiri*, the squirrel monkey, with a similar body size to *Aotus*, is 10% above the predicted value, with other diurnal species near the predicted value (Le Maho *et al.*, 1981; Milton *et al.*, 1979). Nocturnal prosimians, however, have low BMRs similar to owl monkeys, or even lower (Muller, 1979; Hildwein and Goffert, 1975; Chevillard, 1976; McNab and Wright, 1987). It should be noted that the diurnal prosimians studied also have metabolic rates below Kleiber's standard curve (Young *et al.*, 1990).

2. Vision Morphological evidence indicates that *Aotus* is a descendent from a diurnal form. Its eyes vestigially retain many features that characterize diurnal vision in mammals, such as a retinal fovea (Walls, 1942; Polyak, 1957; Ogden, 1975; Noback, 1975).

To see well at low light levels, *Aotus* has evolved large eyeballs, and the lens has a more spherical shape than in diurnal forms, enabling the lens to refract sufficient light to focus an image on the retina. In addition, the posterior placement of its iris diaphragm toward the center of the eye allows an increased pupillary diameter, and thus more light can reach the retina (Noback, 1975). The rod density throughout the retina is several times higher than it is in the human eye (Ogden, 1975).

Aotus, like all anthropoids and *Tarsius,* has no trace of a *tapetum lucidum,* the light-gathering device found behind the retina of almost all nocturnal mammals, including nocturnal strepsirhines (Jones, 1965). Although the *tapetum lucidum* assists vision at minimal light levels, sharpness of focus is sacrificed (Noback, 1975).

The retina of *Aotus* contains both cones and a large number of tightly packed rods. The density of both rods and cones increases continuously from the retinal periphery to the center of the eye, and these are most densely packed in the *area centralis* (Ogden, 1975). In some individuals the retina contains a fovea, but in others there is no clear fovea (Ogden, 1975; Polyak, 1957). The number of cones is much lower than in other primates. In the *area centralis,* cone density reaches about 7300 per mm^2 in *Aotus,* whereas in a comparable part of the retina of the rhesus macaque, the density is 48,000 per mm^2 (Ogden, 1975).

The subcortical relay station of optic nerve fibers, the lateral geniculate body, is large in all nocturnal primates, including *Aotus,* The optic nerve fibers of the rods are fewer, but thicker, with faster conduction speed than the fibers from the cones. The physiological consequence is that impulses from the sensitive rods arrive earlier in the subcortical visual centers than do impulses from the cones (Hassler, 1967). Rod vision is faster, a distinct advantage when catching moving insects or in perceiving substrates while locomoting quickly.

The greatest difference in brain structure seen consistently in nocturnal versus diurnal primates is that the parvocellular layers (perhaps corresponding to fine detail and texture vision receptors) of the lateral geniculate body are greatly expanded in cell number and volume in diurnal primates. The ratio of cell numbers between the magno- and parvocellular layers suggests that the fine texture vision reception in *Aotus* (2.86) is better than that in *Tarsius* (1.81), similar to the nocturnal prosimian *Avahi* (3.09), but appreciably worse than diurnal *Callithrix* (4.86) or *Cebus* (7.53) (Hassler, 1967; Allman and McGuinness, 1988).

Moonlight is rich in the blue-green wavelengths, but light reflected from foliage at night has fewer blue-green wavelengths owing to absorption. The vision of nocturnal prosimians and *Aotus* has maximum sensitivity at 450–500 nm, which is optimal for taking advantage of contrast cues for detecting objects directly luminated by the bluish direct moonlight (Charles-Dominique and Petter, 1980; Jacobs, 1977a).

What does the owl monkey actually perceive with this unique eye? Jacobs (1977b) examined the behavior of owl monkeys in his laboratory and found that the general features of contrast sensitivity functions are the same as in other New World species. However, in owl monkeys the frequency of peak sensitivity and location of the high-frequency cutoff are lower than those in diurnal monkeys. Although humans have three times more spatial resolution than the owl monkey at moderate light levels, the owl monkey has twice the spatial resolution at very low light levels (Jacobs, 1977b).

Jacobs (1977a) also conducted experiments to determine the nature of owl monkey color vision. Owl monkeys have a trichromatic visual system and can perceive all colors of the spectrum, although their color vision is less acute than that of diurnal primates. It is important to note that all of the Old World monkeys studied thus far have color vision similar to normal human trichromats, but the New World monkey data indicate that these species lack normal trichromatic color vision (De Valois and Morgan, 1974; Jacobs, 1981). *Cebus, Saimiri, Callithrix,* and *Aotus* all have weak sensitivity in the red wavelengths, similar to the conditions of humans who are color-blind to red (Jacobs, 1981).

It appears that the *Aotus* eye is faster than that of its diurnal relatives at seeing and following moving objects, particularly at low light levels. It is also better at fine texture vision than the eyes of its diurnal monkey neighbors. Both of these qualities would be assets in capturing insect prey and locomoting in the dark. Although *Aotus* has only weak color vision, nearly all Neotropical monkeys are weak in the red wavelengths, and this does not appear to handicap the activities of this group. Owl monkeys seem to make up for the lack of color vision, because in contrast sensitivity tests at intermediate light levels, they do as well as diurnal monkeys. *Aotus* also has better spatial resolution of objects at low light levels, whereas the day monkeys perceive spatial differences better in bright light (Wright, 1989).

3. Olfaction The *Aotus* nose is equipped with olfactory capabilities similar to those of other New World monkeys. Unlike the Old World monkeys, apes, and humans, the New World monkeys have a functional Jacobsen's organ (vomeronasal) (Stephan and Andy, 1970). The olfactory bulbs in all

New World monkeys are about a third smaller proportionally than those of prosimians. The *Aotus* olfactory bulbs and accessory olfactory bulbs are the largest of all New World monkeys, but only slightly larger than those of tamarins and marmosets (Stephan and Andy, 1970; Stephan *et al.*, 1981).

Scent-marking behaviors seen in lemurs, lorises, tarsiers, and New World monkeys demonstrate the importance of pheromonal olfactory communication to these primates (Epple, 1974; Harcourt, 1981; Milton, 1975, 1985; Wright *et al.*, 1986). *Aotus* uses both urine and glandular secretions to convey olfactory information (Moynihan, 1964; Wright, 1981). Dixson (1982) found in laboratory experiments with anosmic *Aotus* that olfaction plays an important role in sexual recognition and intermale aggression, but has little influence on male sexual behavior. During observations of wild *Aotus* in Paraguay, I found that neither urine washing nor caudal scent rubbing by either sex increased when the female was in estrus. However, both types of olfactory communication increased dramatically during intergroup encounters observed in Peru (Wright, 1989). In contrast, lemurs, lorises, and tarsiers show increases in scent marking before and during estrus (French *et al.*, 1984; Wright *et al.*, 1986).

Scents are stronger and last longer in the humid night atmosphere. Owl monkeys take advantage of this night olfactory environment by communicating aggressive signals chemically and by perceiving ripeness of fruits and edibility of foods.

4. Audition Owl monkeys have a threshold of hearing similar to that of other monkeys, although their auditory bullae are the largest, scaled to body mass, of any of the New World monkeys (Hershkovitz, 1977; Hill, 1960). Peak auditory sensitivity lies between 1 and 8 Hz, although the frequency range extends from below 60 Hz to above 16 kHz (Beecher, 1974).

Long-distance calls emitted by owl monkeys are low in frequency and simple in structure (Wright, 1989). Since low frequencies carry farther in the rain forest (Waser and Waser, 1977) and *Aotus* calls are too low to compete with insect and most frog calls, owl monkey calls are clearly distinguished at night.

II. METHODS

Night in the forest presents similar problems to all nocturnal primates. At night, light levels are low and it is difficult to see to travel, find food, patrol boundaries, visually communicate, and spot predators. Ambient temperatures are lower than in daytime and thermoregulation is also a problem.

To clearly understand the restrictions and adaptive advantages of nocturnal activity, I studied the social behavior, ecology, and diet of owl monkeys in the Peruvian rain forest and in a tropical dry forest in Paraguay (Wright, 1985, 1989). For an additional comparison, I conducted simultaneous research on the behavior and ecology of *Callicebus moloch,* the dusky titi monkey, a diurnal monkey in the rain forest. *Callicebus* and *Aotus* are similar in body weight and social structure, and in fact occupy the same parts of the rain forest, with home ranges overlapping at the Peruvian site (Kinzey, 1981). This comparison between sympatric species was a unique opportunity to study two species using potentially the same food items, but at different times in the 24-hr schedule.

The rain forest study site was located at Cocha Cashu Biological Research Station in the Manu National Park, Peru, at 11°52′S, 17°22′W. This pristine site is located at 400 m elevation, receives 2000 mm of rain annually, and has a high biodiversity, including nearly 1000 species of birds and 13 species of monkeys (Terborgh, 1983; Wright, 1985).

Nine groups of owl monkeys and six groups of titi monkeys were censused periodically during a 15-month period. One focal group of *Aotus* and two focal groups of *Callicebus* were followed throughout their activity cycle for 5 consecutive days or nights each month. Data on activity patterns, social behavior, diet, and range were taken at 5-min intervals. A Varo Noctron V image intensifier and ITT image intensifier with an infrared light source assisted with nocturnal observations.

For a comparison, to understand the ecological constraints on the behavior of *Aotus,* I next studied owl monkeys in a contrasting tropical dry forest habitat in Paraguay. At Golondrina Ranch in the chaco of Paraguay (24°25′S, 58°40′W), both hot and cold temperatures were more extreme, and rainfall was limited to less than 1000 mm each year. The only competing primate species was *Alouatta caraya,* the howling monkey, and the most probable predator was an owl species that does not occur in the rain forest, the great horned owl. The large hawks and eagles of the rain forest were absent in this dry forest habitat. One group of owl monkeys was followed for 5 to 15 days each month for a 4-month period. Often, with the help of a research assistant, the owl monkeys were followed continuously for 24 hr.

III. RESULTS

A. Peruvian Rain Forest

1. Social System and Reproductive Behavior Both *Aotus* and *Callicebus* live in monogamous pairs, and there is some evidence from the wild that this is a long-term bond (Wright, 1989). Monogamy is relatively rare in mammals, but other primates, both diurnal (*Indri* and *Hylobates*) and nocturnal (*Avahi* and *Tarsius*), are monogamous (Kleiman, 1977; Wright, 1984). In the study groups in Peru, pairs with one infant, one juvenile, and one subadult were the rule. In both species, at 3 years of age the subadult left the group, usually during the month of December, presumably to find a mate and establish a new territory. Aggression between parent and offspring was not documented before the dispersal, but the subadults would lag behind the group for the preceding month.

Aotus and *Callicebus* are also unique in having single offspring each year, with fathers assuming primary caretaking responsibilities (Dixson and Fleming, 1981; Wright, 1984; Fragaszy *et al.*, 1982). These parental care behaviors include carrying, guarding, playing with, and food sharing (Wright, 1984). Mothers in the wild offered milk to the infant every 2 to 3 hr, and avoided the father while he was carrying the offspring. Males carrying infants often lagged far behind the rest of the group and arrived at fruit trees 10 to 15 min after other family members had begun feeding. In Peru, the fathers in both *Aotus* and *Callicebus* carried offspring until the infants were 4 months old. Father–offspring conflicts over carrying were heard and observed in both species during the fourth month. In the wild, *Callicebus* mothers continued to nurse offspring until the seventh month of age. During that month, mother–offspring conflicts were often observed when the offspring attempted to suckle. I could not observe *Aotus* suckling bouts in the dark.

In the absence of readily available visual cues, scent and urine marking may provide olfactory cues that indicate the reproductive status of *Aotus* females in estrus, but field observations and laboratory experiments do not support this suggestion (Dixson, 1982; P. C. Wright, personal observation). Most of the diurnal New World monkeys have glandular excretions and urine-marking behaviors that provide reproductive cues. In these species, the frequency of scent-marking behaviors increases when females are in estrus (Milton, 1975, 1985; Goldizen, 1987). However, *Callicebus moloch* was observed breeding in the wild, and neither sex was observed to scent mark before or after each of the three copulations observed. In the case of prosimians, both diurnal and nocturnal lemurs scent mark and/or urine mark to indicate reproductive state (Charles-Dominique and Petter, 1980; Richards, 1978). It appears that neither the social system, the chemical communication of estrus, nor the paternal care of owl monkeys is adapted exclusively to a nocturnal life-style.

2. Ranging Patterns, Territoriality, and Loud-Call Vocalizations The territories of *Callicebus* (mean-6.9 ha, $n = 6$) and *Aotus* (mean-9.2 ha, $n = 9$) were both small compared to those of other sympatric primates (Wright, 1989). For example, the tamarins, which have a smaller body mass, used a territory that was 10 times larger. Similarly, both *Aotus* and *Callicebus* had short daily and nightly path lengths compared to other monkey species at Cocha Cashu. The mean nightly path length of the owl monkey group was 708 m (S.D. = 243, n = 60) and the mean daily path length of *Callicebus* was 671 m (S.D. = 193, n = 63). At Cocha Cashu, the squirrel monkey, smaller than the owl monkey, had a daily path length three times that of owl monkeys (Wright, 1986).

Comparisons of nightly path lengths of *Aotus* showed that distance traveled was associated with phases of the moon. It is clear that available light affects night ranging patterns, and owl monkeys curtailed their traveling and social behaviors on moonless nights. The owl monkeys traveled twice as far during moonlit nights than on nights with no moonlight. On dark nights, including nights with heavy rainfall, travel was slow and not much distance was covered. *Aotus* also used twilight for traveling and the monkeys ranged farther during dawn and dusk than during any other hour of the night (Wright, 1989).

Callicebus moloch defines and defends its territories by loud turkeylike calls given by the male and female pair together in the early morning (Kinzey, 1981). Owl monkeys did not give loud duets to defend territories. The calls of owl monkeys that carry long distances were given by lone individuals and were owllike hoots. These simple calls do not resemble the elaborate *Callicebus* calls that vary in harmonics. Owl monkeys emit one to five single notes that are very low in frequency. These calls were given only during nights with bright moonlight, and hoot calls, which carry for 500 m, were repeated for 1 or 2 hr. The animal calling could be either male or female, but it was always alone. In the main study group, the rest of the group rested together, while a single individual moved alone to the territorial boundary, giving this hootlike call continuously. Observations suggested that the individual was the subadult of the group. The call was answered by individuals in neighboring territories, and often these individuals moved toward the same border. Individuals came within 25 m of the borders, but after several hours they moved back to the center of the respective territories. There is some evidence that this is a mate attraction call, although territory establishment by new pairs has not been observed (Wright, 1984).

Although few incidents of aggression were seen within the owl monkey family group, intergroup aggression at border fruiting trees was seen 15 times within the 15 months. There were no seasonal patterns in intergroup fighting, but all border fights occurred when the moon was full and overhead. A loud vocalization, like a resonant whoop (Moynihan, 1964), preceded the attack. Adults of both sexes in both groups inflated their throat pouches and projected this low, resonating sound that rose in pitch and volume. Piloerection and stiff-legged jumping accompanied the whoops. Chases, more aggressive whoops, and wrestles could be seen from the ground. Fights were usually terminated within 10 min, where upon each group retreated into its own territory. In captivity it has been observed that males fight males, and females fight females (Wright, 1985), but in the wild it was difficult to distinguish sexes in the moonlight.

3. Playing and Grooming Behavior Playing behavior was observed between the subadult, juvenile, and father within the study group, on moonlit nights from October through May. During these months abundant fruit was available, and in other months the group may have been conserving energy. In captivity the mother rarely plays, but it was difficult to confirm this at night in the rain forest (Wright, 1984).

No instances of grooming were observed in owl monkeys in the rain forest. However, all members of the *Callicebus* monkey family group groomed throughout rest periods and before sleeping in the evening. Moynihan (1964) describes owl monkeys as grooming only before copulation, but owl monkey copulations were not seen in Peru.

4. Diet A comparison of the diet between these diurnal and nocturnal monkeys, which had range overlap and the same availability of foods, showed that they ate fruits, flowers, young leaves, and insects, but with the following differences. *Callicebus* trap-lined many small-crowned fruit trees of a single species each month, but *Aotus* preferred feeding in larger-canopied fruit trees (Wright, 1989). At all times of year, *Callicebus* ate more vegetation than *Aotus,* and fecal samples indicated that *Aotus* obtained more protein from insects. Nevertheless, diets were not dramatically different during the season when food was abundant, although diets tended to differentiate during the season when food was scarce. *Callicebus* spent 25% of their feeding minutes on bamboo and vine leaves in the time of abundant fruits, and 40% on vegetation during the season of scarce fruits. During the latter season, *Aotus* concentrated on nectar, figs, and insects.

During this period of scarce resources, other diurnal species of monkey also fed on nectar and figs (Terborgh, 1983). However, during the day the small-bodied monkeys with small group size, such as *Callicebus,* were chased from the big trees by many groups of each of the species of large monkey, namely, *Cebus apella, Cebus albifrons,* and *Ateles paniscus.* When denied access to the figs and nectar resources, the dusky titi monkeys resorted to leaf eating. Owl monkeys did not have to compete with these large, diurnal monkeys at night. Nocturnal competitors for food have either small body size or small group size. Owl monkeys were observed feeding on nectar and fig trees with kinkajous and marsupials with no aggression.

From these data it appears that owl monkeys gain certain feeding advantages by being nocturnal. First, interference competition for preferred foods is minimal: They can feed on figs and nectar resources at all times of the year. In addition, insects, especially large orthopterans and edible lepidopterans, are active and calling at night, facilitating location of this high-protein food.

5. Predation Pressures, Sleep Trees, and Activity Patterns A different set of predators is active at night. Although predation events are rare occurrences, they are important selection pressures. To measure the differences between day and night predation, I inferred information from indirect associations in differences in (1) population densities, (2) sleeping sites and timing of entering and exiting, and (3) height used in the forest.

In the rain forests of South America, diurnal raptors are the main predator of monkeys (Fowler and Cope, 1964; Rettig, 1978; Terborgh, 1983). Nocturnal predators such as snakes, owls and felids rarely eat monkeys (Emmons, 1987, 1990). These differences might result in different behavioral choices for diurnal and nocturnal monkeys. To avoid the harpy and crested eagles, an owl monkey might choose a safe sleeping site where the raptors could not detect them during the day, whereas titi monkeys would be less selective about sleep trees, since nocturnal predators would not be a threat. *Aotus* might travel and feed while the diurnal raptors were sleeping and end activities before the raptors awakened. *Callicebus,* to avoid the constant danger of the harpy and crested eagles, might spend more time hidden in the lower levels of the rain forest. Group size and population densities might be lower in the species that was most preyed upon.

Indeed, population densities of the owl monkeys were twice that of the titi monkeys, despite the fact that territory sizes were not that different. An infant or juvenile was absent from three out of six groups in *Callicebus,* but the only owl monkey group seen without a full complement of offspring was a newly formed pair. Because infants were seen before they disappeared, I attributed this difference in reproductive success to predation rather than to competition for food that would result in a reduced reproductive rate. The only observed incidence of predation was a crested eagle carrying off a yearling *Callicebus*.

As predicted, there was a marked difference in the selection of sleeping sites between the nocturnal and diurnal species regardless of season. *Aotus* group I slept in five different trees during the year, choosing one tree on 57% of the nights. The effect of such high selectivity meant that path length was often circular, with the group reentering the same tree at dawn that it had left at dusk. Sleeping sites were protected by being either in tree holes or in vine tangles where the animals could not be seen.

Callicebus, however, slept in tree sites where they were clearly visible. One group chose 38 different sleeping trees during the year that they were studied. There were favorite sites that they chose, but there was no pattern. During the 5-day observation periods, a group would never use less than three trees. *Callicebus* never slept in tree holes or in any of the trees chosen by *Aotus*.

A striking difference between the two species was the regularity with which *Aotus* exited its sleeping site, and the irregularity with which *Callicebus* exited. *Aotus* chose its exit times according to sunrise and sunset, entering 15 min before sunrise and exiting 15 min after. *Callicebus* varied its entrance and exit according to temperature and weather conditions. In the cold months of June, July, and August, *Callicebus* would become active 4 hr later than in hot season months (Wright, 1989).

Titi monkeys spent their active hours low in the rain forest, often in the shadows, 50% of the time below 10 m in a 40 m high forest. In contrast, the larger, diurnal species, who are less vulnerable to aerial predators, used upper canopy levels (Fleagle and Mittermeier, 1980; Terborgh, 1983). Owl monkeys spent most of their time at higher canopy levels and only 28% of their time at 10 m or lower. Behaviors that drew attention to owl monkeys, such as playing, intergroup fighting, and loud calling, always occurred high in the canopy during bright moonlit nights. The same behaviors by *Callicebus* occurred in vine tangles or below 10 m.

B. Comparative Study in Tropical Dry Forest

To better understand the ecological variables of predation pressure, interference competition, and environmental temperature extremes, I also studied owl monkeys in the tropical dry forest of Paraguay. At this site, large diurnal monkeys are absent, except for the folivorous black howler (*Alouatta caraya*). Large eagles like the harpy and crested eagles are rare, but the great horned owl, a nocturnal raptor with females weighing twice as much as an owl monkey, is common. Temperatures were extreme, ranging from −5°C to 40°C (Kreig, 1930), and the annual rainfall was half that received at the Manu National Park.

1. Social System and Reproductive Activity The social system of *Aotus* was no different from that in the rain forest and all groups sighted were in pairs. In a more extensive census in the Argentina chaco, Rathbun (1980) also found *Aotus* only in family groups. It appears that the monogamous social system does not change with changes in predation pressure. Since owl monkeys feed on fruits, leaves, and flowers found in small-canopied trees, it is possible that resource size may be selecting for this monogamous social system (Wright, 1986).

For the first time in the wild, copulations were seen. On August 4, 1982 a male *Aotus* approached a female during the day, mounted her, and thrust nine times. There were no observations of grooming before the copulation, the pair rested together for 5 min afterward, and then they resumed foraging. In this seasonal environment, it would be a good strategy to have offspring in December, the season of high fruit production. The gestation length of *Aotus* is 4 months (Hunter *et al.*, 1979).

2. Ranging Behaviors and Vocalization Home range size was only 5 ha, half the territory size in the rain forest, and path length was shorter, 619 m on average. A smaller group size of two adults with no offspring was observed. During the bright moonlit nights, the loud hooting calls were heard, as in the rain forest. However, in contrast to the continuous calling pattern heard in the rain forest, the calls were given more intermittently, and the calling owl monkey traveled farther between calls in the dry forest.

3. Diet The diet of Paraguayan owl monkeys remained omnivorous, including fruits, leaves, insects, and flowers, but these were consumed in different proportions. Leaf eating was high (46% of feeding minutes) in cold, winter months when fruits were scarce. Insect eating and flower eating increased dramatically during the spring, when these resources became abundant. Interference competition for food items was not a factor. The black howler monkeys fed on leaves from large trees in the winter, whereas the owl monkeys restricted their leaf eating to vines and small trees found primarily on the edges of the forest. In the spring, black howler monkeys and owl monkeys were observed feeding on flowers in the same tree in bright sunshine, without any obvious interaction between them. Other mammals, such as two species of marsupials, did not interfere with owl monkey feeding.

4. Predation Pressures, Sleep Trees, and Activity Patterns In the dry forest, there was a change in owl monkey sleeping patterns to random sleep tree choice, in contrast to the rain forest pattern of choosing habitual sleep trees. In the 4 months of this study, 43 different trees were slept in by one pair of owl monkeys, and most of these sites were in the open where the animals were visible. It could not be predicted where the owl monkeys would sleep.

As previously reported by Kreig (1930) and Rathbun (1980), the owl monkeys studied did not adhere strictly to a nocturnal schedule. Instead, the owl monkeys from the chaco could be seen feeding any time of the day, even at noon in bright sunshine. During 24-hr observation periods, the owl monkeys studied spent 1 to 3 hr foraging during daylight hours during the winter months. However, more than half of their active hours were always at night.

There was some indication that owl monkeys were conscious of the activity of the pair of great horned owls that lived in the forest. Traveling at night was quieter in this forest than in the rain forest, and the pair did not give the contact calls that they gave in the rain forest. Owl monkey sleeping sites were at least 200 m from the great horned owl's nest. Casts of the pair of owls were examined each morning to determine their diet. No owl monkey bones were found; the only evidence we have that owls take monkeys is indirect. There were no offspring in the pairs observed and in a census of 25 groups of *Aotus* in the Argentine chaco, Rathbun (1980) found that the mean group size was 2.9, with a juvenile or infant lacking from each group. These figures suggest that offspring mortality is high owing to poor nutrition or higher predation.

These results from the chaco show that when environmental factors change (predators, competitiors, food availability, and temperature extremes), some aspects of the behavior of owl monkeys change, including diet proportions, activity rhythms, and sleep tree choice. But other behaviors, such as the monogamous social system, primarily nocturnal activity, and the long, loud calls, do not change. It should be noted that the pattern of calling, giving calls intermittently with long travel bouts between calls, did change in Paraguay.

IV. DISCUSSION

The problems that all nocturnal mammals must overcome to be night-active are (1) colder temperatures, (2) low light levels, (3) differences in predators and competitors, and (4) difference in availability of certain food items. By integrating knowledge gained from the laboratory on physiology and neurobiology with information from field studies of *Aotus*, we can better understand the nocturnal monkey niche and the adaptations that better equip this monkey to live at night.

The comparative approach to the behavioral ecology of the owl monkey has revealed several important differences between the diurnal *Callicebus* and the nocturnal *Aotus*. In contrast to the day monkey, ranging patterns were often circular and chosen routes were habitual and rigid, and one or two sleep trees were used for many years. These habitually traveled and scent-marked routes would make orientation easier in the dark. Sleep sites were hidden in tree holes or dense vine tangles. This suggested that owl monkeys needed to hide from daytime predators, but were not threatened by nighttime predators who could easily memorize habitual travel routes. The owl monkeys traveled more than twice as far during twilight and full moon, suggesting that low light levels may be restrictive. Social behavior, including intragroup calling, playing, and intergroup fighting, occurred when the moon was bright and overhead. This monthly rhythm of activity may be a consequence of the restrictions imposed by darkness during parts of each month (Wright, 1989,1994).

Advantages of being active at night in the New World include the increase in availability of active and calling insects, which are a good protein source. Indeed, *Aotus* relied on insects more than leaves for protein. Avoidance of the major predators, the day-active hawks and eagles, and the main food competitors, the larger diurnal monkeys, was also an advantage to night activity. When these latter two selective pressures were eliminated, as seen in the site in Paraguay, the owl monkeys changed activity patterns by spending 1 to 3 hr feeding during daylight. The large nocturnal predator (great horned owl) found in Paraguay may have influenced the second major change, the fact that habitual routes were not used and sleep tree sites were randomly chosen in the forest. Other behaviors, including the monogamous breeding system and loud calls, did not change in this contrasting habitat.

ACKNOWLEDGMENTS

The Peruvian Ministry of Agriculture and the Peruvian Forestry Institute gave me permission to do research in the Manu National Park in Peru. Many thanks to Diane and Anthony Espinoza, the U.S. Peace Corps, J. J. and A. Earhart, D. Norman, C. H. Janson, J. W. Terborgh, W. G. Kinzey, J. G. Fleagle, R. B. Foster, J. F. Oates, L. H. Emmons, J. M. Allman, P. S. Daniels, A. W. Goldizen, A. Wright, and L. Martin. This research was supported by NSF Grants BNS 81-15368 and BNS 77-24921, the World Wildlife Fund-US Primate Program, and the Wenner-Gren Foundation for Anthropological Research Grant 4282.

REFERENCES

Allman, J., and McGuinness, E. (1988). The visual cortex in primates. *In* "Comparative Primate Biology" (H. P. Steklis, ed.), Vol. 3. Liss, New York.

Beecher, M. D. (1974). Audition in the owl monkey (*Aotus trivirgatus*). *J. Comp. Physiol. Psychol.* **86,** 898–901.

Charles-Dominique, P., and Petter, J. J. (1980). Ecology and social life of *Phaner furcifer*. *In* "Nocturnal Malagasy Primates" (P. Charles-Dominique, ed.), pp. 75–85. Academic Press, New York.

Chevillard, M. C. (1976). Capacites thermoregulatrices d'un Lemurien malgache, *Microcebus murinus* (Miller 1777). Thesis, University of Paris VII.

Crompton, A. W., Taylor, C. R., and Jagger, J. J. (1978). Evolution of homeothermy in mammals. *Nature (London)* **272,** 333–336.

De Valois, R. L., and Morgan, H. C. (1974). Psychophysical studies on monkey vision. Vol. II. Squirrel monkey wavelength and saturation discrimination. *Vision Res.* **14,** 69–73.

Dixson, A. F. (1982). The owl monkey (*Aotus trivirgatus*). *In* "Reproduction in New World Primates" (J. P. Hearn, ed.), pp. 69–114. MTP Press, Boston.

Dixson, A. F., and Fleming, D. (1981). Parental behaviour and infant development in owl monkeys (*Aotus trivirgatus griseimembra*). *J. Zool. Soc. London* **194,** 25–39.

Emmons, L. (1987). Comparative feeding ecology of felids in a neotropical rain forest. *Behav. Ecol. Sociobiol.* **20,** 271–283.

Emmons, L. (1990). "Neotropical Rainforest Mammals: A Field Guide." Univ. of Chicago Press, Chicago.

Epple, G. (1974). Olfactory communication in South American primates. *Ann. N. Y. Acad. Sci.* **237,** 261–278.

Fleagle, J. G., and Mittermeier, R. A. (1980). Locomotor behavior, body size and comparative ecology of seven Surinam monkeys. *Am. J. Phys. Anthropol.* **52,** 301–314.

Ford, S. M. (1988). Systematics and biogeography of *Aotus*. *Int. J. Primatol.* **8,** 741.

Fowler, J. M., and Cope, J. B. (1964). Notes on the harpy eagle in British Guiana. *Auk* **81,** 257–273.

Fragaszy, D. M., Schwartz, S., and Shimosaka, D. (1982). Longitudinal observation of care and development of infant titi monkeys (*Callicebus moloch*). *Am. J. Primatol.* **2,** 191–200.

French, J. A., Abbott, D. H., and Snowdon, C. T. (1984). The effect of social environment on estrogen excretion, scent marking and sociosexual behavior in tamarins (*Saguinus oedipus*). *Am. J. Primatol.* **6,** 155–167.

Galbreath, G. J. (1983). Karyotypic evolution in *Aotus. Am. J. Primatol.* **4**(3), 245–251.

Goldizen, A. W. (1987). Facultative polyandry and the role of infant-carrying in wild saddle-back tamarins (*Saguinus fuscicollis*). *Behav. Ecol. Sociobiol.* **20,** 99–109.

Harcourt, C. S. (1981). An examination of the function of urine washing in *Galago senegalensis. Z. Tierpsychol.* **55,** 119–128.

Hassler, R. (1967). Comparative anatomy of the central visual systems in day and night active primates. *In* "Evolution of the Forebrain" (R. Hassler and H. Stephan, eds.), pp. 419–434. Plenum, New York.

Hershkovitz, P. (1977). "Living New World Primates (Platyrrhini)." Univ. of Chicago Press, Chicago.

Hershkovitz, P. (1983). Two new species of night monkeys, genus *Aotus* (Cebidae, Primates): A preliminary report on *Aotus* taxonomy. *Am. J. Primatol.* **4,** 209–243.

Hildwein, G., and Goffert, M. (1975). Standard metabolism and thermoregulation in a prosimian *Perodicticus potto. Comp. Biochem. Physiol. A* **50A,** 201–213.

Hill, W. C. O. (1960). "Cebidae," Part A. Edinburgh Univ. Press, Edinburgh.

Hunter, J., Martin, R. S., Dixson, A. F., and Rudder, B. C. C. (1979). Gestation and inter birth intervals in the owl monkey. (*Aotus trivirgatus griseimembra*). *Folia Primatol.* **31,** 165–175.

Jacobs, G. H. (1977a). Visual capacities of the owl monkey (*Aotus trivirgatus*) I. Spectral sensitivity and color vision. *Vision Res.* **17,** 811–820.

Jacobs, G. H. (1977b). Visual capacities of the owl monkey (*Aotus trivirgatus*)—II. Spatial contrast sensitivity. *Vision Res.* **17,** 821–825.

Jacobs, G. H. (1981). "Comparative Color Vision." Academic Press, New York.

Jones, A. E. (1965). The retinal structure of the owl monkey (*Aotus trivirgatus*). *J. Comp. Neurol.* **125,** 19–27.

Kinzey, W. G. (1981). The titi monkeys, genus *Callicebus. In* "Ecology and Behavior of Neotropical Primates," (A. F. Coimbra-Filho and R. A. Mittermeier, eds.), Vol. 1, pp. 241–276. Acad. Bras. Cienc., Rio de Janeiro.

Kleiman, D. G. (1977), Monogamy in mammals. *Q. Rev. Biol.* **52,** 39–69.

Kreig, H. (1930). Die affen der Gran Chaco und seiner grenzbegeite. *Z. Morphol. Oekol.* **4,** 760–785.

Le Maho, Y., Geffert, M., Rochas, A., Felbabel, H., and Chatonnet, J. (1981). Thermoregulation in the nocturnal simian: The night monkey *Aotus trivirgatus. J. Physiol.* (*London*) R156–R165.

Ma, N. S. F. (1981). Chromosome evolution in the owl monkey, *Aotus. Am. J. Phys. Anthropol.* **54,** 293–303.

Ma, N. S. F., Aquino, R., and Collins, N. E. (1985). The new karyotypes in the Peruvian owl monkey (*Aotus trivirgatus*). *Am. J. Primatol.* **9,** 333–341.

McNab, B., and Wright, P. C. (1987). Temperature regulation and oxygen consumption in the Philippine tarsier *Tarsius syrichta. Physiol. Zool.* **60,** 596–600.

Milton, K. (1975). Urine rubbing behavior in the mantled howler monkey (*Alouatta palliata*). *Folia Primatol.* **23,** 105–112.

Milton, K. (1985). Urine washing behavior in the woolly spider monkey (*Brachyteles arachnoides*). *Z. Tierpsychol.* **67,** 154–160.

Milton, K., Casey, T. M., and Casey, K. K. (1979). The basal metabolism of mantled howler monkeys (*Alouatta palliata*). *J. Mammal.* **60,** 373–376.

Moynihan, M. A. (1964). Some behavior patterns of platyrrhine monkeys. 1. The night monkey (*Aotus trivirgatus*). *Smithson. Misc. Collec.* **146,** 1–84.

Muller, E. F. (1979). Energy metabolism, thermoregulation and water budget in the slow loris (*Nycticebus coucang* Boddaert 1785). *Comp. Biochem. Physiol.* **64,** 109–119.

Noback, C. R. (1975). The visual system of primates in phylogenetic studies. *In* "Phylogeny of the Primates" (P. Luckett and F. S. Szalay, eds.), pp. 199–218. Plenum, New York.
Ogden, T. E. (1975). The receptor mosaic of *Aotus trivirgatus:* Distribution of rods and cones. *J. Comp. Neurol.* **163,** 193–202.
Polyak, S. (1957). "The Vertebrate Visual System." Univ. of Chicago Press, Chicago.
Rathbun, G. P. (1980). Ecological survey of the night monkey, *Aotus trivirgatus,* in Formosa Province in Argentina. *Primates* **21,** 211–219.
Rettig, N. (1978). Breeding behavior of the harpy eagle (*Harpia harpyja*). *Auk* **95,** 629–643.
Richards, A. F. (1978). "Behavioral Variation: Case Study of a Malagasy Lemur." Bucknell Univer. Press, Lewisburg, PA.
Setoguchi, T., and Rosenberger, A. L. (1987). A fossil owl monkey from La Venta, Colombia. *Nature (London)* **326,** 692–694.
Stephan, H., and Andy, O. J. (1970). The allocortex in primates. *In* "The Primate Brain" (C. R. Noback and W. Montagna, eds.), pp. 109–135. Appleton-Century-Croft, New York.
Stephan, H., Frahm, H., and Baron G. (1981). New and revised data of volumes of brain structures in insectivores and primates. *Folia Primatol.* **35,** 1–29.
Terborgh, J. W. (1983). "Five New World Primates: A Study in Comparative Ecology." Princeton Univ. Press, Princeton, NJ.
Thorington, R. W., Jr., and Vorek, R. E. (1976). Observations on the geographic variation and skeletal development of *Aotus. Lab Anim. Sci.* **26**(6), 1006–1021.
Walls, G. L. (1942). "The Vertebrate Eye and Its Adaptive Radiation." Cranbrook Inst. Sci. Bloomfield Hills, MI.
Waser, P. M., and Waser, M. S. (1977). Experimental studies of primate vocalization: Specializations for long distance propagation. *Z. Tierpsychol.* **43,** 239–263.
Wright, P. C. (1981). The night monkeys, genus *Aotus. In* "Ecology and Behavior of Neotropical Primates." (A. F. Coimbra-Filho and R. A. Mittermeier, eds.). Vol. 1, pp. 211–244. Acad. Bras. Cienc. Rio de Janeiro.
Wright, P. C. (1984). Biparental care in *Aotus trivirgatus* and *Callicebus moloch. In* "Female Primates: Studies by Women Primatologists" (M. E. Small, ed.), pp. 59–75. Liss, New York.
Wright, P. C. (1985). The costs and benefits of nocturnality in *Aotus trivirgatus* (the night monkey). Ph. D. Dissertation, City University of New York, New York.
Wright, P. C. (1986). Ecological correlates of monogamy in *Aotus* and *Callicebus moloch. In* "Primate Ecology and Conservation" (J. Else and P. C. Lee, eds.), pp. 159–167. Cambridge Univ. Press, Cambridge, UK.
Wright, P. C. (1989). The nocturnal monkey niche in the New World. *J. Hum. Evol.* **18,** 635–658.
Wright, P. C. (1994). Night watch on the Amazon. *Nat. Hist.* **103,** 44–51.
Wright, P. C., Toyama, L. M., and Simons, E. L. (1986). Courtship and copulation in *Tarsius bancanus. Folia Primat.* **46,** 142–148.
Young, A. L., Richard, A. F., and Aiello, L. C. (1990). Female dominance and maternal investment in strepsirhine primates. *Am. Nat.* **135**(4), 473–488.
Yunis, E., Torres de Caballero, O. M., and Ramiriz, C. (1977). Genus *Aotus* Q and G band karyotypes and natural hybrids. *Folia Primatol.* **27,** 165–177.

4

Reproductive Biology of the Owl Monkey

Allen F. Dixson
Medical Research Council Group on the Development and Integration of Behavior
Sub-Department of Animal Behavior
University of Cambridge
Madingley, Cambridge CB3 8AA, United Kingdom

I. INTRODUCTION

Traditionally, owl monkeys were considered to be delicate and difficult to breed in captivity. Most laboratories maintained these animals for research on malaria or cancer, however, and little attempt was made to study their reproductive biology. The Colombian form of the owl monkey, now classified as a distinct species (*Aotus lemurinus;* Hershkovitz, 1983), has been particularly useful for studies of human malaria, since it can be infected experimentally with *Plasmodium falciparum* (Geiman and Meagher, 1967). Because 150 to 200 million people are estimated to be infected with *Plasmodium* at any given time and more than a million children under 5 years of age die of malaria each year (Dobson, 1992), the use of Colombian owl monkeys for research on *P. falciparum* is understandable. However, the pace of rain forest destruction in Colombia, coupled with the removal of many thousands of monkeys for biomedical research, led to the realization that captive breeding programs were essential. Accordingly, a number of studies of reproductive physiology and behavior in *Aotus* were undertaken during the 1970s and 1980s (e.g., Elliott *et al.*, 1976; Merritt, 1976; Cicmanec and Campbell, 1977; Dixson, 1982, 1983a, 1987), as well as several valuable field studies of behavioral ecology (e.g., Wright, 1978, 1981; Rathbun, 1980; Aquino and Encarnación, 1986).

This chapter will review current knowledge of the reproductive physiology and reproductive behavior of Colombian owl monkeys in captivity. We will present information on gonadal function, sexual behavior, gestation, interbirth intervals, and parental behavior. We will also attempt to identify the most important questions for future research and to suggest how observations of captive owl monkeys may relate to their reproductive biology under natural conditions.

II. TESTICULAR FUNCTION

Perhaps the most remarkable feature of reproduction in owl monkeys concerns their testicular physiology. Hunt *et al.* (1975) and King (1975) originally described a condition resembling partially arrested spermatogenesis in captive adult male *Aotus,* which they attributed to vitamin E deficiency. Subsequently, histological studies of whole testes, or testicular biopsies, from 19 adult *A. lemurinus* confirmed that very few spermatozoa were present. Furthermore, many of the spermatocytes in the seminiferous tubules appeared degenerate, and deposits of an orange-brown pigment were seen in the intertubular tissues. Yet there is no doubt that males that exhibit this testicular condition are fertile. All specimens that I was able to examine exhibited the same testicular structure, yet they sired offspring (Dixson, 1983a). Nor is it likely that vitamin E deficiency in captivity is responsible for testicular anomalies in *Aotus,* or that the condition is necessarily pathological. Treatment of one male with vitamin E (total dose of 1940 mg) between the ages of 72 to 577 days had no effect on testicular histology. How these monkeys, which copulate infrequently and produce so few spermatozoa, are able to impregnate their female partners remains a mystery. Andrologists concerned with the effects of diminishing sperm counts in human males might consider this problem.

There have been no reports published on testicular structure in wild owl monkeys. Examination of a single specimen collected in the Canal Zone revealed that its testicular histology resembled the captives' condition. However, it is not known whether this male was obtained directly from the forest.

Levels of plasma testosterone in captive adult male owl monkeys are apparently normal and are not subject to seasonal changes in animals maintained under artificial day/night conditions. A diurnal rhythm of plasma testosterone occurs (Fig. 1), with highest levels seen during the illuminated phase (mean ± S.E.M., 24.8 ± 5.3 ng/ml) and a nadir (4.7 ± 1.2 ng/ml) during periods of darkness (Dixson and Gardner, 1981). These daily changes in circulating testosterone are the reverse of those reported in some diurnal primate species (e.g., rhesus monkey, Goodman *et al.*, 1974; ringtail lemur, Van Horn *et al.*, 1976). However, since the owl monkey is nocturnal, its plasma testosterone rhythm corresponds to that of other primates in the sense that the highest hormone levels occur during periods of rest or sleep and the lowest levels during periods of wakefulness and activity.

Figure 2 shows changes in circulating testosterone in six young male owl monkeys during growth and throughout puberty. Plasma testosterone first increases when males are 211 to 337 days old (median, 313 days), with larger rises at 300 to 400 days (Dixson *et al.*, 1980). Males enter puberty at a surprisingly early age, particularly when it is considered that in another

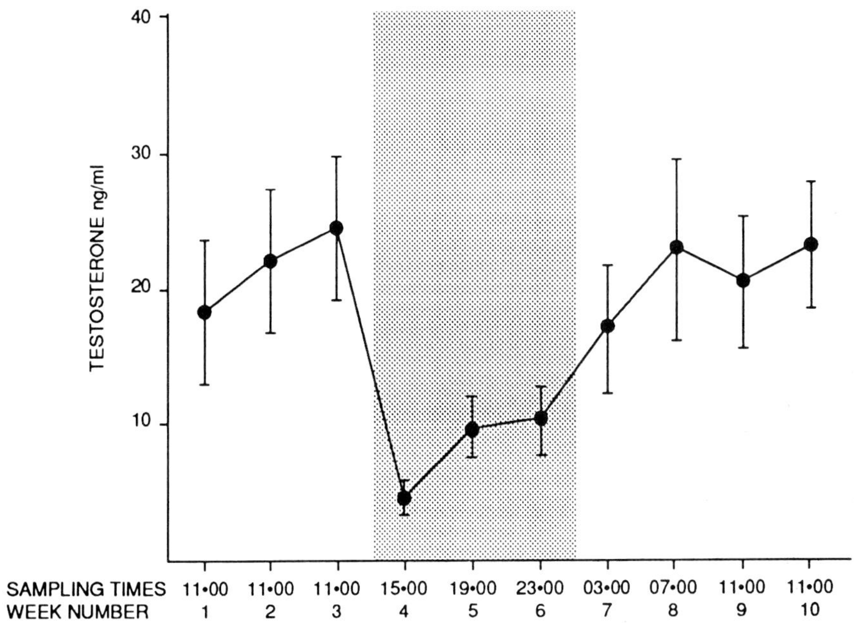

FIGURE 1 Concentrations of plasma testosterone (mean ± S.E.M) in adult male owl monkeys, sampled once weekly at different times of day during 10 consecutive weeks. Animals were maintained under a reversed lighting schedule (lights on from 0200 to 1400 hr; lights off from 1400 to 0200 hr). Stippled area = period of darkness. Number of monkeys sampled = 6 (Weeks 1 to 3) or 5 (Weeks 4 to 10).

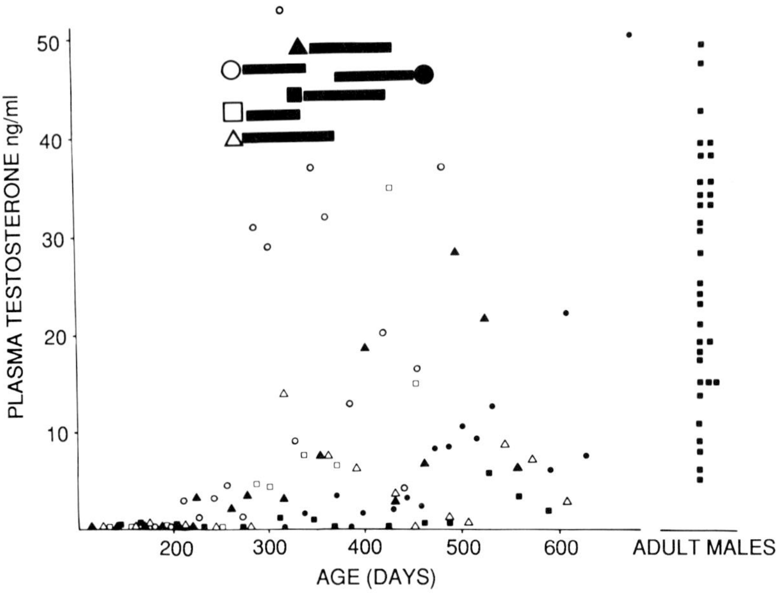

FIGURE 2 Plasma testosterone levels and growth of subcaudal gland in six male owl monkeys aged betwen 114 and 711 days; male 2 (▲), male 3 (△), male 5 (□), male 7 (■), male 4829 (○), and male 4465 (●). Values for adult males are shown on the right for comparison. Horizontal bars indicate time elapsed between appearance and complete development of specialized subcaudal hairs in each male.

New World species, the common marmoset (*Callithrix jacchus*), which weighs half as much as *Aotus*, pubertal increases in plasma testosterone occur between 200 and 250 days (Dixson, 1986). In human males, the first pubertal increases in plasma testosterone are measured in nocturnal blood samples (Judd *et al.*, 1974). All the data shown in Fig. 2 derive from blood samples collected during the *illuminated phase* of the lighting cycle when owl monkeys were resting or asleep. It is probable that the earliest rises in pubertal testosterone were measured under these conditions rather than the later general increases measurable in plasma from males during the nocturnal phase, when owl monkeys are active and foraging.

Changes in mean testis volume during puberty in male owl monkeys are shown in Fig. 3, along with data on testicular volume in adults. Even in the adult, the testes are very small (range, 378 to 673 mm^3; median, 514 mm^3; Dixson *et al.*, 1980). Small errors of measurement can lead to incorrect calculations of testicular volumes during puberty. This observation may account for the fact that, although testis volume increased in all subjects, consecutive measurements fluctuated considerably and obscured changes that occurred at puberty. Given the unusual testicular histology of adult

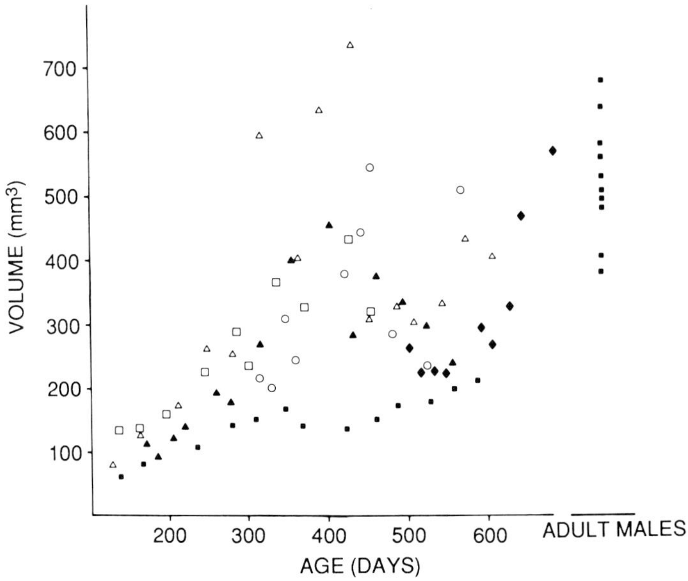

FIGURE 3 Mean testis volume in six male owl monkeys aged between 114 and 711 days. Symbols as in Fig. 2. Individual values for adult males are shown on the right.

males, it would be valuable to conduct studies on histological changes during puberty in owl monkeys. No such studies have been reported.

A useful external marker of testicular activity during puberty in owl monkeys is provided by growth of the subcaudal scent-marking gland. This glandular complex, which is situated under the base of the tail in both sexes, consists of sebaceous and apocrine elements covered by specialized hairs (Hill *et al.*, 1959). The subcaudal hairs begin to stiffen and to turn brown during puberty. In six young males, these changes began between 282 and 370 days of age (median, 316 days) and were complete between 336 and 442 days (median, 397 days). As can be seen in Fig. 2, development of the subcaudal complex correlates with increased levels of circulating testosterone at puberty. Indeed, treatment of one juvenile male with testosterone caused precocious development of the gland, so that it had attained an adult appearance by 278 days of age (Dixson, 1983a).

III. THE OVARIAN CYCLE

The female owl monkey exhibits no changes in its external genitalia or in vaginal cornification throughout the ovarian cycle. Menstruation does not occur, and erythrocytes are never present in appreciable numbers in vaginal smears (Dixson, 1983a). Vaginal smears are occasionally tinged with an orange-brown material, referred to as "smegma" by Hill (1960). In fact, this substance is produced not by the genital tract but by the subcaudal cutaneous glands. These secretions are sometimes introduced into the vagina if the female sits with the subcaudal gland pressed against the genitalia.

Measurements of circulating or urinary steroids provide a reliable means of monitoring the course of the ovarian cycle in *Aotus* (Bonney *et al.*, 1979, 1980). Concentrations of circulating estradiol 17β, estrone, progesterone, and testosterone during ovarian cycles in two females are shown in Fig. 4. The follicular phase of the cycle lasts for 6 days and the luteal phase for 10 days, on average. Basal levels of estradiol 17β average 0.1 to 0.2 ng/ml, and peak levels (0.68 ± 0.05 ng/ml) occur at intervals of 15.5 days. By contrast, progesterone reaches much higher levels in the circulation, averaging 250 ng/ml during the midluteal phase of the cycle. This finding, which

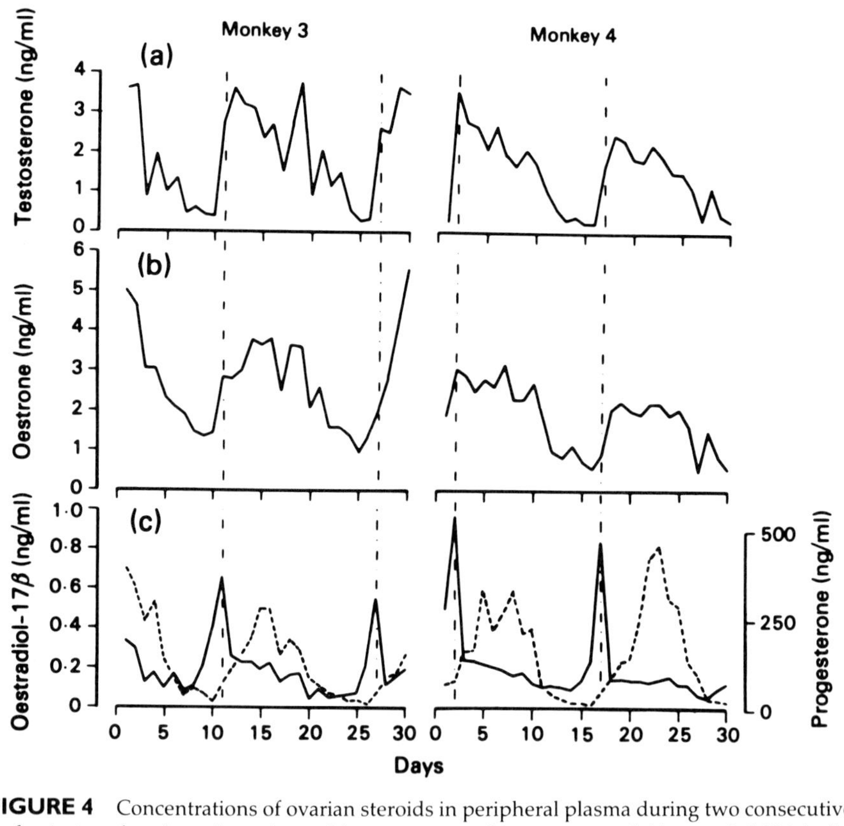

FIGURE 4 Concentrations of ovarian steroids in peripheral plasma during two consecutive cycles in two female owl monkeys: (a) testosterone, (b) estrone, and (c) estradiol 17β (solid line) and progesterone (dashed line). Vertical lines indicate estradiol peaks.

also applies to a number of other New World monkeys, such as the marmosets and tamarins, the squirrel monkey, and *Cebus,* contrasts with the situation found in catarrhine primates. The ovaries of New world monkeys are larger than those of Old World species relative to body weight, mainly due to large amounts of interstitial glandular tissue (Koering, 1974, 1987). Extensive luteinization of the ovaries in some New World monkeys may be responsible for the high levels of progesterone measurable in their plasma (Hertig *et al.*, 1976).

Measurements of steroids excreted in the urine of unmanipulated female owl monkeys confirm that the ovarian cycle lasts for approximately 16 days (Bonney *et al.*, 1979). Sedation using ketamine hydrochloride in order to collect blood samples had no effect on the characteristics of the cycle in habituated subjects, as was also the case in rhesus monkeys (Channing *et al.*, 1977) and baboons (Koyama *et al.*, 1977). This point is significant because, in several other New World primates, variability in cycle lengths has probably been caused by the methods used to handle monkeys in order to collect blood samples or vaginal smears (e.g., in *Callithrix jacchus, Saguinus oedipus,* and *Saimiri sciureus;* Dixson, 1983b).

The major estrogen excreted in the urine of owl monkeys is estrone, as in some other New World primates (*Saguinus fuscicollis, Saimiri sciureus, Cebus albifrons,* and *Ateles fusciceps;* Hodges *et al.*, 1981). The owl monkey excretes progesterone mainly as 6β-hydroxypregnanolone (Bonney and Setchell, 1980), also the major urinary metabolite in marmosets (Shackleton, 1974, 1975). Pregnanediol and 16-hydroxypregnanolone also occur in the urine of the owl monkey; radioimmunoassay of pregnanediol-3α-glucuronide provides a reliable indicator of progesterone excretion during the ovarian cycle.

IV. SEXUAL BEHAVIOR

Owl monkeys in the wild usually live in small groups consisting of a mated pair and younger animals, which are presumed to be their offspring. The mating system is believed to be monogamous (Wright, 1978, 1981, 1989). One characteristic of monogamous primates is the relatively low frequency of copulatory behavior in established pairs (Kleiman, 1977). Captive owl monkeys are rarely observed mating; long-term observations on five captive groups revealed that the males mounted just 19 times during a period of 278 hr (Table I). Given that females in these groups become pregnant, it follows that copulatory activity is coordinated with the female's periovulatory period. Yet detailed studies of sexual activity do not support any assumption that females are sexually receptive or "in estrus" only during the periovulatory phase. Thus, if female owl monkeys are "pair-tested" daily with the same male partners throughout the ovarian cycle, copulations occur much more frequently than anticipated by their monogamous behavior (Table I) and at all stages of the cycle.

Results of one such experiment are shown in Fig. 5. This female was transferred to her male partner's home cage for a 15-min behavioral test on each of 33 consecutive days. Observations were made, under dim red illumination, from behind a one-way-vision mirror to minimize disturbance to the monkeys. Urine samples were collected each day from the floor of the female's cage to measure concentrations of estrone and pregnanediol-3α-glucuronide. The female exhibited two complete ovarian cycles during the course of the experiment, and the frequency of her male partner's sexual and associated behavior, including anogenital inspections, mount attempts, and mounts with pelvic thrusting, did not vary in a predictable fashion during these cycles. Nor were there any changes in mount latencies or mount durations. Copulation is brief in the owl monkey and it is probable that the male ejaculates during a single intromission with pelvic thrusting. However, males do not exhibit the body tremor or pause in thrusting movements that accompany ejaculation in many primate species (Dixson, 1983b). Hence it is not certain whether ejaculations occurred during pair-tests, although mounts and intromissions certainly took place at all stages of the ovarian cycle (Fig. 5).

Results similar to those shown in Fig. 5 were obtained from studies of two further pairs of owl monkeys in which females showed regular ovarian cycles. In a fourth pair, the female conceived during the first few days of the study but continued to accept the male regularly during the first 3 weeks of pregnancy. The fifth female studied failed to show cyclic changes in urinary steroids, but despite the absence of a normal ovarian cycle, she copulated with her partner during most tests.

TABLE I Hourly Frequencies of Selected Behavioral Patterns Observed in Family Groups of Owl Monkeys or during Daily, Opposite-Sex, Pair-tests

Behavioral patterns	Family groups[a] (N = 5)		Pairs[b] (N = 5)	
	Male and female		*Male and female*	
Arching displays	0		2.44	
Copulations	0.07		3.87	
	Male	*Female*	*Male*	*Female*
"Urine-washing"	8.3	5.3	3.93	3.93
Tail-rubbing	7.4	0.56	3.13	7.25

[a] 278 hr of observations

[b] 150 tests of 15 min duration.

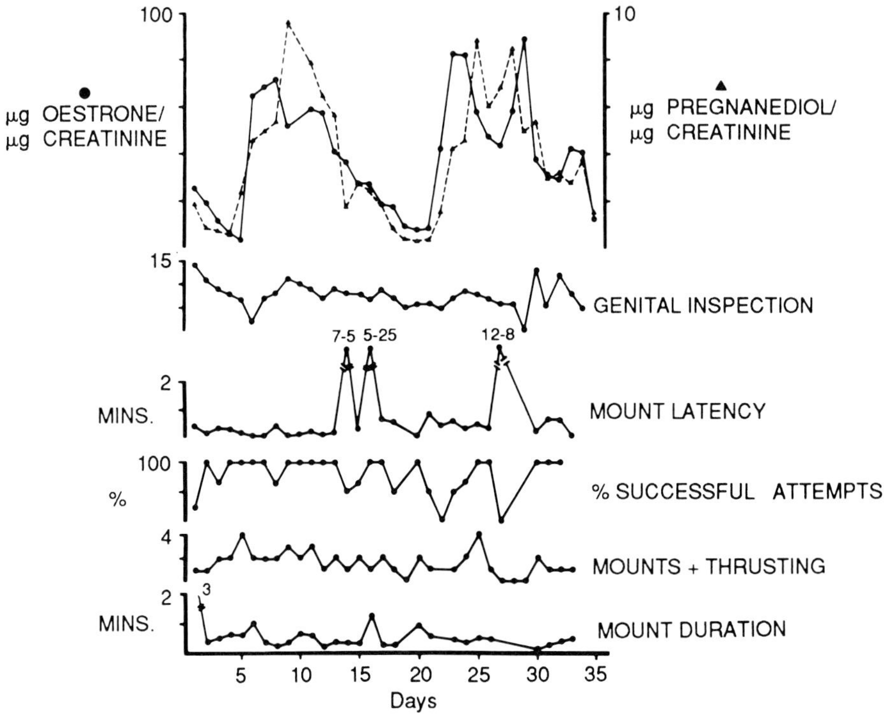

FIGURE 5 Sexual interactions during two consecutive ovarian cycles of a female owl monkey paired daily with the same male. Upper graph shows urinary level of estrone (●) and pregnanediol -3α-glucuronide (▲). Other graphs show measurements of male partner's sexual activity: genital inspections, latency to onset of mounting, percentage of male's mount attempts that were successful, number of mounts with pelvic thrusting, and duration of mounts with thrusting.

These experiments indicate that, under the artificial conditions of a "pair-test," female owl monkeys were sexually receptive (and attractive) to their male partners throughout the ovarian cycle, or even in the absence of a regular cycle. Sexual presentations and other proceptive patterns (Beach, 1976) were never observed during these tests and, as far as could be ascertained, all copulations were initiated by males. Although both sexes exhibited scent-marking patterns (rubbing with subcaudal gland and "urine-washing" displays, see Table I), their frequency bore no relationship to the stages of the ovarian cycle.

In long-term associated groups of owl monkeys, the mechanisms that coordinate sexual activity and ensure fertilization are unknown. It seems likely, however, that the infrequent copulations in captive groups (Table I) are cued primarily by chemical stimuli that enhance female attractiveness.

Chemical communication is of considerable importance in New world primates, many of which possess a vomeronasal organ and accessory olfactory bulbs as well as receive inputs via the main olfactory system (Hunter *et al.*, 1984; Abbott *et al.*, 1993; Epple *et al.*, 1993). Owl monkeys are known to employ olfactory cues in sexual and aggressive contexts (Hunter, 1981; Hunter and Dixson, 1983). It seems likely, therefore, that such cues might play an important role in synchronizing mating behavior within the social group. This is not to deny the fact that tail-rubbing and urine-washing patterns play an important role in communication between owl monkey groups and increase in frequency during intergroup encounters in the wild (Wright, 1989). In captivity, when animals are isolated and the sexes placed together only for brief pair-tests, the male partners are sufficiently aroused to initiate copulation at any stage of the ovarian cycle. Both sexes also scent-mark under these conditions, and females show particularly high frequencies of tail-marking (Table I), perhaps because they are less familiar with their male partners and react in a similar fashion as during intergroup encounters in the wild.

V. GESTATION AND INTERBIRTH INTERVALS

On the basis of a single timed mating, Hunter *et al.* (1979) found that the gestation period of the Colombian owl monkey is 133 days. This is more accurate than previous estimates, which were based on the duration of interbirth intervals (149 to 159 days, Elliott *et al.*, 1976; 5 months, Merritt, 1976). D. A. Merritt, Jr. (personal communication, cited by Hunter *et al.*, 1979) has also recorded that one female gave birth 138 days after first being caged permanently with a male. Interbirth intervals for owl monkeys, shown in Fig. 6, range from 166 to 419 days in 36 cases where no stillbirths, premature births, or abortions were involved (Dixson, 1983a). Thus, in captivity, female owl monkeys will produce one offspring every 8 months on average. There is no obvious pattern in the distribution of interbirth intervals, such as would be expected if a postpartum "estrus" had occurred. Detailed behavioral observations on eight family groups of owl monkeys, for 4 months after the birth of an infant, confirm that there was no resumption of sexual activity afterward. Nor was there any seasonality in the frequency of births, at least under conditions when heating and illumination were controlled (Fig. 6). Whether seasonal changes occur in the wild remains unknown.

VI. PARENTAL BEHAVIOR AND DEVELOPMENT OF OFFSPRING

Newborn owl monkeys weigh between 90 and 105 g (average weight of four newborns, 96.5 g). The pelage is well developed at birth, except that

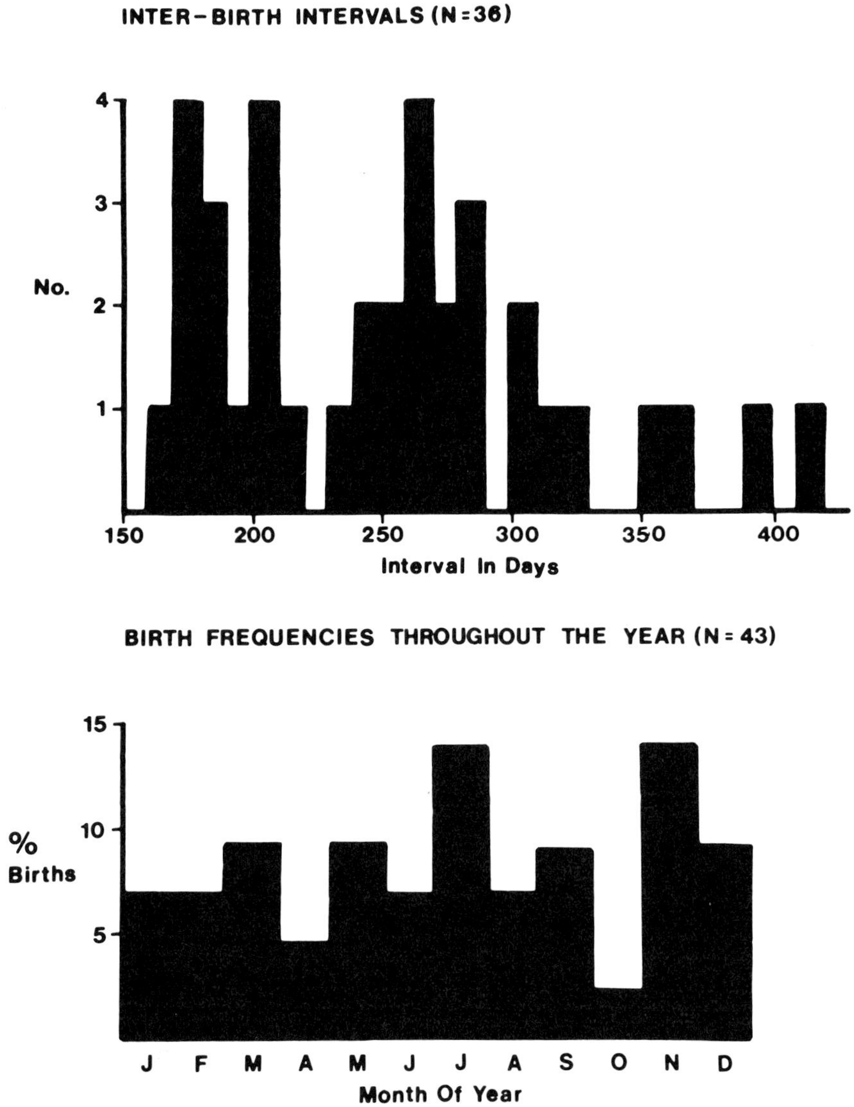

FIGURE 6 Interbirth intervals and birth frequency throughout the year in a captive colony of Colombian owl monkeys. All data refer to animals kept under a reversed lighting regime (12 hr artificial light and 12 hr darkness each day) and at temperatures ranging from 20° to 25°C.

the abdomen and inner surfaces of the limbs are sparsely covered with fine hair. The black triradiate markings on the infant's head join to form a single stripe, which extends across the scalp and down the center of the back. The newborn clings to the mother in a distinctive ventrolateral posture, resting in the flexure of her thigh with its head pointing outward. This positional behavior is either innate or is rapidly learned within 24 hr of birth. Thus, in experiments in which mothers were tranquilized on the day of parturition and restrained in a seated position, their infants were still able to move into a ventrolateral clinging position without assistance (Fig. 7).

Male owl monkeys, as well as females, play a major role in carrying and caring for their offspring. Detailed observations of parental behavior and infant development in 13 captive groups of *Aotus lemurinus* were reported some years ago (Dixson and Fleming, 1981). Some results of those studies are summarized in Table II and Figs. 8–10.

The ages of first observed occurrences of selected behavioral patterns are shown in Table II. Development is rapid, and the offspring may transfer from the female to the male parent on the day of birth in some groups. Dorsal clinging may also be observed on Day 1, but it is not until 3 to 4 weeks of age that dorsal clinging replaces ventrolateral clinging as the major position in which the infant rests on the parent. The offspring first gets off the parents between 22 and 46 days and begins to eat solid food between

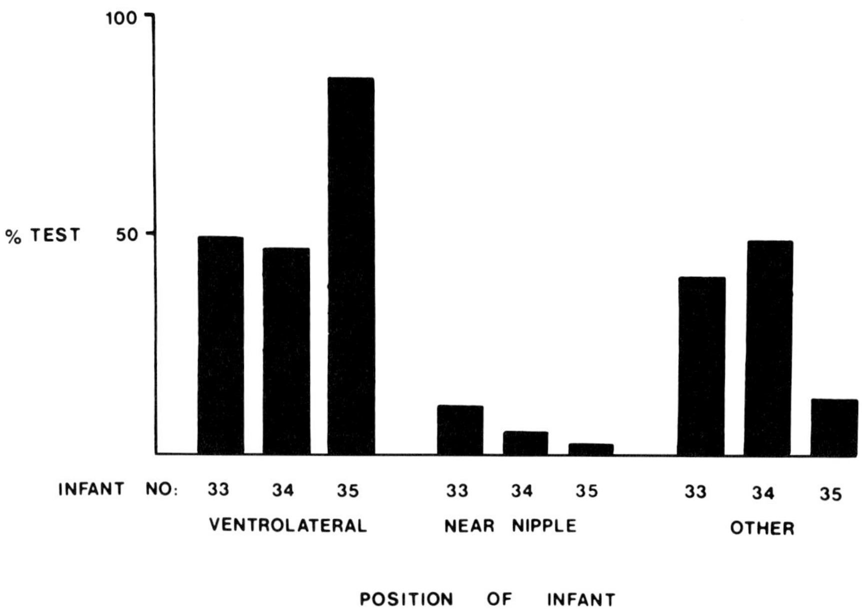

FIGURE 7 Positional behavior by newborn infant owl monkeys after tranquilization of their mothers. Three mother–infant pairs were studied for 4 or 5 trials, each of 5 min duration. At the start of each trial, the infant was removed from the mother's body and immediately replaced at shoulder level. All infants spent the majority of the trials resting in a "ventrolateral" position in the flexure of the mother's thigh.

TABLE II Age of First Observed Occurrence of Selected Behaviors in 10 Infant Owl Monkeys

Infant no.	Sex	Day of first observation (days)	Transfers from male to female parent	Clings dorsally	Gets off	Eats solid food	"Urine-washes"[a]
2	M	5	5	5	40	45	101
3	M	6	6	6	35	35	70
4	F	1	3	3	22	35	63
5	M	1	2	2	26	48	105
12	M	2	2	2	23	37	62
13	M	1	1	1	37	47	109
14	M	2	2	2	27	48	109
15	M	1	2	1	26	39	76
CZ_1	M	3	3	6	46	60	141
CZ_2	F	1	1	1	34	42	48
	Range		1–6	1–6	22–46	35–60	48–141
	Median		2	2	30.5	43.5	88.5

[a] A scent marking behavior.

35 and 60 days. The growth of independence is a complex process, however, in which the parents play different roles. The infant is carried by the mother for most of the first postnatal week, but thereafter the male parent performs this function, and the infant transfers back to its mother mainly to suckle (Fig. 8). Time spent on the male peaks during Week 3, then shows a gradual decline. This correlates with the fact that from Week 4 or 5 onward, infants spend increasing amounts of time off their parents (Fig. 8). The number of "time-off" episodes is greatest between Weeks 7 and 11, then declines as the infant spends progressively longer periods without returning to its parents. By the 18th week, the offspring moves independently most of the time, but should any disturbance occur, it invariably returns to the adult male and attempts to reestablish contact.

Female owl monkeys suckle their offspring for periods lasting from 7 to 160 sec (in 146 cases, mean duration = 27 sec). As in other New World primates, the nipples are axillary in position, and the female usually lifts her arm slightly to facilitate the infant's attempts to suckle. The infant makes rapid side-to-side head movements as it seeks the nipple. It may shift from one nipple to the other a number of times during each suckling episode (range, 1 to 9 shifts/suckling episode; mean, 1.57). Owl monkeys rarely remain on the nipple except while suckling, in contrast to many Old World Monkeys (e.g., rhesus monkeys, Hinde and Spencer-Booth, 1968; baboons, Rowell *et al.*, 1968). Figure 9 shows that suckling frequency declines from an average of 1.0 occurrence/hr in Weeks 1 to 3 to a mean of

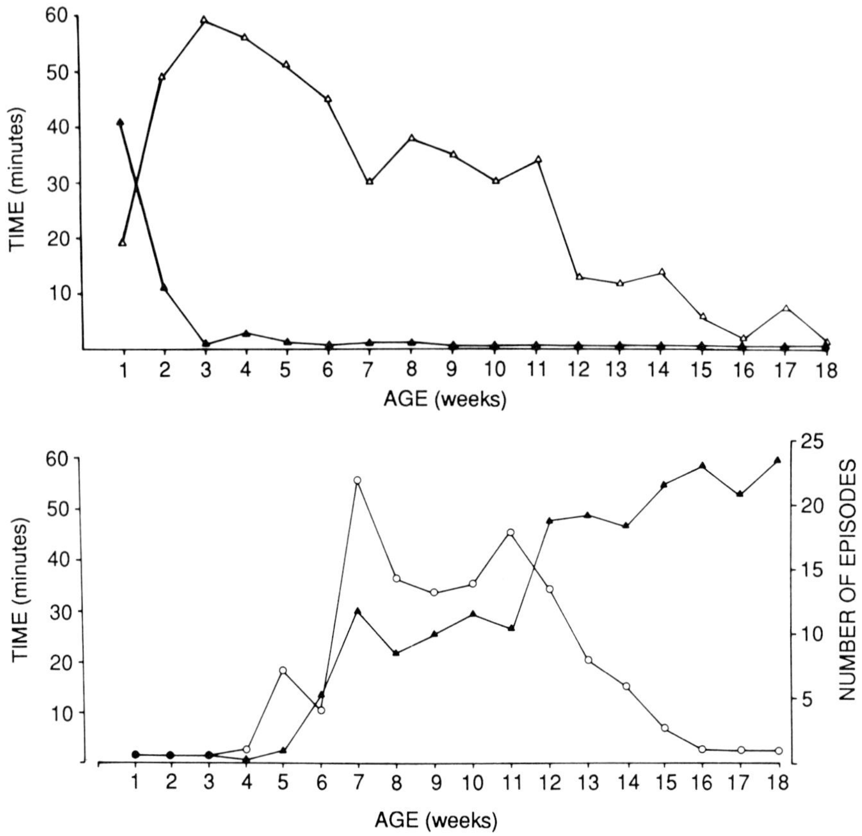

FIGURE 8 Upper graph: Time spent by infant owl monkeys on mother (▲) and on father (△). Lower graph: Time spent "off" parents by infant owl monkeys (▲) and number of time-off episodes (○). Data, expressed as mean/hr/week, refer to seven family groups, each containing a single offspring.

0.27/hr during Weeks 16–18. Forty of the 146 observed suckling episodes (i.e., 27%) were followed immediately by maternal rejection, which consisted of biting the infant's tail, hands, and feet. The infant promptly transferred to the male parent after 90% of such rejections. The pattern of postsuckling rejection varied according to the infant's age and was highest (65%) between 4 and 6 weeks of age (Fig. 9). This is the period when the offspring begins to take solid food, so it is possible that the mother's punitive behavior plays some role in the weaning process. It is also intriguing to consider whether some link might exist between the neuroendocrine reflex of milk ejection and subsequent infant rejection by female owl monkeys, given that the infant is rejected so promptly after it has suckled (Dixson and Fleming, 1981).

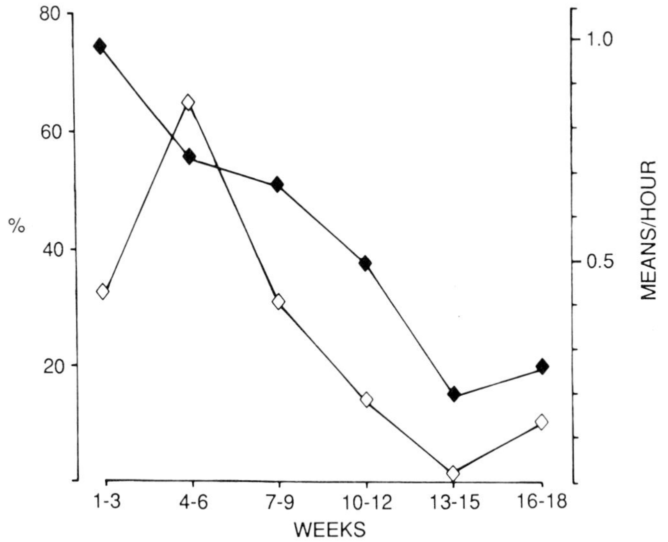

FIGURE 9 Frequencies of suckling (◆) and of maternal rejections after suckling (◇) in owl monkeys. Data are for 10 infants in 7 family groups. Suckling episodes expressed as mean frequencies/hr/week. Rejections after suckling are expressed as percentages.

In Harlow's (1971) studies of maternal–infant attachment in rhesus monkeys, one classic experiment provided infants with a choice between two artificial "surrogate mothers" that offered either tactile comfort or a supply of milk. Infants spent more time clinging to the surrogates that were covered with soft material rather than to wire mesh surrogates equipped with a milk supply. In a sense, the parenting system of owl monkeys provides a natural parallel to this situation. Female owl monkeys provide their offspring with milk but reject them more frequently than males when they are between 2 and 8 weeks of age (Fig. 10). The infant then transfers actively to the male parent, which provides tactile comfort, warmth, etc., and moves to the mother only to suckle. Males rarely reject their offspring before they are 8 weeks old, but then frequencies of paternal rejections increase and exceed the maternal levels (Fig. 10). Many rejections occur in response to the infant's attempts to reestablish contact with the male after a "time-off" episode. The male's behavior may be important in encouraging the offspring to spend longer periods on its own and hence may contribute to the growth of its independence.

Hinde and Spencer-Booth (1968) and Hinde (1971) have employed a method of measuring the maintenance of proximity between parent and offspring that is useful when applied to the owl monkey. The percentage of approaches (AP) and "leavings" (L) that are originated by the infant

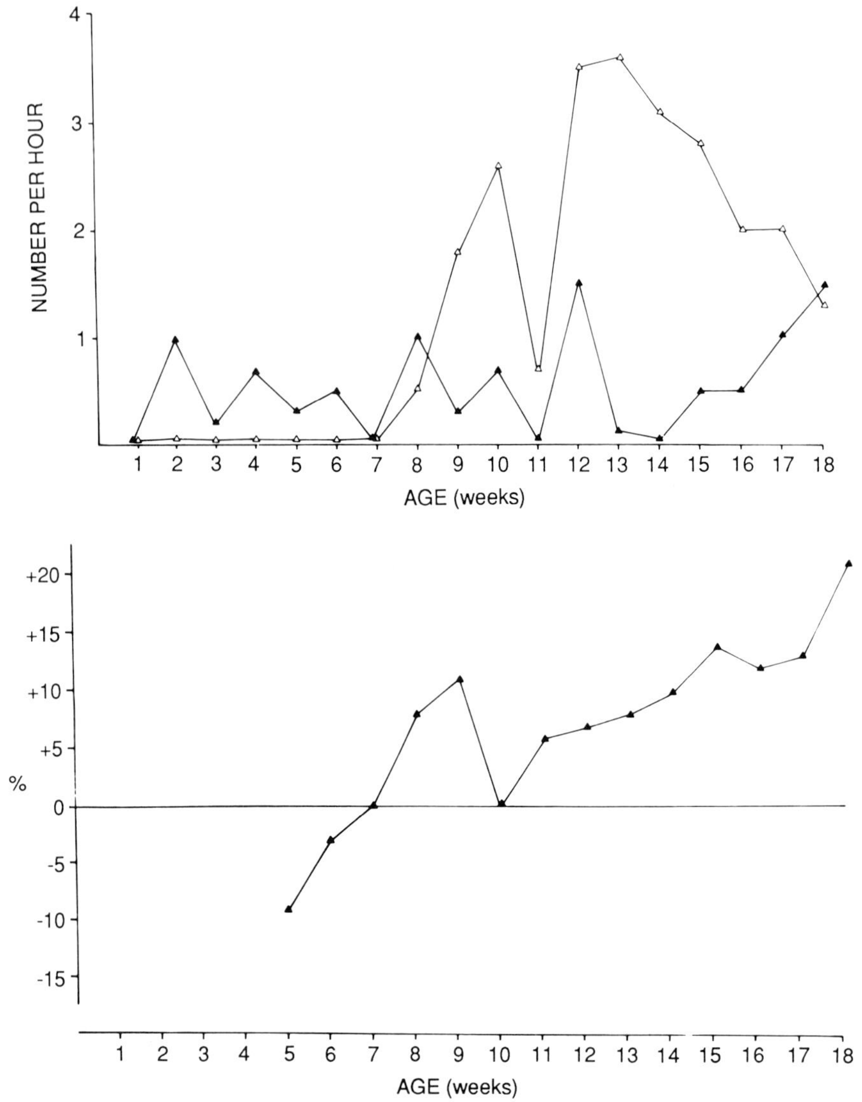

FIGURE 10 Upper graph: Rejection of infant owl monkeys by mother (▲) and by father (△). Lower graph: Percentage of approaches (AP) and "leavings" (L) originated by infant owl monkey, rather than male parent, expressed by the formula $\%AP - \%L_{inf}$. Data are mean frequencies/hr/week for seven family groups, each containing a single offspring.

rather than its parent are represented by the formula $\%AP - \%L_{inf}$. The data in Fig. 10 show that during Weeks 5 and 6 (when there are negative values for the equation), the male parent is principally responsible for maintaining proximity between itself and the offspring. From Week 8 onward the situation is reversed: maintenance of proximity is due mainly to the infant, as indicated by positive values for the equation.

VII. DISCUSSION

Much has been learned about the reproductive biology of owl monkeys during the last two decades; there is now sufficient information to manage captive breeding programs. However, there are several major gaps in our knowledge of this most interesting genus. It is not known whether seasonal factors influence reproduction in the wild. Although owl monkeys breed throughout the year in captivity, it is well to remember that the nine species of the genus *Aotus* occupy an enormous distribution range (Hershkovitz, 1983). All of these species are nocturnal but they may nonetheless, experience a wide variety of environmental pressures, including seasonal changes in temperature, rainfall, and diet. It is quite possible that seasonal changes in birth frequencies occur in some populations.

Owl monkeys live in small groups and are believed to be monogamous (Wright, 1981, 1989). Studies of their sexual and parental behavior in captivity are certainly consistent with this view. The high degree of care shown by male owl monkeys toward their offspring accords with theories of parental investment (Trivers, 1972). Their small testicular size resembles the situation found in monogamous primates in which sperm competition is minimal compared to that in certain polygamous primates (Harcourt *et al.*, 1981). However, detailed studies of many "monogamous" avian species have demonstrated that extra-pair copulations and "cuckoldry" are not uncommon (Birkhead and Moller, 1992). The application of DNA fingerprinting techniques is currently producing valuable insights into the mating systems of nonhuman primates (Martin *et al.*, 1992). It would be useful to apply such techniques to studies of genetic relatedness and reproductive success in a monogamous primate such as *Aotus*. Such an approach might provide information concerning the formation of new social groups of owl monkeys. In captivity, maturing offspring are tolerated by their parents; there is no increase in agonistic behavior between pubertal owl monkeys and their parents (Dixson, 1983a). However, serious fights occasionally occur in some groups, between postpubertal offspring and their parents, as the former approach maturity, at between 18 and 24 months of age. The reasons for these sudden aggressive outbursts are not known. Perhaps, under normal conditions, mature offspring emigrate actively from their natal groups before displays of aggression to their parents reach such high levels.

In the laboratory, many questions remain to be answered concerning the reproductive physiology of the owl monkey. For example, we still know very little about endocrine changes during pregnancy. Most puzzling to this author, however, is the question of testicular function in captive owl monkeys. Why do males produce so few spermatozoa, and how is fertilization accomplished under these circumstances? The small numbers of sperm produced must possess remarkable qualities! There is also the question of how the small numbers of sperm are transported within the female's reproductive tract. Before the testicular condition found in captive male owl monkeys is assumed to be pathological, it is essential to examine feral males. It might be possible, for instance, to collect testicular biopsies from live feral animals. One hopes that an opportunity to undertake these studies may arise in the future.

ACKNOWLEDGMENTS

The original research on Colombian *Aotus* described here received generous financial support from the Wellcome Trust (U.K.), the World Health Organization, and the Zoological Society of London.

REFERENCES

Abbott, D. H., Barrett, J., and George, L. M. (1993). Comparative aspects of the social suppression of reproduction in female marmosets and tamarins. *In* "Marmosets and Tamarins: Systematics, Behavior and Ecology" (A. B. Rylands, ed.), pp. 152–163. Oxford Univ. Press, Oxford.

Aquino, R., and Encarnación, F. (1986). Population structure of *Aotus nancymae* (Cebidae: Primates) in Peruvian Amazon lowland forests. *Am. J. Primatol.* **11,** 1–7.

Beach, F. A. (1976). Sexual attractivity, proceptivity and receptivity in female mammals. *Horm. Behav.* **7,** 105–138.

Birkhead, T., and Moller, A. P. (1992). "Sperm Competition In Birds." Academic Press, London and New York.

Bonney, R. C., and Setchell, K. D. R. (1980). The excretion of gonadal steroids during the reproductive cycle of the owl monkey (*Aotus trivirgatus*). *J. Steroid Biochem.* **12,** 417–421.

Bonney, R. C., Dixson, A. F., and Fleming, D. (1979). Cyclic changes in the circulating and urinary levels of ovarian steroids in adult owl monkeys. *J. Reprod. Fertil.* **56,** 271–280.

Bonney, R. C., Dixson, A. F., and Fleming, D. (1980). Plasma concentrations of oestradiol 17B, oestrone, progesterone and testosterone during the ovarian cycle of the owl monkey (*Aotus trivirgatus*). *J. Reprod. Fertil.* **60,** 101–107.

Channing, C. P., Fowler, S., Engel, B., and Vitek, K. (1977). Failure of daily injections of ketamine HCl to adversely alter menstrual cycle length, blood estrogen and progesterone levels in the rhesus monkey. *Proc. Soc. Exp. Biol. Med.* **155,** 615–619.

Cicmanec, J. C., and Campbell, A. K. (1977). Breeding the owl monkey (*Aotus trivirgatus*) in a laboratory environment. *Lab. Anim. Sci.* **27,** 512–517.

Dixson, A. F. (1982). Some observations on the reproductive physiology and behavior of the owl monkey in captivity. *Int. Zoo Yearb.* **22,** 115–119.

Dixson, A. F. (1983a). The owl monkey. *In* "Reproduction in New World Primates" (J. P. Hearn, ed.), pp. 69–113. MTP Press, Lancaster, UK.

Dixson, A. F (1983b). The hormonal control of sexual behavior in primates. *Oxford Rev. Reprod. Biol.* **5,** 131–219.

Dixson, A. F. (1986). Plasma testosterone concentrations during postnatal development in the male common marmoset. *Folia Primatol.* **47,** 166–170.

Dixson, A. F. (1987). The owl monkey. *In* "Universities Federation of Animal Welfare Handbook" (T. Poole, ed.), Chapter 38. Longmans, UK.

Dixson, A. F., and Fleming, D. (1981). Parental behavior and infant development in owl monkeys (*Aotus trivirgatus griseimembra*). *J. Zool.* **194,** 25–39.

Dixson, A. F., and Gardner, J. S. (1981). Diurnal variations in plasma testosterone in a male nocturnal primate, the owl monkey. *J. Reprod. Fertil.* **62,** 83–86.

Dixson, A. F., Gardner, J. S., and Bonney, R. C. (1980). Puberty in the male owl monkey, a study of physical and hormonal development. *Int. J. Primatol.* **1,** 129–139.

Dobson, A. (1992). People and disease. *In* "The Cambridge Encyclopedia of Human Evolution" (S. Jones, R. Martin, and D. Pilbeam, eds.), pp. 411–420. Cambridge Univ. Press, Cambridge, UK.

Elliott, M. W., Sehgal, P. K., and Chalifoux, L. V. (1976). Management and breeding of *Aotus trivirgatus*. *Lab. Anim. Sci.* **26,** 1037–1040.

Epple, G., Belcher, A. M., Kuderling, I., Zeller, U., Scolnick, L., Greenfield, K. L., and Smith, A. B. (1993). Making sense out of scents: Species differences in scent glands, scent-marking behavior and scent-mark composition in the Callitrichidae. *In* "Marmosets and Tamarins: Systematics, Behavior and Ecology" (A. B. Rylands, ed.), pp. 123–151. Oxford Univ. Press, Oxford.

Geiman, Q. M., and Meagher, M. J. (1967). Susceptibility of a New World monkey to *Plasmodium falciparum* from man. *Nature (London)* **215,** 437.

Goodman, R. L., Hotchkiss, J., Karsch, F. J., and Knobil, E. (1974). Diurnal variations in serum testosterone concentrations in the adult male rhesus monkey. *Biol. Reprod.* **11,** 624–630.

Harcourt, A. H., Harvey, P. H., Larson, S. G., and Short, R. V. (1981). Testis weight, body weight and breeding system in primates. *Nature (London)* **293,** 55–57.

Harlow, H. F. (1971). "Learning to Love." Ballantine Books, New York.

Hershkovitz, P. (1983). Two new species of night monkeys, genus *Aotus* (Cebidae, Platyrrhini): A preliminary report on *Aotus* taxonomy. *Am. J. Primatol.* **4,** 209–243.

Hertig, A. T., Barton, B. R., and Mackey, J. J. (1976). The female reproductive tract of the owl monkey (*Aotus trivirgatus*) with special reference to the ovary. *Lab. Anim. Sci.* **26,** 1041–1067.

Hill, W. C. O. (1960). "Primates: Comparative Anatomy and Taxonomy," Vol. 4, Part B. Edinburgh Univ. Press, Edinburgh.

Hill, W. C. O., Appleyard, H. M., and Auber, L. (1959). The specialized area of skin glands in *Aotus* Humboldt (Simiae: Platyrrhini). *Trans.—R. Soc. Edinburgh* **63,** 535–551.

Hinde, R. A. (1971). Development of social behavior. *In* "Behavior of Non-human Primates" (A. M. Schrier and F. Stollnitz, eds.), Vol. 3, pp. 1–68. Academic Press, New York and London.

Hinde, R. A., and Spencer-Booth, Y. (1968). The study of mother–infant interaction in captive, group-living rhesus monkeys. *Pro. R. Soc. London, Ser. B* **169,** 177–201.

Hodges, J. K., Gulick, B. A., Czekala, N. M., and Lasley, B. L. (1981). Comparison of urinary oestrogen excretion in South American primates. *J. Reprod. Fertil.* **61,** 83–90.

Hunt, R. D., Chalifoux, L. V., King, N. W., and Trum, B. F. (1975). Arrested spermatogenesis in owl monkeys (*Aotus trivirgatus*). *J. Med. Primatol.* **4,** 399–400.

Hunter, A. J. (1981). Chemical communication, aggression and sexual behavior in the owl monkey (*Aotus trivirgatus griseimembra*) Ph.D. Thesis, University of London, London.

Hunter, A. J., and Dixson, A. F. (1983). Anosmia and aggression in male owl monkeys (*Aotus trivirgatus*). *Physiol. Behav.* **30,** 875–879.

Hunter, A. J., Martin, R. D., Dixson, A. F., and Rudder, B. C. C. (1979). Gestation and interbirth intervals in the owl monkey (*Aotus trivirgatus griseimembra*). *Folia Primatol.* **31,** 165–175.

Hunter, A. J., Fleming, D., and Dixson, A. F. (1984). The structure of the vomeronasal organ and nasopalatine ducts in *Aotus trivirgatus* and some other primate species. *J. Anat.* **138,** 217–225.

Judd, H. L., Parker, D. C., Siller, T. M., and Yen, S. S. C. (1974). Nocturnal rise of plasma testosterone in pubertal boys. *J. Clin. Endocrinol. Metab.* **38,** 710.

King, N. (1975). Morphological observations on the owl monkey testicles. *Primate Rec.* **6,** 13.

Kleiman, D. G. (1977). Monogamy in mammals. *Oxford Rev. Biol.* **52,** 36–69.

Koering, M. J. (1974). Comparative morphology of the primate ovary. *Contrib. Primatol.* **3,** 38–81.

Koering, M. J. (1987). Follicle maturation and atresia: Morphological correlates. *In* "the Primate Ovary" (R. L. Stouffer, ed.), pp. 3–23. Plenum, New York.

Koyama, T., de la Peria, A., and Hagino, N. (1977). Plasma estrogen, progestin and luteinizing hormone during the normal menstrual cycle in the baboon: Role of luteinizing hormone. *Am. J. Obstet. Gynecol.* **127,** 67–72.

Martin, R. D., Dixson, A. F., and Wickings, E. J. (1992). "Parternity in Primates: Genetic Tests and Theories." Karger, Basel.

Merritt, D. A., Jr. (1976). The owl monkey, *Aotus trivirgatus;* Husbandry, behavior and breeding. *Proc. Natl. Conf. American Association of Zoological Parks and Aquaria,* pp. 107–123.

Rathbun, G. B. (1980). Ecological survey of the night monkey, *Aotus trivirgatus,* in Formosa Province, Argentina. *Primates* **21,** 211–229.

Rowell, T. E., Din., N. A., and Omar, A. (1968). The social development of baboons in their first three months. *J. Zool.* **155,** 461–483.

Shackleton, C. H. L. (1974). Progesterone and oestrogen metabolism in the pregnant marmoset (*Callithrix jacchus*). *J. Steroid Biochem.* **5,** 597–600.

Shackleton, C. H. L. (1975). The excretion of steroids by the adult marmoset monkey (*Callithrix jacchus*). *J. Steroid Biochem.* **6,** 1429–1432.

Trivers, R. L. (1972). Parental investment and sexual selection. *In* "Sexual Selection and the Descent of Man" (B. Campbell, ed.), pp. 136–179. Heinemann, London.

Van Horn, R. N., Beamer, N., and Dixson, A. F. (1976). Diurnal variations of plasma testosterone in two prosimian primates, *Galago crassicaudatus crassicaudatus* and *Lemur catta. Biol. Reprod.* **15,** 523–528.

Wright, P. C. (1978). Home range, activity pattern and agonistic encounters of a group of night monkeys (*Aotus trivirgatus*) in Peru. *Folia Primatol.* **29,** 43–55.

Wright, P. C. (1981). The night monkeys, genus *Aotus. In* "Ecology and Behavior of Neotropical Primates" (A. E. Coimbra-Filho and R. A. Mittermeier, eds.), Vol. 1, pp. 211–240. Acad. Bras. Cienc., Rio de Janeiro.

Wright, P. C. (1989). The nocturnal primate niche in the New World. *J. Hum. Evol.* **18,** 635–658.

5

Husbandry and Medical Management of the Owl Monkey

Janet F. Baer
California Institute of Technology
Division of Biology
Pasadena, California 91125

I. INTRODUCTION

Members of the genus *Aotus,* commonly referred to as owl monkeys, night monkeys, or douroucouli, have been used historically in biomedical research to address a diverse array of scientific questions. These moderately small, nocturnal Neotropical primates possess several unique features that make them ideal for use as an animal model of human disease. These features include susceptibility to infection with *Plasmodium* spp., large eyes with easily visualized retinas, adaptability to captivity, and relative ease of handling.

Although they have been maintained in zoological gardens and biomedical research facilities for more than 30 years, little has been written on the management and veterinary care of captive owl monkeys. The purpose of this chapter is to provide the reader with a general understanding of the husbandry and medical management of the owl monkey.

II. COLONY HUSBANDRY

A. Caging Systems

Owl monkeys adapt readily to standard laboratory primate caging systems constructed of stainless steel, aluminum alloys, or heavy plastic-coated wire mesh (Fig. 1). A variety of cage designs have been used in zoological settings. Ease of cleaning, durability, resistance to rust and corrosion, and animal safety and well-being must be considered when evaluating cage design and materials. The wire mesh of the cage floor should be wide enough to allow urine and feces to pass through but narrow enough to provide a substantial surface for locomotion, confine infants, and prevent limb injuries due to accidental entrapment.

The use of woodchips as floor litter has been reported to increase activity and time spent on the ground and decrease aggression in a number of primate species, including the primarily arboreal marmosets and tamarins (Chamove *et al.*, 1982). Contrary to expectation, woodchip litter was found to inhibit growth of *Salmonella typhimurium* and coliform bacteria over time (Chamove *et al.*, 1982). Additional benefits include a decrease in overall odor, cleaner-appearing animals and walls, and provision of a surface suitable for foraging opportunities. McKenzie *et al.* (1986) reported similar findings when using shredded paper as litter for two arboreal callitrichid primates (*Callithrix jacchus* and *Saguinus oedipus*). Extrapolated to *Aotus,* these data suggest that the use of floor litter may serve as a viable environmental enrichment resource.

The Animal Welfare Act (1992) specifies that "the minimum space that must be provided to each nonhuman primate, whether housed individually

FIGURE 1 Caging system with removable partition for cage expansion.

or with other nonhuman primates, will be determined by the typical weight of animals of its species." Owl monkeys, because they are similar in size to the squirrel monkey, are classified as Group 2 species. Minimum cage dimensions for Group 2 species include a 3-ft^2 floor area and cage height of 30 inches. Group-housed animals must be provided with the sum of the floor space required for each individual animal with the exception of infants under 6 months of age.

The Animal Welfare Act (1992) also mandates that "a physical environment adequate to promote the psychological well-being of primates" must be provided for all captive nonhuman primates. At a minimum, the cage environment should include a perch, a food container, and a water device or system (i.e., bowl, bottle, or automatic water dispenser). The physical environment of the cage may be enhanced by incorporation of a multilevel perch system, a nest box, a cage partition, and enrichment devices such as novel objects or simple food puzzles. Multilevel perches increase the amount of effective cage space. Nest boxes provide visual barriers that may be used by an animal to avoid direct eye contact with conspecifics or a human observer. Cage partitions can be removed, allowing expansion of overall cage size to accommodate a large family group or provide additional space during the introduction of new animals. Environmental enrichment devices, such as novel objects or simple food puzzles, may be designed to provide an opportunity for expression of species-typical activities. Items utilized for environmental enrichment should be easy to sanitize, durable, or replaceable, and pose no threat to animal safety.

Owl monkeys are strictly arboreal in the wild and prefer to remain in the upper horizontal portion of their environment when given the option in captivity. If provided with elevated perches, owl monkeys rarely move to the cage floor during periods of bright illumination but can be seen utilizing a greater amount of their cage space, including the floor, during observation under red light illumination. A similar preference for elevated perches has been clearly documented for the squirrel monkey (*Saimiri boliviensis*), an arboreal species similar in size to the owl monkey (Williams *et al.*, 1988). Perches and other structures within the cage should be placed according to the animal's tendency to maintain an arboreal status within the captive environment.

In the wild, owl monkeys do not build nests but are very selective in their choice of sleeping sites (Aquino and Encarnación, 1986b). Characteristics of preferred sleeping sites include protection from predators, easy access, shelter from environmental parameters, and provision of sufficient space to accommodate an entire family group. A tree hole with a cylindrical internal cavity and circular access hole appears to be the most preferred sleeping site (Aquino and Encarnación, 1986b). Owl monkeys in captivity prefer to remain protected from full view when a human observer is present. When approached by a human other than a familiar caretaker, many owl monkeys will enter a nest box if given the opportunity. Captive-born, hand-reared animals may be an exception to this generalization. Even in the absence of a human observer, nest boxes serve as preferred sleeping sites for owl monkeys in captivity.

Nest boxes should be constructed to withstand routine cleaning and normal wear and tear. Small animal airline transport boxes made of plastic can be modified to serve as nest boxes. This type of nest box may be placed within or attached to the side of the cage. Modifications of the transport box include removal of the front door, replacement of a minimum of half the floor with wire mesh securely attached to the bottom of the transport box, and placement of a perch above the box floor. Replacement of the floor and installation of a perch above the floor reduce the risk of infection due to an animal sitting in a grossly soiled nest box; the modified design allows fecal material and urine to fall through the bottom of the nest box and the perch allows the animal to remain off the soiled nest box floor (Fig. 2).

A second nest box design consists of an 8-inch-diameter, 18-inch length of PVC pipe with a circular 4-inch "peep hole" cut out nondiagonally in the upper and lower thirds of the pipe. A horizontal perch constructed of 1.5-inch-diameter PVC pipe may be placed approximately two-thirds of the

FIGURE 2 Nest box created from a small animal transport box.

distance from the bottom of the larger PVC pipe. This type of nest box may be placed on the bottom of the cage floor or may be suspended within the cage (Figs. 3 and 4).

B. Sanitation

Routine cleaning and sanitation of the caging system, including perches, nest boxes, food and water containers, and any additional items within the cage environment, are essential to remove potential pathogens and reduce waste that may attract insects or rodents. Urine, feces, and uneaten food should be removed daily. In most suspended-floor caging systems, daily hosing of the cage floor with water is adequate. Care should be taken to ensure that the animals remain dry during the daily cleaning routine as a wet animal may be especially prone to hypothermia. Food and water containers should be sanitized daily. Cages and nest boxes must be thoroughly sanitized a minimum of once every 2 weeks by using a cage washer with water temperatures of at least 82.8°C (180°F) or by hand scrubbing with an appropriate disinfectant.

Owl monkeys possess a subcaudal scent gland that they rub in a back and forth motion on tree limbs during agonistic encounters with conspecifics in

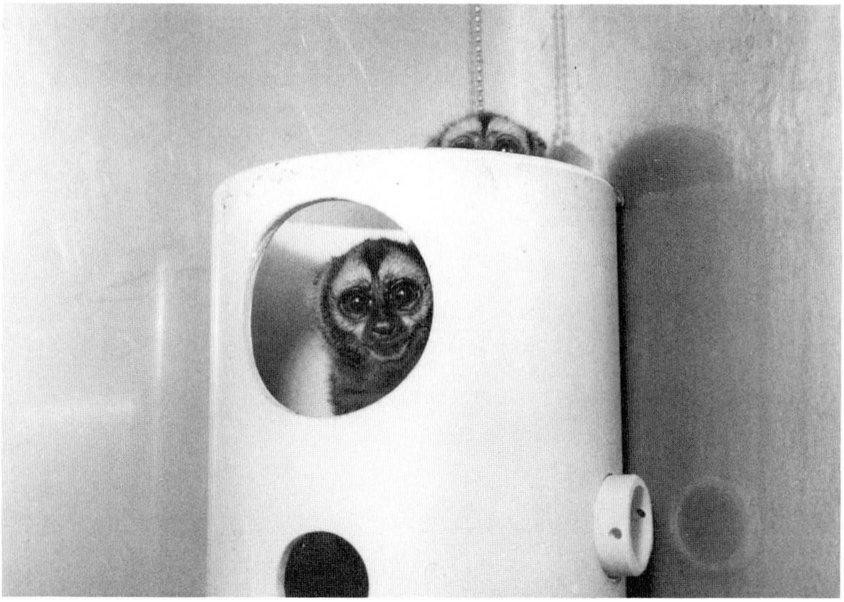

FIGURE 3 Nest box created from PVC pipe material.

the wild (Wright, 1978; Aquino and Encarnación, 1986a). This scent gland is located on the ventral surface of the base of the tail. Additionally, a large aggregate of apocrine glands is located on the chest of males and a smaller aggregate of similar glandular tissue on female and juvenile owl monkeys (Chapman *et al.*, 1985; Aquino and Encarnación, 1986a). The function of this chest gland is unknown but may be related to territorial marking.

Owl monkeys appear less stressed during cage washing and nest box changes if these items are cleaned on alternate weeks. Sanitizing these items on an alternating cycle provides stability of olfactory cues within the animal's environment. Ideally, owl monkeys should be trained to enter their scent-marked nest box, which can then be secured and transferred to a clean cage. This system precludes having to stress the animals by individually capturing each one, and allows the transfer of olfactory cues through transfer of the scent-marked nest box. Nest boxes may simply be removed for cleaning while the animals are secured within their scent-marked cage.

Items that are transferable between cages, including food and water containers and environmental enrichment devices, should be sanitized prior to transfer. Items used to handle numerous animals sequentially (i.e., transport boxes, gloves, nets, etc.) should be sanitized as frequently as possible to minimize the possibility of transfer of an infectious agent. Where

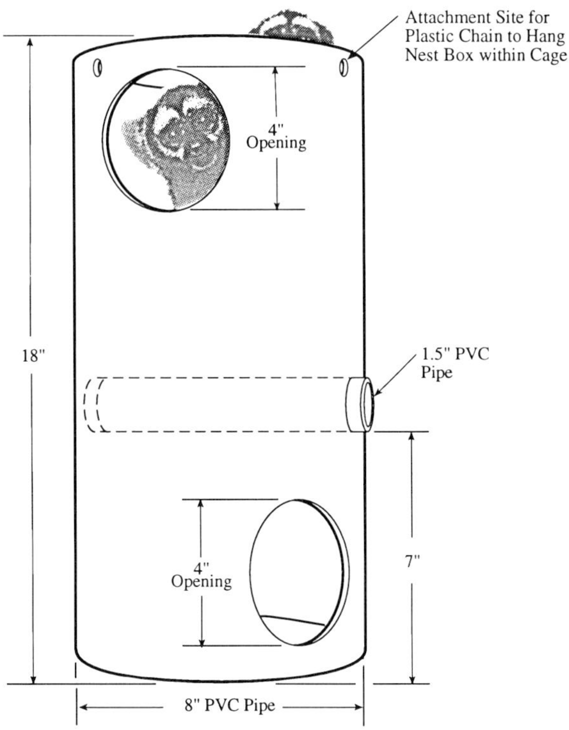

FIGURE 4 Schematic of nest box created from PVC pipe.

appropriate these items should be restricted for use within defined groups of animals. All items that come in contact with a sick animal should be either discarded or appropriately disinfected prior to reuse.

C. Environmental Conditions

Like many tropical species, owl monkeys exhibit a narrow thermoneutral zone, with critical lower and upper temperatures at 28° and 30°C, respectively (Le Maho *et al.*, 1981). Behavioral and physical adaptations to low ambient temperatures include high body insulation provided by dense pelage, body heat generated by nocturnal activity, reduced total body surface area through resting in a ball-like posture, and shivering during exposure to temperatures below 28°C. Adaptation to elevated ambient temperatures includes resting during the hottest period of the day and panting at ambient temperatures above 30°C (Le Maho *et al.*, 1981).

The recommended ambient temperature for owl monkeys in captivity is 24–27°C (Cicmanec and Campbell, 1977; Lehner, 1984; Umana *et al.*, 1984;

Weller *et al.*, 1991a). Sick or debilitated animals may require temperatures 2–5°C warmer.

The recommended relative humidity for housing owl monkeys is 30–60% (Cicmanec and Campbell, 1977; Lehner, 1984; Weller *et al.*, 1991a). An increased incidence of respiratory disease and keratoconjunctivitis may be observed in animals maintained at low humidity levels, especially in conjunction with poor air circulation (Umana *et al.*, 1984). A minimum of 10 to 15 room air changes per hour are recommended to supply an adequate amount of fresh air and reduce objectionable odors.

Wild owl monkeys, the only nocturnal simian primates, exit their sleeping sites at dusk and return around dawn (Wright, 1978; Aquino and Encarnación, 1986b; Garcia and Braza, 1987). Nocturnal behavior continues in captivity with periods of greatest activity occurring during the dark phase of the light/dark cycle (Erkert, 1976; Kavanau and Peters, 1976). Measurement of electrophysiologic activity has shown that rapid eye movement sleep occurs most frequently during the light phase of the light/dark cycle whereas wakefulness occurs most frequently during the dark phase (Perachio, 1971).

The recommended light cycle for captive owl monkeys is 12 hr of bright illumination followed by 12 hr of dim red illumination. Twenty-five to 100 foot-candles (fc) of illumination at a distance of 1 foot above the floor is adequate for bright illumination (Weller *et al.*, 1991a). Red light bulbs or red plastic covers over existing fluorescent light fixtures can be used to provide dim red illumination. The light cycle should be coordinated to provide bright illumination for performance of routine daily husbandry duties and animal observation. Institution of a reverse light cycle, with the dim red light period beginning in the early afternoon, creates an opportunity to observe the animals' behavior during their most active period.

Based on experimental data, dim red illumination is recommended in contrast to total darkness during the dark phase of the light/dark cycle. Erkert (1989) simulated average activity patterns similar to those observed under natural conditions by providing a nocturnal lighting pattern in the laboratory that corresponded with the four phases of the moon. Activity patterns in the laboratory were markedly influenced by illumination intensity during the dark phase of the light/dark cycle. A marked decrease in activity was observed during conditions of total darkness. Maximal nocturnal activity was demonstrated at 0.1 lx (9.3×10^{-3} fc), an illumination that approximates that of the full moon (Erkert, 1976). Recommendations based on these data suggest that optimal lighting conditions for owl monkeys are 100–1000 lx (9.3–93.0 fc) during the light phase and 0.01–0.5 lx (9.3×10^{-4}–4.65×10^{-2} fc) during the dark phase. Additional data are provided by Jacobs (1977), who demonstrated that when given a choice between two lights of low to moderate illuminance, owl monkeys demonstrate a clear preference for the brighter of the two lights.

III. DIETARY REQUIREMENTS

Owl monkeys are primarily frugivorous in the wild, selecting a variety of fruits ranging in size from 1 to 5 cm in diameter (Wright, 1978). Their diet is supplemented by ingestion of insects, particularly when fruit is scarce. In captivity they appear to adapt readily to nontraditional foods (Moynihan, 1964).

Little is known about the exact nutritional requirements of owl monkeys; dietarily they are assumed to be similar to other New World primates. The vitamin and mineral allowances for New World primates published by Ausman *et al.* (1985) and presented in Tables I and II, respectively, provide a general guideline that may be useful in making dietary decisions in relation to owl monkeys. New World primates require dietary supplementation of vitamin D3 from animal sources or exposure to an ultraviolet light source. New World primates also require dietary supplementation with vitamin C. A nutritionally balanced owl monkey diet must therefore contain or be supplemented with vitamin D3 and ascorbic acid.

A number of commercially prepared, nutritionally balanced New World

TABLE I Vitamin Allowances for New World Monkeys[a]

Vitamin	Vitamin content	Recommended allowance[b]
	(mg/kg diet)	
Thiamin hydrochloride	4.0	4–6
Riboflavin	8.0	8–10
Pyridoxine hydrochloride	4.0	4–6
Calcium pantothenate	25.0	11[c]
Niacinamide	40.0	40–50
Folic acid	4.0	4–6
Biotin	0.2	0.2
Cyanocobalamin	0.15	0.004[d]
Ascorbic acid	612.5	500–600
Menadione	5.0	0.800[e]
	(IU/kg diet)	
dl-Alpha-tocopheryl acetate (500 IU/g)	50.0	20–100
Vitamin A acetate (500,000 IU/g)	12,500	12,500
Vitamin D_3 (200,000 IU/g)	1,250	1,250

[a] From Ausman *et al.* (1985).
[b] Taken from Nicolosi and Hunt (1979).
[c] Calculated using a U.S. intake of 7 mg/kg, a 2500 kcal/day intake, and a diet density of 4 kcal/g.
[d] Calculated using the human RDA of 3 μg/day, a 2500 kcal intake, and a diet density of 4 kcal/g.
[e] Calculated using the usual intake of 500 μg/day (U.S. RDA), a 2500 kcal/day intake, and a diet density of 4 kcal/g.

TABLE II Mineral Allowances for New World Monkeys[a]

Element	Content in Ausman–Hayes mineral mix	Content in diet[b]	Recommended daily allowance[c]
	(g/kg)	(g/kg diet)	(g/kg diet)
Ca	133.2	6.660	6–8
K	147.2	7.360	2.4–10.9
P	71.40	3.570	3–4
Mg	29.06	1.453	1–1.5
S	13.11	0.656	–
Na	63.89	3.194	2.2–4.4
Cl	98.98	4.949	2.7–6.2
	(mg/kg)	(mg/kg diet)	(mg/kg diet)
Fe	2220.0	111	100
I	59.1	2.95	2
Mn	396.9	19.84	20
Zn	438.8	21.94	20
Cu	73.8	3.69	2
Cr	10.05	0.502	0.5
Se	1.96	0.098	0.1
F	10.5	0.525	0.5

[a] From Ausman *et al.* (1985).
[b] When incorporated into the diet at 5 g/100 g.
[c] Derived from Nicolosi and Hunt (1979).

primate diets are available. Many of these diets were formulated based on data gathered from nutritional studies on squirrel monkeys. New World primate diets typically have an increased protein and fat content as compared to Old World primate diets. Protein and fat requirements of owl monkeys have not been determined; therefore, the suitability of the protein and fat content of commercially available New World primate diets for owl monkeys is unknown. Exposure to heat during transportation or storage may lead to instability of the vitamin C component of commercially prepared diets, thus decreasing the product's shelf life. Care must be taken to avoid exposure to excessive heat in order to preserve the vitamin C component of the diet.

The texture, size, palatability, hardness, and availability of a commercial diet should be evaluated in the selection of an appropriate diet for a group of animals. Young, sick or debilitated animals, or animals with poor dentition may have difficulty eating large or hard, dry biscuits. These animals often benefit from softened or liquified (blended) diets, which are easier to ingest. Water added to the diet may result in rapid deterioration of vitamin C; therefore, unsweetened fruit juice is recommended for softening the biscuits.

In the wild, owl monkeys are reported to spend 31 to 53% of their time foraging (Wright, 1978; Garcia and Braza, 1987). Feeding occurs at night during the active period, with several long rest periods interspersed throughout the night. Captive management implications of these data suggest that an optimal feeding regime for owl monkeys would consist of provision of food during the beginning of the dark portion of the light/dark cycle.

Although most commercial New World primate diets typically supply essential dietary needs, fresh food items may be added to the diet to provide variety and serve as a form of environmental enrichment. Evidence from a feeding study performed on lion-tailed macaques (*Macaca silenus*) suggests that presentation of whole foods, rather than prepared, chopped foods, results in increased dietary diversity, time spent foraging, and total amount of food consumed (Smith *et al.*, 1989). Feeding whole foods as a supplement to commercial primate diets may contribute to psychological well-being by providing captive primates the opportunity to forage and actively participate in food preparation. Certain whole foods may be too large or otherwise difficult for owl monkeys to manipulate successfully. Partial preparation of these food items should be performed to allow further preparation and ingestion by the animals themselves.

Fresh fruits (i.e., apples, bananas, and oranges), vegetables (i.e., sweet potatoes, peas, and carrots), mealworms, and yogurt are usually well received by owl monkeys. These foods should be offered only as supplements as they do not provide a balanced diet by themselves. Care must be taken to ensure that animals eat a nutritionally balanced diet in addition to supplements. One method of ensuring consumption of the nutritionally balanced diet is to offer the diet during the beginning of the animals' most active period. Supplements may be given separately a few hours later. As a rule, supplements should not exceed 15% of the total daily dietary intake.

IV. COLONY RECORDS

The importance of accurate and thorough colony records cannot be overemphasized in relation to their potential impact on colony and individual animal management. Decisions regarding preventive veterinary care, reproductive pairing, and research assignment will draw upon data maintained in the colony records. The following is a brief listing of suggested data files that should be maintained for each animal.

1. Basic Animal History—This file should contain date of birth or arrival into the facility, tattoo or permanent identification number (ID), parent ID and/or source if from another facility or wild caught, karyotype, sex, age (if known), and final disposition.

2. Medical Record—This file should contain a record of medical problems, diagnoses, and treatments performed. Results of routine physical examinations, TB tests, fecal parasitology, bloodwork, and any additional diagnostic procedures should be recorded in this file.

3. Reproductive History—This data base is essential for maximizing reproductive performance. The file should contain the date the animal was first paired, the ID and karyotype of the mate, breeding dates and gestation length when known, date and method used to determine pregnancy (i.e., palpation, ultrasound), outcome of pregnancies (i.e., livebirth or abortion) and the corresponding dates, the ID and sex of infants born, and the date(s) of infant weaning.

4. Research History—This file should contain specific data on the animal's experimental history. The date of protocol initiation and termination and a description of the experimental protocol should be included.

5. Body Weight Data—This file should contain historical and current body weight data for each animal in the colony. Each animal should be weighed routinely. In addition, each animal should be weighed if anesthetized for other procedures. Timely evaluation of body weight data may assist in early diagnosis and treatment of disease.

6. Necropsy Data—This record should contain information regarding the death of the animal, including date of death, relevant history (i.e., chronic illness, traumatic injury, termination for experimental purposes), gross and histopathological findings, morphological diagnosis, etiology, and a cause of death statement that is consistent with both clinical and necropsy findings.

7. Cage Location History—In a large colony this file provides a method for rapid location of an individual animal, tracking animals in case of an infectious disease outbreak, and monitoring population stability within a defined area (i.e., cage or room). It should contain information on the location of each animal and should be updated each time an animal's location is changed.

The use of a computerized data base that provides compilation and search features is highly recommended for ease of data collection, assimilation, and use (Kuehl and Dukelow, 1977). Data files should be modified to reflect the needs of the colony. The data base should be constructed in such a way that output is readily available for an individual animal or for a particular topic involving a specific group of animals. Preprinted data sheets or computer screens help assure complete data collection and assist in ease of data tracking at later dates.

V. SOCIAL GROUPING

Owl monkeys in the wild live in family groups consisting of an adult male and female with their recent offspring. The average number of offspring per family group ranges from two to three (Aquino and Encarnación, 1986a; Aquino *et al.*, 1990). The long-term association between offspring and parents provides an opportunity for the offspring to observe parental behavior and gain experience in the caretaking of a new infant.

In captivity, adult owl monkeys are typically housed as a family group with a male–female pair of similar karyotype and their offspring. This grouping mimics the social structure observed under natural conditions and thus promotes psychological well-being by providing an environment in which species-typical social behavior may be expressed. Subadults are typically removed as their younger sibling approaches 6–8 months of age. In the wild, nonreproductive adult *A. nancymae* offspring, of either sex but not both simultaneously, are apparently tolerated within the family group, whereas only adult male offspring appear to be tolerated in *A. vociferans* family groups (Aquino *et al.*, 1990). This tolerance of adult male offspring is interesting in light of the observation that in captivity, owl monkey infants are carried for a greater percentage of time by their male siblings than by their female siblings (Dixson and Fleming, 1981).

Establishment of breeding pairs in which both members of the pair are the same karyotype results in improved reproductive success (Cicmanec and Campbell, 1977). Mated pairs of different karyotypes can breed and produce hybrid offspring. However, the reproductive success of these pairs is frequently less than that of karyotypically similar pairs and the reproductive success of the hybrid offspring produced is questionable. Moreover, a further reason to discourage the mating of adults with dissimilar karyotypes is the potentially confounding effect of the genetic background of a hybrid animal on research results.

Severe aggression, occasionally resulting in fatalities, may be observed during the formation of a male–female pair. The adult male owl monkey is most often the aggressor in these attacks although instances of female aggression toward a male have been observed. Recently introduced adults should be closely observed on a daily basis for the first month following the introduction for evidence of aggression (i.e., bite wounds, agonistic behavior). Bite wounds are usually inflicted around the head and on the arms and hands (Aquino and Encarnación, 1986a; also personal observation). Fighting between mating pairs may erupt as long as 2–3 years following an introduction; therefore, daily visual examination of mated pairs should be routinely performed. In addition to overt signs of aggression such as fighting, less obvious signs may occur. A dominant animal may limit the access of a subordinant animal to food and water through aggressive behavior. Thus, daily observation of hydration status and general body condition of each member of the pair is important.

Introductions should be made in a recently sanitized, neutral cage that has not been scent marked by either animal. Removal of all scent marks from the cage environment denies either animal a territorial advantage. The cage partition should be partially removed to allow both animals access to either cage. Two nest boxes, feed containers, and water sources should be made available to prevent competition for resources. Additional visual barriers and escape options should be added if available. Prior to introductions it is advisable to acclimate the animals to one another by caging them separately in adjacent cages.

Provision of mate choice may play an integral role in success or failure of breeding programs involving monogamous primate species (Tilson, 1986). At Battelle Northwest Laboratories (BNW), juvenile animals were typically housed in groups of four animals following removal from their familial groups. When possible the four animals consisted of two members of each sex. Groups were housed in two adjoining cages with the cage partition removed. In this way, each group of four animals had access to two nest boxes, two food containers, and two water sources. With time pairs would form, with each male–female pair sleeping separately in their own nest box. Once pairs were formed and appeared stable, the cage partition was replaced, dividing the large cage into two smaller cages. Little aggression was noted when pair formation followed this introduction pattern for young pairs. This form of introduction may also be applicable to adult pairs, where an adult male is placed in a large cage with two females and a choice of nest boxes. However, introduction of two adult males and two adult females in this fashion may result in severe aggression between the two adult males and should not be done.

Adult females can be paired if no suitable mate is available. Adult male owl monkeys, if paired together, will fight and may fatally injure one another, therefore pairing of adult males is strongly discouraged.

VI. PREVENTIVE MEDICINE

A. Quarantine

All imported animals should be quarantined upon arrival in accordance with the Centers for Disease Control (CDC) (1990) guidelines for newly imported nonhuman primates. Newly arrived animals pose a major health risk to an established colony through potential introduction of infectious agents acquired prior to or during transport. Introduction of an infectious agent into a captive population can have a devastating impact on an established colony as a result of the high stocking density and potential for transfer of an infectious organism from one animal to another.

Because of its unknown health history, an animal originating from a wild population potentially poses a greater health risk to an established colony than an animal arriving from a captive population. Animals arriving directly from wild populations should be examined thoroughly and quarantined for a longer period than is necessary for animals originating from established, disease-free colonies. Animals from disease-free captive populations should be quarantined for a minimum of 31 days. Animals from wild populations should be quarantined for a minimum of 60 days. The quarantine period should be extended if clinical signs of a potentially infectious disease are observed in any of the incoming animals.

The quarantine area should be physically separated from the established colony. Quarantine procedures designed to prevent transmission of disease between incoming animals and the existing colony should be established in conjunction with existing CDC guidelines (CDC, 1990) prior to receipt of new animals. Only personnel trained on the quarantine procedures should be allowed in the quarantine area. Personnel entering the area should be kept to the minimum essential to care for the animals. The use of protective clothing, masks, caps, gloves, and shoe covers should be required of all individuals entering the quarantine area. Protective clothing should be changed prior to exiting the quarantine area. To reduce the risk of transfer of an infectious agent, work schedules should be arranged such that personnel working in the quarantine area are not required to enter the established animal colony; if entry into the established colony is essential then personnel should work in the established colony prior to going into the quarantine area. Equipment necessary for conducting husbandry duties in the quarantine area should be dedicated for use only in that area. The quarantine area, including all equipment, must be thoroughly cleaned and disinfected between each group of incoming animals.

B. Conditioning

Each animal should be weighed and given a physical examination upon arrival at the new facility. Injuries sustained during transit should be treated as soon as possible. Hyperthermia, hypothermia, dehydration, and/or hypoglycemia should be treated as appropriate (see Chapter 7). Determination of sex and assignment of an identification number should be performed prior to establishing mated pairs. Every attempt should be made to keep previously compatible mated pairs together.

Daily observation of each animal during the quarantine/conditioning period is critical for the early detection of disease. As is the case with many nondomestic animals, owl monkeys may not exhibit behavioral signs of disease in the presence of a human observer. This can result in detection of disease late in its course when therapeutic intervention may be less successful. At a minimum, daily observation should consist of noting food and water consumption, hydration status, quantity and character of feces and urine, and any behavioral or other abnormalities. Animals demonstrating clinical signs of disease should be examined by a veterinarian as soon as possible and isolated if there is any indication of infectious disease.

A 3- to 7-day acclimatization period is recommended prior to handling animals after initial arrival. This period of time allows the animals to begin to recover from the stress associated with shipping and to become familiar with their new environment. The percentage difference in individual body weights obtained prior to shipping and upon arrival at a new facility provides a physiological indicator of shipping stress (Malaga *et al.*, 1991b). Malaga *et al.* found that shipping resulted in significant weigh loss in all groups regardless of age, sex, or transit time. Weight loss in subadults was greater than that in adults. However, subadults regained the weight lost during shipping and gained additional weight over the 30-day observation period immediately following shipping. In contrast, adult animals failed to regain the weight lost during shipping over the course of the 30-day observation period following shipping (Malaga *et al.*, 1991b).

Medical evaluation of each animal should be performed following the initial acclimatization period. Physical examination, fecal examination for internal parasites and culture for enteric pathogens, and identification by tattooing (if applicable) should be performed. In addition, three consecutive negative tuberculin tests should be performed at 2-week intervals for animals with unknown health histories. Two consecutive negative TB tests are adequate for animals that have been tested and found negative prior to shipping.

Permanent identification may be provided through application of a tattoo to the medial aspect of the thigh. Following induction of general anesthesia, hair removal, and cleansing of the skin, the tattoo should be applied in large, legible numbers and/or letters using commercially available tattoo equipment. The skin of the ventral thorax and abdomen provide an alternate site for application of a tattoo; however, a tattoo applied in this area may be difficult to visualize once the animal's fur has regrown. The medial aspect of the thigh is the preferred site because fur covering this area is more sparse and the area is routinely shaved prior to venipuncture, allowing easy visualization of the tattoo.

C. Isolation

An area used to isolate an animal or animals with a potentially infectious disease should be available in each primate facility. As in the quarantine area, designated isolation area procedures should be written and personnel trained prior to use of this area. Equipment should be dedicated for use solely within the isolation area or should be thoroughly sanitized prior to use in the established colony.

D. Physical Examination

A brief physical examination may be performed on a well-restrained animal; however, chemical immobilization is generally required to perform a thorough physical examination, venipuncture, electrocardiography, or cardiac ultrasound.

At a minimum, physical examination of an owl monkey should include: evaluation of coat condition; observation for external parasites, skin lesions, wounds, and hydration status; examination of the musculoskeletal system; evaluation of the oral cavity, noting mucous membrane color, capillary refill time, and the presence of dental tartar, wear, and any dental abnormalities; evaluation of ophthalmologic structures and function; thoracic auscultation; palpation of lymph nodes and abdominal structures; and evaluation of reproductive and urogenital structures. In addition, collection and evaluation of a baseline electrocardiogram and cardiac ultrasound is recommended because of the relatively high incidence of cardiac hypertrophy observed in the owl monkey (Rodger *et al.*, 1986; Jones *et al.*, 1992). Collection of a blood sample and evaluation of a complete blood count and clinical chemistry are also recommended to complete the physical examination. Fecal parasitology and culture for enteric pathogens should be performed on newly arrived animals, especially if they have recently arrived from the wild.

E. Preventive Medicine

A physical examination should be performed on each animal at least annually. This examination may be scheduled to coincide with restraint necessary for a research protocol. Problems noted during the physical examination should be recorded in the animal's medical record and appropriate therapy initiated. The annual examination should include TB testing, physical examination, and, if necessary, dental prophylaxis. Monthly evaluation of body weight may be used to evaluate general health. A loss of body weight may signal a health or husbandry problem.

F. Tuberculin Testing

Although owl monkeys appear to be relatively resistant to infection with *Mycobacterium* spp. (Bone and Soave, 1979), routine tuberculin (TB) testing should be performed at least annually in an established colony (CDC, 1993). Animals known to be previously exposed to Freund's Complete Adjuvant

(FCA) should be excluded from TB testing as a misleading positive TB reaction usually occurs due to the presence of killed *Mycobacterium tuberculosis* cells in FCA (Pierce and Dukelow, 1988). Physical restraint is necessary prior to performing a TB test. Chemical immobilization is recommended to reduce the risk of causing trauma to the eyelid or adjacent ocular structures and to reduce the risk of injury to personnel administering the test. Injury to the eyelid may result in a false positive reaction.

Using a tuberculin syringe and a 25- to 27-gauge 0.5-inch needle, 0.1 ml of a mammalian old tuberculin is administered intradermally into the skin of the eyelid. The eyelid is then visually examined at 24, 48 and 72 hr for signs of hyperemia, edema, induration, and ulceration. The severity of the reaction is graded, with a (−) representing no visible reaction and a (5+) representing a severe reaction. Erythema without swelling is usually interpreted as a negative reaction at 24 hr. Palpebral swelling with or without erythema is interpreted as a positive reaction if no history of trauma is associated with administration of the test.

An animal with a positive TB test should be considered infected and immediately isolated and retested in the opposite eyelid and on the abdominal skin, and a thoracic radiograph obtained. Extreme caution must be used when working with a TB-positive animal to minimize potential exposure to animal care personnel and to other animals in the colony. If an animal is negative upon retest, it should be tested again in 1 week and have five consecutive negative TB tests, each 2 weeks apart, prior to reintroduction to the existing colony. The decision regarding final disposition of an animal with a positive reaction should be based on the intrapalpebral TB test result, the experience of the person administering and evaluating the test, and the medical history of the colony. An animal testing positive on retest with no known history of treatment with FCA should be destroyed and a necropsy performed by a qualified veterinarian under animal biosafety level 3 conditions (CDC, 1993).

VII. VETERINARY PROCEDURES

A. Procedures

Because of similarities in size and anatomy, the procedures described for use in the squirrel monkey (Abee, 1985) may be adapted for use in the owl monkey. Intravenous access is most commonly obtained in the owl monkey via the femoral vein (Fig. 5). Blood samples can be collected from this site by skilled personnel without the use of anesthesia; however, any laceration to the vein or closely adjacent artery may result in the formation of a large hematoma. For this reason, and also to reduce the stress associated with manual restraint, chemical restraint should be used unless both the

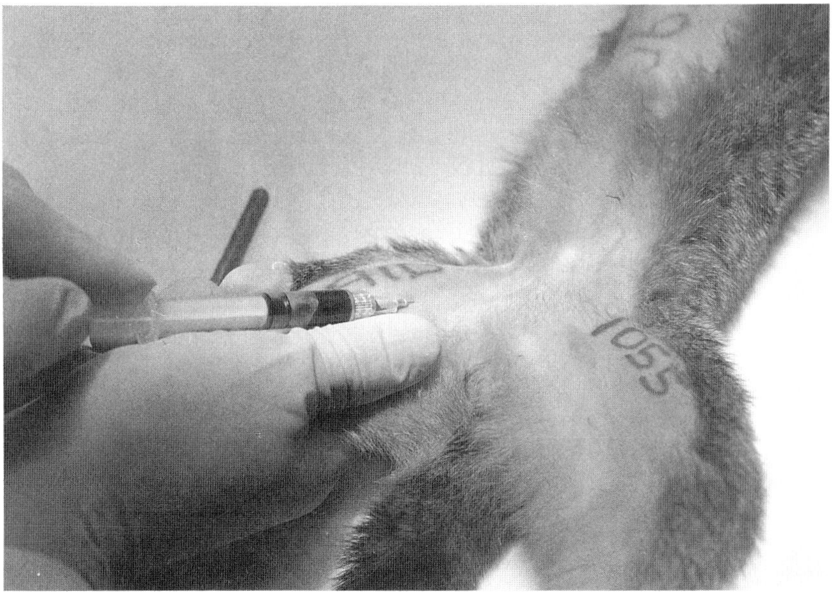

FIGURE 5 Venipuncture utilizing the femoral vein in the owl monkey.

person obtaining the blood sample and the person restraining the animal are highly skilled.

In severely debilitated animals, where venous access is impeded owing to hypotension, intraosseous infusion may be used to rapidly administer large volumes of fluid. The trochanteric fossa of the femur is the most accessible site for intraosseous infusion in the owl monkey. Following aseptic preparation and local anesthesia of the skin and periosteum with 1% lidocaine, a 20-gauge needle is aseptically placed into the medullary cavity through the trochanteric fossa. When properly placed, the needle will be stable and fluid may be delivered through a standard intravenous administration set. The needle may be secured by placing a tape butterfly around the needle hub and suturing this to the skin. The catheter must be maintained under aseptic conditions while in place. Unlike a vein, the bone marrow cavity cannot expand to accommodate increased flow rates; therefore, maximal flow rates with intraosseous infusion are limited. Use of this route of fluid administration is limited to severely disabled or restrained animals; active, nonsedated animals may dislodge the intraosseous needle.

B. Restraint

Owl monkeys can be trained to enter a nest box or a restraining device (Moseley and Davis, 1989). A nest box attached to the side of the cage with a guillotine door is ideal for frequent capture of animals. Confining an animal to a smaller space such as a nest box prior to capture increases the ease of capture and decreases the risk of injury to the animal. Short-handled fishing nets made of a fine, durable mesh cloth can be used to capture an animal in a room or cage. Care must be taken not to injure the animal with the rigid metal rim of the net. Leather primate handling gloves should be worn when handling owl monkeys to protect the handler from bite wounds and scratches. The animal should be caught as quickly and gently as possible either by hand or with a net. Once caught, the animal may be restrained by placing one hand gently but securely around the neck and chest and then gently grasping and extending the legs with the other hand (Fig. 6). Care must be taken not to place pressure upon the trachea during manual restraint.

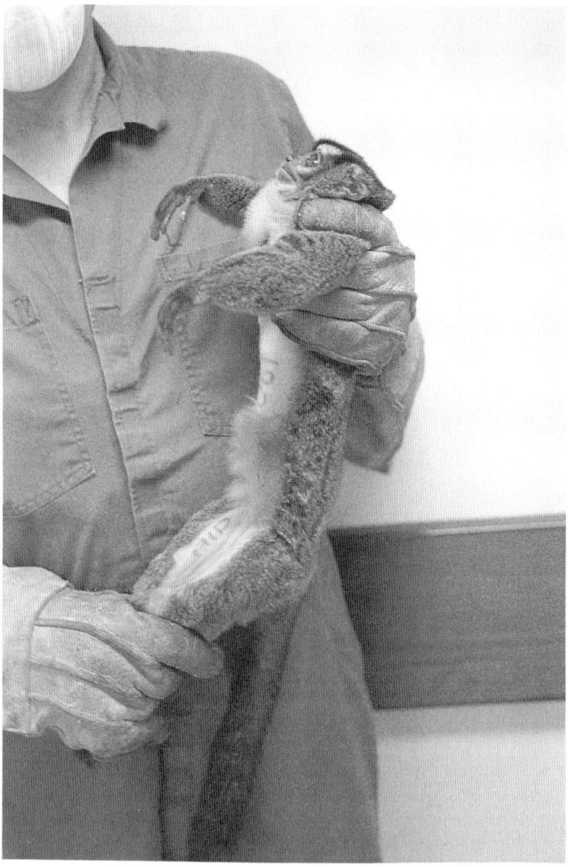

FIGURE 6 Manual restraint of the owl monkey.

HCl (10–25 mg/kg) administered intramuscularly or subcutaneously. Xylazine (1.2 mg/kg) may be administered in conjunction with ketamine HCl to prolong anesthesia and increase muscle relaxation.

VIII. Normative Physiological Data

A. Hematological, Clinical Chemistry, and Urinalysis Data

Normative hematological and clinical chemistry data for owl monkeys have been published by several authors (Porter, 1969; Wellde *et al.*, 1971, 1972; Sehgal *et al.*, 1980; Chalifoux *et al.*, 1981; Klein *et al.*, 1985; Beland *et al.*, 1981; Mrema *et al.*, 1988; Malaga *et al.*, 1990). Differences among the data from different sources are apparent; whether these differences are due to variation in sample handling, assay techniques, karyotype, husbandry, or other environmental factors is unknown. Hematological and clinical chemistry values for adult, clinically healthy owl monkeys maintained at BNW are presented in Tables III and IV, respectively. The method of handling and analysis of these samples have been previously reported (Malaga *et al.*, 1990). A comparison of hematological reference values reported for *Aotus* spp. is presented in Table V.

Sex differences in hematological and clinical chemistry values have been reported for *A. nancymae* (Malaga *et al.*, 1990). Significant differences in erythrocyte, hemoglobin, mean corposcular volume, and platelet values

TABLE III Hematological Data for Owl Monkeys of Four Karyotypes[a]

	K-I (*N* = 254)		K-II (*N* = 62)		K-III (*N* = 55)		K-V (*N* = 35)	
Analyte	*X*	S.D.	*X*	S.D.	*X*	S.D.	*X*	S.D.
RBCs (10^6/ml)	6.2	0.6	5.3	0.5	5.3	0.9	6.4	0.5
Hgb (g/dl)	16.3	1.6	14.3	1.5	14.0	2.3	17.1	1.3
VPRC (%)	49.7	4.4	43.6	4.2	43.6	7.0	52.2	3.7
Plate (10^3/ml)	431	119	342	88	333	117	295	90
MCV (μm^3)	79.8	3.5	82.6	3.0	82.3	4.0	81.3	4.1
MCH (pg)	26.2	1.6	27.0	1.4	26.5	1.6	26.6	1.6
MCHC (%)	32.8	1.1	32.8	0.9	32.2	0.8	32.8	0.6
WBC (10^3/ml)	10.1	3.2	12.2	4.4	14.2	11.2	8.8	3.8
Segs (10^3/ml)	3.0	2.0	3.4	1.8	4.3	2.2	2.6	1.2
Lymphs (10^3/ml)	6.1	2.6	6.9	2.2	8.1	2.5	5.5	1.5
Monos (10^3/ml)	0.4	0.3	0.3	0.3	0.3	0.2	0.3	0.3
Eosin (10^3/ml)	0.4	0.5	1.4	1.1	1.4	1.5	1.6	1.1
Basos (10^3/ml)	0.2	0.2	0.2	0.2	0.1	0.2	0.1	0.1

[a] Data collected at Battelle Pacific Northwest Laboratories.

TABLE IV Serum Chemistry Data for Owl Monkeys of Four Karyotypes[a]

	K-I (N = 254)		K-II (N = 57)		K-III (N = 53)		K-V (N = 35)	
Analyte	X	S.D.	X	S.D.	X	S.D.	X	S.D.
T bili (mg/dl)	0.8	0.4	0.5	0.2	0.5	0.2	0.7	0.2
Choles (mg/dl)	150	46	91	34	111	44	99	34
Creatin (mg/dl)	1.0	0.4	1.0	0.2	1.1	0.4	1.0	0.3
Glucose (mg/dl)	139	35	153	39	150	47	172	40
Calcium (mg/dl)	10.4	1.0	9.3	0.9	9.2	0.9	9.6	0.7
Alk phos (IU/l)	494	469	183	151	143	87	364	381
Phosph (mg/dl)	4.0	1.5	4.4	1.5	4.6	1.5	4.8	1.5
ALT (IU/l)	47	37	44	34	49	35	59	34
SUN (mg/dl)	15	5.4	15	6.7	17	11	15	8.9
T prot (g/dl)	8.3	0.9	8.0	0.7	8.1	1.1	8.2	0.5
Albumin (g/dl)	4.4	0.5	3.8	0.5	3.7	0.5	4.6	0.4
GGT (IU/l)	17	14	20	14	23	18	26	21
Sodium (mEq/l)	152	5	156	9	154	6	148	3
Potass (mEq/l)	3.8	0.7	4.6	1.6	4.8	2.0	3.3	0.7

[a] Data collected at Battelle Pacific Northwest Laboratories.

between male and female animals were observed. In addition, significant differences in serum cholesterol, calcium, and albumin values were noted. The authors suggest that, even though these differences are statistically significant, they are unlikely to be clinically important owing to the small amount of the differences. No consistently significant sex differences in hematological and clinical chemistry values were reported by Porter (1969).

Minimal data have been published on the effect of captivity on hematological values in the owl monkey. Porter (1969) compared hematological values obtained from a population of newly acquired, wild-caught animals to values obtained from the same population during a 2-year period in captivity. Leukocyte values increased significantly over time in the captive population.

Significant diurnal variation in several blood parameters, including total leukocytes, eosinophils, erythrocytes, and hemoglobin, serum protein, and iron concentrations, has been reported by Klein *et al.* (1985). These data are consistent with data obtained from other species and provide a rationale for obtaining blood samples at consistent time periods, especially when samples are collected as an integral part of an experimental procedure.

Variation in hematological values among different *Aotus* karyotypes has been reported (Beland *et al.*, 1981). Macrocytic normochromic hemolytic anemia observed in karyotype II, III, and IV animals was not present in karyotype I and V animals. Anemic animals also had a leukocytosis characterized by a band neutrophilia. Idiopathic eosinophilia has been reported

TABLE V Comparison of Hematic Reference Intervals Reported for *Aotus* spp.[a]

	Battelle[b]		Iquitos[c]		Wellde[d]		Sehgal[e]		Umana[f]	
	Mean	S.D.	Mean	S.D.	Mean	S.D.	Mean	S.D.	Mean	S.D.
Erythrocytes (10^6/ml)	6.2	0.6	4.5	0.6	5.2	0.8	5.7	0.7	6.5	0.2
Packed cell volume (%)	49.7	4.4	36.9	4.4	42.0	5.4	42.3	4.9	52.8	5.6
Hemoglobin (g/dl)	16.3	1.6	14.5	1.2	14.3	1.1[b]	14.2	1.7	18.7	1.5
MCV (μm^3)	79.8	3.5	82.5	8.6	82.4	11.9	74.9	5.6	80.2	5.1
MCH (pg)	26.2	1.6	32.8	4.2	26.9	3.1[b]	25.2	2.1	n/a	n/a
MCHC (%)	32.8	1.1	39.9	4.4	34.1	3.1[b]	33.5	1.2	n/a	n/a
Platelets (10^3/ml)	431	119	n/a	n/a	397	109	n/a	n/a	n/a	n/a
Leukocytes (10^3/ml)	10.1	3.2	7.2	2.1	12.7	4.7	12.4	4.3	14.1	3.8
Neutrophils (%)	29.1	15.3	38.8	7.7	55.4	7.6	45.9	17.1	38.4	12.3
Eosinophils (%)	4.4	4.6	6.0	3.5	9.5	9.2	6.1	6.2	7.0	4.3
Basophils (%)	1.7	1.7	0.0		0.0		1.0	1.4	0.0	0.1
Lymphocytes (%)	61.2	15.9	52.4	7.4	35.5	18.3	43.8	17.4	51.1	13.0
Monocytes (%)	3.6	2.5	2.0	1.0	0.0		2.5	2.3	3.7	2.3

[a] From Malaga *et al.* (1990).
[b] *Aotus nancymae*, $n = 254$.
[c] *Aotus nancymae*, $n = 32$.
[d] *Aotus* sp., $n = 157$, (*b*) $n = 60$.
[e] *Aotus* sp., $n = 390$.
[f] *Aotus lemurinus griseimembra*, $n = 57$.

in karyotype IV but not karyotype I animals; eosinophilia was variable among other karyotypes examined (Beland *et al.*, 1981).

Monoclonal-antibody-defined peripheral blood lymphocyte subsets have been evaluated in owl monkeys using anti-human monoclonal antibodies. Many of the monoclonal antibodies used to identify pan T, helper/inducer, suppressor/cytotoxic, pan B, and NK cell subsets of lymphocytes in humans exhibited positive reaction with owl monkey lymphocytes (Caldwell *et al.*, 1987; Smith *et al.*, 1992).

Although spontaneous and experimentally induced renal disease is common in owl monkeys (Hunt *et al.*, 1976; Chalifoux *et al.*, 1981; Aikawa *et al.*, 1988), information on normative renal function parameters and clinical interpretation of renal data is limited. Nephrotic syndrome, glomerulonephropathy, and interstitial nephritis have all been documented in owl monkeys. However, the lack of normative renal data for the owl monkey has made it difficult to diagnose and determine the clinical extent of renal disease in individual animals. Urinalysis and indices of renal function have been reported for owl monkeys with and without evidence of renal disease (Weller *et al.*, 1991b, 1992, 1994). Those studies indicated that normal renal function and urine parameters of owl monkeys are not significantly different from those of other mammalian species (Weller *et al.*, 1992). Urine parameters obtained from 98 adult clinically healthy owl monkeys maintained at BNW are summarized in Table VI.

B. Body and Organ Weights

Mean adult body weights for karyotype I, II, III, and V owl monkeys maintained by BNW are listed by sex in Table VII. Male karyotype I and V owl monkeys weighed significantly more than females of the same karyotype ($p < 0.05$). Sex differences in body weights of karyotype II and III animals were not observed. Karyotype I males weighed significantly more than males of karyotypes II, III, and V ($p < 0.05$). Karyotype II and III males weighed significantly more than karyotype V males ($p < 0.05$). No significant difference was found between body weights of karyotype I, II, and III females, however, these karyotypes did weigh significantly more than karyotype V females ($p < 0.005$).

Weight gain data collected from a total of 82 BNW parent-reared owl monkeys ranging in age from 1 to 23 months are presented in Fig. 7 and 8. The growth curves for male and female owl monkeys were derived by fitting the data for each animal using a cubic interpolating spline technique. Percentiles smoothed by the cubic spline technique were then plotted to obtain the growth curves represented by Fig. 7 and 8. On average, male and female owl monkeys reached a mature body weight of 750 g by 14 months of age. Little difference in the rate of weight gain was observed between male and female infants. These weight gain data correspond with

TABLE VI Normal Urine Parameters in Owl Monkeys

Urine	Mean ± S.E.M.	Range
Specific gravity	1.010 ± 0.001	1.002–1.023
Volume (ml/kg/day)	81.9 ± 7.5	9–303
Urinalysis—semiquantitative values		
Protein multistix	0–3+	
Glucose	0–1+	
Ketones	0–2+	
Bilirubin	0	
Urobilinogen	0	
Occult blood	0–4+	
pH	6–8.5	
Appearance	Light–dark yellow; clear–cloudy	
Microscopic urinalysis		
Red blood cells	Rare–few (3–5/hpf)	
White blood cells	Rare–occasional	
Epithelial cells	Rare–occasional squamous	
Casts	Occasional granular	
Crystals	Few–moderate triple phosphate; rare calcium oxalate	
Miscellaneous	Occasional renal tubular epithelial cells	

previously published weight gain data for laboratory-reared owl monkeys (Cicmanec *et al.*, 1979).

Mean organ weights, obtained at necropsy from 62 animals of various karyotypes that died from nonexperimental causes at BNW, are listed in Table VIII.

C. Electrocardiographic Parameters

Mean electrocardiographic parameters obtained from 45 clinically healthy adult owl monkeys anesthetized with ketamine HCl (10 mg/kg) and

TABLE VII Mean Adult Owl Monkey Body Weight (g, Mean ± 1 S.D.) Listed by Karyotype and Sex[a]

Karyotype	Male	*N*	Female	*N*	Range
I	957 ± 124	130	885 ± 120	124	567–1232
II	880 ± 98	55	864 ± 111	48	531–1281
III	898 ± 109	52	907 ± 99	37	529–1237
V	838 ± 118	18	751 ± 91	17	572–1077

[a] Data collected at Battelle Pacific Northwest Laboratories.

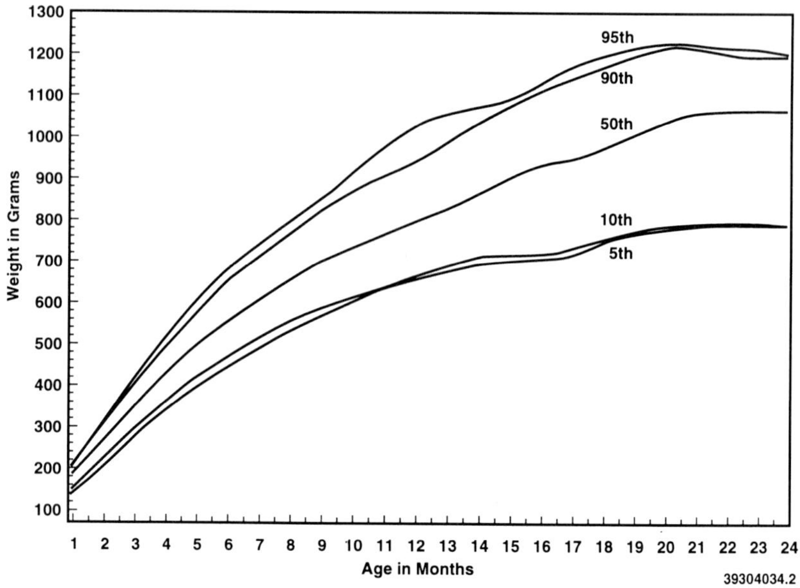

FIGURE 7 Weight gain data for captive-born male owl monkeys.

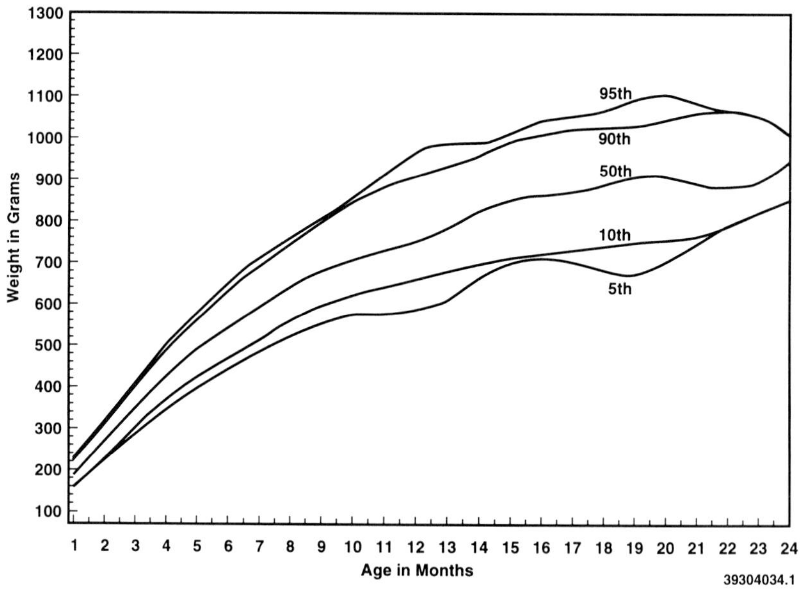

FIGURE 8 Weight gain data for captive-born female owl monkeys.

TABLE VIII Mean Organ Weights (g) for Adult Owl Monkeys ($n = 62$)

Organ	Mean weight ± 1 S.D.	Range
Body weight	739 ± 164	420–1020
Heart	7.46 ± 2.74	2.65–13.84
Liver	37.63 ± 15.15	16.22–85.17
Lung	14.06 ± 7.51	4.83–29.07
Pancreas	1.99 ± 0.82	0.75–3.89
Spleen	1.94 ± 1.70	0.22–7.89
Kidney		
Left	4.53 ± 1.96	1.76–8.68
Right	4.58 ± 1.95	1.73–8.98
Adrenal gland		
Left	0.15 ± 0.06	0.08–0.31
Right	0.14 ± 0.04	0.07–0.22
Thyroid gland		
Left	0.05 ± 0.02	0.03–0.09
Right	0.05 ± 0.03	0.02–0.15
Pituitary gland	0.02 ± 0.01	0.01–0.03
Brain	16.80 ± 1.44	14.39–19.57

screened for the absence of cardiac hypertrophy by echocardiographic and radiologic evaluation are presented in Table IX. These data are similar to data reported for nine other nonhuman primate species (Gonder *et al.*, 1980), the exception being the QRS duration, which appears to be prolonged in the owl monkey. The range of means reported for the QRS duration for nine nonhuman primate species was 0.01 to 0.04 sec (Gonder *et al.*, 1980). The mean QRS duration in the owl monkey is 0.044 ± 0.005 sec.

TABLE IX Mean Electrocardiographic Parameters for Clinically Healthy Adult Owl Monkeys ($n = 45$) Obtained under Ketamine Sedation

ECG parameter	
P-wave (sec)	0.036 ± 0.005
P–Q interval (sec)	0.075 ± 0.015
QRS duration (sec)	0.044 ± 0.005
Q–T interval (sec)	0.136 ± 0.036
P-wave (mV)	0.29 ± 0.074
R-wave lead II (mV)	1.15 ± 0.694
Heart rate (min^{-1})	258 ± 63
R-wave lead I (mV)	0.487 ± 0.357
R-wave lead aVf (mV)	0.823 ± 0.71
Mean frontal axis (deg)	62.41 ± 21.4

D. Miscellaneous Physiological Data

Chronologic age in the owl monkey may be estimated by radiologic evaluation of bone maturation (Cleveland *et al.*, 1986). Femoral length and initial epiphyseal ossification or completed fusion of various secondary ossification centers may be used to predict chronologic age in the developing owl monkey. Age of wild-caught animals may be estimated based on overall dental wear, canine length, eruption of permanent molars, development of the subcaudal and pectoral glands, and pigmentation of the external genitalia (Tables X and XI) (Aquino and Encarnación, 1986a). Limited longevity data suggest that in captivity the life span of an owl monkey may exceed 20 years in some cases (Atiya Hakeem, personal communication, California Institute of Technology).

The heme precursor, protoporphyrin, is elevated in owl monkeys as compared to other species (Heltne *et al.*, 1981). Unlike many species in which protoporphyrin is bound to zinc, in owl monkeys it is found in the unbound state as the free base protoporphyrin IX. Protoporphyrinemia has been reported in the five *Aotus* karyotypes examined. A relationship between several major disease states of owl monkeys and protoporphyrinemia has been suggested (Heltne *et al.*, 1981).

Albumin polymorphism, conforming to a codominant allele of transmis-

TABLE X Mean Eruption Time for Deciduous Teeth of 28 Owl Monkeys[a]

Tooth	Age (weeks)			% of times observed erupting first[b]
	Mean	1 S.D.	Range	
I_1 MX	2.3	0.7	1.0–3.5	26
MD	2.3	0.8	0.5–3.5	26
I_2 MX	3.5	0.9	1.5–4.5	0
MD	2.9	0.8	1.0–4.5	70
C MX	4.7	1.0	1.5–6.0	27
MD	4.6	1.2	2.0–6.5	23
P_1 MX	3.5	0.7	1.5–5.0	13
MD	3.4	0.7	2.0–5.0	27
P_2 MX	3.8	0.8	2.0–5.5	21
MD	3.8	0.9	2.0–5.0	18
P_3 MX	5.6	1.2	3.0–8.0	13
MD	5.5	1.1	3.0–7.5	27

[a] From Hall *et al.* (1979).

[b] Percentage of observations in which the maxillary or mandibular tooth erupted before its ipsilateral counterpart. The difference between the sum of MX + MD and 100% equals the percentage of observations in which the maxillary and mandibular teeth erupted together. MX = maxillary tooth, MD = mandibular tooth.

TABLE XI Mean Eruption Time for Permanent Teeth of Owl Monkeys[a]

Tooth		No. of observations[b]	Age (months)			% of times observed erupting first[c]
			Mean	1 S.D.	Range	
I_1	MX	24	9.4	0.8	8.0–11.0	55
	MD	18	9.6	0.6	9.5–11.0	9
I_2	MX	13	10.8	0.6	10.0–12.0	64
	MD	8	10.6	0.4	10.0–11.0	0
C	MX	18	15.0	1.3	13.5–18.0	29
	MD	15	14.0	1.0	12.5–16.0	43
P_1	MX	12	12.4	1.1	11.0–14.5	33
	MD	13	12.4	0.9	11.5–14.0	17
P_2	MX	12	11.9	1.0	10.5–14.0	67
	MD	13	12.3	0.8	11.5–14.0	0
P_3	MX	12	11.5	0.9	10.0–13.0	13
	MD	16	11.0	0.7	10.0–12.0	50
M_1	MX	36	4.9	0.6	4.0–6.0	0
	MD	38	4.3	0.5	3.5–5.0	84
M_2	MX	28	7.3	0.6	6.5–8.5	0
	MD	32	6.4	0.8	5.0–7.5	100
M_3	MX	17	11.2	1.1	10.0–13.0	0
	MD	18	9.9	1.2	8.5–11.5	91

[a] From Hall *et al.* (1979).
[b] Observation is the eruption of a single tooth.
[c] Percentage of observations in which the maxillary or mandibular tooth erupted before its ipsilateral counterpart. The difference between the sum of MX + MD and 100% equals the percentage of observations in which the maxillary and mandibular teeth erupted together. MX = maxillary tooth, MD = mandibular tooth.

sion, has been reported in owl monkeys (Reardon *et al.*, 1979). Differences in three phenotypic forms of albumin, designated as fast, double, and slow, were reported among the different owl monkey karyotypes examined. A unique alpha globulin, which appears to be a dominant characteristic, has been identified in karyotype I and VII owl monkeys (Reardon *et al.*, 1979).

Based on calculations for a mammal of its size, the owl monkey has been characterized as a hypometabolic animal (Schölander *et al.*, 1950). Consistent with this hypometabolic rate, decreased serum triiodothyronine (T3) and tetraiodothyronine (T4) levels have been reported in owl monkeys (Whittow *et al.*, 1979). In a comparison between owl monkeys and cynomolgus macaques (*Macaca fasicularis*), T3, T4, and growth hormone levels were shown to be significantly lower in owl monkeys (Whittow *et al.*, 1979). Thyroid-stimulating hormone (TSH) levels of these two species were found to be comparable, suggesting a decrease in thyroid sensitivity to TSH in the owl monkey. Owl monkey T3 levels were increased at night, corresponding with the period of increased behavioral activity of this species.

REFERENCES

Abee, C. R. (1985). Medical care and management of the squirrel monkey. *In* "Handbook of Squirrel Monkey Research" (L. A. Rosenblum and L. C. Coe, eds.), 1st ed., pp. 447–488. Plenum, New York and London.

Aikawa, M., Broderson, J. R., Igarashi, I., Jacobs, G., Pappaioanou, M., Collins, W. E., and Campbell, C. C. (1988). "An Atlas of Renal Disease in *Aotus* Monkeys with Experimental Plasmodial Infection." Am. Inst. Biol. Sci., Washington, DC.

Animal Welfare Act (1992). Code of Federal Regulations, Title 9, Subchapter A, Part 3, p. 74.

Aquino, R., and Encarnación, F. (1986a). Population structure of *Aotus nancymae* (Cebidae:Primates) in Peruvian Amazon lowland forest. *Am. J. Primatol.* **11,** 1–7.

Aquino, R., and Encarnación, F. (1986b). Characteristics and use of sleeping sites in *Aotus* (Cebidae:Primates) in the Amazon lowlands of Peru. *Am. J. Primatol.* **11,** 319–331.

Aquino, R., Puertas, P., and Encarnación, F. (1990). Supplemental notes on population parameters of northeastern Peruvian night monkeys, genus *Aotus* (*Cebidae*). *Am. J. Primatol.* **21,** 215–221.

Ausman, L. M., Gallina, D. L., and Nicolosi, R. J. (1985). Nutrition and metabolism of the squirrel monkey. *In* "Handbook of Squirrel Monkey Research" (L. A. Rosenblum and L. C. Coe, eds.), pp. 349–378. Plenum, New York.

Beland, M. F., Bronson, R. T., Ma, N. S. F., McIntyre, K. W., Sehgal, P. K., and Keadle, T. L. (1981). Karyotypic variation in susceptibility to hemolytic anemia and idiopathic eosinophilia of owl monkeys, *Aotus trivirgatus. Primates* **22,** 551–556.

Bone, J. F., and Soave, O. A. (1970). Experimental tuberculosis in owl monkeys (*Aotus trivirgatus*). *Lab. Anim. Care* **20,** 946–948.

Caldwell, C. W., Mrema, J. E. K., Taylor, H. M., and Green, T. J. (1987). Reference values for peripheral blood lymphocytes from *Aotus lemurinus* spp. *griseimembra* (owl monkey). *J. Med. Primatol.* **16,** 261–271.

Centers for Disease Control (CDC) (1990a). Update: Ebola-related filovirus infection in nonhuman primates and interim guidelines for handling nonhuman primates during transit and quarantine. *Morbid. Moral. Wkly. Rep.* **39,** 22–24, 29–30.

Centers for Disease Control (CDC) (1993). Interim guidelines for tuberculin skin testing of nonhuman primates during quarantine. *Morbid. Mortal. Wkly. Rep.* **42,** 575.

Chalifoux, L. V., Bronson, P., Sehgal, P., Blake, B. J., and King, N. W. (1981). Nephritis and hemolytic anemia in owl monkeys (*Aotus trivirgatus*). *Vet. Pathol.* **18,** 23–27.

Chamove, A. S., Anderson, J. R., Morgan-Jones, S. C., and Jones, S. P. (1982). Deep woodchip litter: Hygiene, feeding, and behavioral enhancement in eight primate species. *Int. J. Stud. Anim. Prob.* **3,** 27–37.

Chapman, W. L., Hanson, W. L., Hayre, M. D., and Harrison, D. P. (1985). An enlarged aggregate of apocrine glands on the chest of karyotype I owl monkeys. *Lab. Anim. Sci.* **35,** 491–492.

Cicmanec, J. C., and Campbell, A. K. (1977). Breeding the owl monkey (*Aotus trivirgatus*) in a laboratory environment. *Lab. Anim. Sci.* **27,** 512–517.

Cicmanec, J. C., Hernadez, D. M., Jenkins, S. R., Campbell, A. K., and Smith, J. A. (1979). Hand-rearing infant Callitrichids (*Saguinus* spp and *Callithrix jacchus*), owl monkeys (*Aotus trivirgatus*), and capuchins (*Cebus albifrons*). *In* "Nursery Care of Nonhuman Primates" (G. C. Ruppenthal and D. J. Reese, eds.), 1st ed., pp. 307–312. Plenum, New York and London.

Cleveland, R. H., Hultsch E., Kushner, D. C., Correia, J. A., and Peters, M. E. (1986). Radiographic determination of bone maturation as an aid in predicting chronologic age in the owl monkey (*Aotus trivirgatus*). *J. Med. Primatol.* **15,** 147–167.

Dixson, A. F., and Fleming, D. (1981). Parental behaviour and infant development in owl monkeys (*Aotus trivirgatus griseimembra*). *J. Zool.* **194,** 25–39.

Erkert, H. G. (1976). Beleuchtungsabhangiges aktivitatsoptimum bei nachtaffen (*Aotus trivirgatus*). *Folia Primatol.* **25,** 186–192.
Erkert, H. G. (1989). Lighting requirements of nocturnal primates in captivity: A chronobiological approach. *Zoo Biol.* **8,** 179–191.
Garcia, J. E., and Braza, F. (1987). Activity rhythms and use of space of a group of *Aotus azare* in Bolivia during the rainy season. *Primates* **28,** 337–343.
Gonder, J. C., Gard, E. A., and Lott, N. E. (1980). Electrocardiograms of nine species of nonhuman primates sedated with ketamine. *Am. J. Vet. Res.* **41,** 972–975.
Hall, R. D., Beattie, R. J., and Wyckoff, G. H. (1979). Weight gains and sequence of dental eruptions in infant owl monkeys (*Aotus trivirgatus*). *In* "Nursery Care of Nonhuman Primates" (G. C. Ruppenthal and D. J. Reese, eds.), 1st ed., pp. 321–328. Plenum, New York and London.
Heltne, P. G., Chisolm, J. J., Wojcik, J. F., Hall, R. D., and Loew, F. M. (1981). Erythrocyte protoporphyrin levels in *Aotus trivirgatus*. *J. Med. Primatol.* **10,** 240–254.
Hunt, R. D., Van Zwieten, M. J., Baggs, R. B., Sehgal, P. K., King, N. W., and Wickum, M. (1976). Glomerulonephritis in the owl monkey (*Aotus trivirgatus*). *Lab. Anim. Sci.* **26,** 1088–1092.
Jacobs, G. H. (1977). Brightness preference in nocturnal and diurnal South American monkeys. *Folia Prematol.* **28,** 231–240.
Jones, C. L., Blatt, H., and Perri, S. (1992). Diagnostic ultrasonogram findings on a colony of owl monkeys (*Aotus trivirgatus*). *Contemp. Top. Lab. Anim. Sci.* **31,** 26.
Kavanau, J. L., and Peters, C. R. (1976). Activity of nocturnal primates: Influences of twilight zeitgebers and weather. *Science* **191,** 83–86.
Klein, R., Bleiholder, B., Jung, A., and Erkert, H. G. (1985). Diurnal variation of several blood parameters in the owl monkey *Aotus trivirgatus griseimembra*. *Folia Primatol.* **45,** 195–203.
Kuehl, T. J., and Dukelow, T. J. (1977). Computer program for maintenance of individual animal records in a nonhuman primate colony. *Lab. Anim. Sci.* **27,** 391–395.
Lehner, N. D. (1984). Biology and diseases of Cebidae. *In* "Laboratory Animal Medicine" (J. G. Fox, B. J. Cohen, and F. M. Loew, eds.), pp. 321–353. Academic Press, Orlando, FL.
Le Maho, Y., Goffart, M., Rochas, A., Felbabel, H., and Chatonnet, J. (1981). Thermoregulation in the only nocturnal simian: The night monkey *Aotus trivirgatus*. *Am. J. Physiol.* **240,** R156–R165.
Malaga, C. A., Weller, R. E., Buschbom, R. L., and Ragan, H. A. (1990). Hematology of the wild caught karyotype 1 owl monkey (*Aotus nancymae*). *Lab. Anim. Sci.* **40,** 204–206.
Malaga, C. A., Weller, R. E., Buschbom, R. L., and Ragan, H. A. (1991a). Serum chemistry of the wild caught karyotype 1 night monkey (*Aotus nancymae*). *Lab. Anim. Sci.* **41,** 143–145.
Malaga, C. A., Weller, R. E., Montoya, E., Moro, J., and Buschbom, R. L. (1991b). Mortality and body weight changes in *Aotus nancymae* shipped from Iquitos, Peru to Richland, Washington. *J. Med. Primatol.* **20,** 6–11.
McKenzie, S. M., Chamove, A. S., and Feistner, A. T. C. (1986). Floor-coverings and hanging screens alter arboreal monkey behavior. *Zoo Biol.* **5,** 339–348.
Moseley, J. R., and Davis, J. A. (1989). Psychological enrichment techniques and New World monkey restraint device reduce colony management time. *Lab. Anim.* **18,** 31–33.
Moynihan, M. (1964). Some behavior patterns of platyrrhine monkeys I. The night monkey (*Aotus trivirgatus*). *Smithson. Misc. Collect.* **146,** 1–84.
Mrema, J. E. K., Johnson, G. S., Kelley, S. T., and Green, T. J. (1984). Activated partial thromboplastin time of owl monkey (*Aotus trivirgatus*) plasma. *Lab. Anim. Sci.* **34,** 295–298.
Perachio, A. A. (1971). Sleep in the nocturnal primate, *Aotus trivirgatus*. *Proc. Int. Congr. Primatol., 3rd, 1970,* Zurich, Vol. 2, pp. 54–60.
Pierce, D. L., and Dukelow, W. R. (1988). Misleading positive tuberculin reactions in a squirrel monkey colony. *Lab. Anim. Sci.* **38,** 729–730.

Porter, J. A. (1969). Hematology of the night monkey, *Aotus trivirgatus*. *Lab. Anim. Sci.* **19,** 470–472.

Reardon, M. J., Hall, R. D., and Davidson, C. E. (1979). Serum electrophoretic patterns of karyotypically defined owl monkeys (*Aotus trivirgatus*). *Lab. Anim. Sci.* **29,** 617–620.

Rodger, R. F., Hartley, L. H., Ringler, D. J., and Nicolosi, R. J. (1986). Hypertrophic cardiomyopathy in owl monkeys (*Aotus trivirgatus*): Clinical diagnosis and clinicopathological correlations. *Lab. Anim. Sci.* **36,** 561.

Schölander, P. F., Hock, R., Walters, V., and Irving, L. (1950). Adaptation to cold in arctic and tropical mammals and birds in relation to body temperature, insulation, and basal metabolic rate. *Biol. Bull (Woods Hole, Mass.)* **99,** 259–271.

Sehgal, P. K., Bronson, R. T., Brady, P. S., McIntyre, K. W., and Elliott, M. W. (1980). Therapeutic efficacy of vitamin E and selenium in treating hemolytic anemia of owl monkeys (*Aotus trivirgatus*). *Lab. Anim. Sci.* **30,** 92–98.

Smith, A., Lindburg, D. G., and Vehrencamp, S. (1989). Effect of food preparation on feeding behavior of lion-tailed macaques. *Zoo Biol.* **8,** 57–65.

Smith, S., Morris, J. E., Hoff, C. A., Davis, W. E., and Weller, R. E. (1992). Immunophenotypic characterization of owl monkey peripheral blood mononuclear cells. *Ann. N. Y. Acad. Sci.* **653,** 146–153.

Tilson, R. L. (1986). Primate mating systems and their consequences for captive management. *In* "Primates; the Road to Self-Sustaining Populations" (K. Benirschke, ed.), pp. 361–373. Springer-Verlag, New York.

Umana, J. A., Ramierz, J. C., Espinal, C. A., and Sabogal, E. S. (1984). Establishment of a colony of nonhuman primates (*Aotus lemurinus griseimembra*) in Colombia. *Pan Am. Health Organ. Bull.* **18,** 221–229.

Wellde, B. T., Johnson, A. J., Williams, J. S., Langbehn, H. R., and Sadun, E. H. (1971). Hematologic, biochemical, and parasitologic parameters of the night monkey (*Aotus trivirgatus*). *Lab Anim. Sci.* **21,** 575–580.

Wellde, B. T., Johnson, A. J., Williams, J. S., and Sadun, E. H. (1972). Experimental infection with *Plasmodium falciparum* in *Aotus* monkeys. I. Parasitology, hematologic and serum biochemical determination. *Am. J. Trop. Med. Hyg.* **21,** 260–271.

Weller, R. E., Wierman, E. L., Malaga, C. A., Baer, J. F., and LeMieux, T. P. (1991a). Battelle primate facility. *J. Med. Primatol.* **20,** 133–137.

Weller, R. E., Málaga, C. A., Buschbom, R. L., Baer, J. F., and Ragan, H. A. (1991b). Protein concentration in urine of normal owl monkeys. *J. Med. Primatol.* **20,** 365–369.

Weller, R. E., Baer, J. F., Málaga, C. A., Buschbom, R. L., and Ragan, H. A. (1992). Renal clearance and excretion of endogenous substances in the owl monkey. *Am. J. Primatol.* **28,** 115–123.

Weller, R. E., Collins, W. E., Buschbom, R. L., Málaga, C. A., and Ragan, H. A. (1994). Impaired renal function in owl monkeys (*Aotus nancymae*) infected with *Plasmodium falciparum*. *Mem. Inst. Oswaldo Cruz* (in press).

Whittow, G. C., Guernsey, D. L., and Morishige, W. K. (1979). Thyroid activity in a hypometabolic primate, the owl monkey (*Aotus trivirgatus*). *Arch. Int. Physiol. Biochim.* **87,** 963–967.

Williams, L. E., Abee, C. R., Barnes, S. R., and Ricker, R. B. (1988). Cage design and configuration for an arboreal species of primate. *Lab. Anim. Sci.* **38,** 289–291.

Wright, P. C. (1978). Home range, activity pattern, and agonistic encounters of a group of night monkeys (*Aotus trivirgatus*) in Peru. *Folia Primatol.* **29,** 43–55.

6

Handrearing the Owl Monkey

Carlos A. Málaga
Animal Resources Center
Medical Sciences Campus
University of Puerto Rico
San Juan, Puerto Rico 00936

I. INTRODUCTION

In any breeding colony it is occasionally necessary to handrear offspring. The most frequent reasons for this are perinatal death of the mother, lack of strength of the infant, inadequate nursing, maternal inexperience, or an aggressive or indifferent mother.

As with all medical conditions, early detection is desirable. However, in the case of rejected infants, it is of critical importance. Though daily rounds of the colony are usually adequate, an infant left alone at the bottom of the cage, or injured by its parents, has little chance of survival after only a few hours. The greatest chance for success depends on a strong infant and the early detection of clinical problems.

Personnel in contact with the animals should be aware of which females are pregnant and should check them frequently as they near parturition. This procedure is simplified by clearly identifying cages with pregnant animals.

Owl monkeys are usually shy and remain in the nest box when in the presence of humans. However, they ordinarily do not get overly excited and do come out of the nest box when it is tapped gently, thus allowing visual inspection. Care should be exercised with animals known to be nervous, as they can react violently to any disturbance. These animals should be inspected with extreme care and from a distance to minimize the risk of trauma to the fetus or to the newborn.

Parents with a poor infant-rearing history or with a history of injuring infants should be closely monitored around the expected time of parturition.

II. HANDS-OFF INSPECTION OF THE INFANT

A health record should be initiated for the newborn animal with pertinent history and clinical facts, including any evidence of nursing. The infant should be visually examined for physical defects and/or wounds.

The newborn should hold on firmly to one of the parents during the first weeks after birth. Because owl monkeys do not actively support their newborn, the infant must cling to ensure its survival (Dixson and Fleming, 1981; Cicmanec and Campbell, 1977; Dixson *et al.*, 1980). Although the female will initially carry the newborn, the infant can transfer to the male soon after birth (within 24 hr). As the infant matures the male will progressively replace the female as primary bearer of the infant, who will only transfer to the female for nursing (Dixson and Fleming, 1981; Dixson *et al.*, 1980). The infant will also increasingly transfer to older siblings in the cage (Dixson *et al.*, 1980; Málaga *et al.*, 1991) as it matures, with whom it can spend the majority of its active hours.

Upon birth and thereafter the infant should look strong and hold on to the parent's fur with both hands and feet. The head should be held firmly over the shoulders. A drooping head or one held at an abnormal angle are indications of a weak infant, as the infant is not using all limbs to hold on to the parent.

Owl monkeys are usually, but not always, born with a healthy hair coat. Small infants—those weighing less than 60 g—and premature animals, identifiable by their sparse hair coat, should be observed closely. However, experienced mothers frequently will nurture them through weaning without outside intervention.

III. DECISION TO INTERVENE

Careful observation of the infant during the first few days of life has proven critical to achieving the survival of rejected or weak infants. The following are essential guidelines:

- If the infant is found alone within the first week after birth, the animal should be removed for handrearing, even if one of the adults picks the infant up when approached by a human.
- An infant showing indications of having been injured must be removed immediately.
- Failure to grasp by one or more extremities, as well as holding of the head in an unnatural position, indicates a lack of strength in the infant that can be caused from either insufficient milk production by the mother (agalactia) or the inability of the infant to nurse properly. These animals should be monitored closely and removed for human care if the condition deteriorates. In this case it is better to act hastily than to allow the infant to become hypothermic, hypoglycemic, or dehydrated.

IV. EVALUATION OF THE INFANT AFTER INTERVENTION

Before attempting handrearing, the desirability and potential of successful handrearing must be assessed for the individual infant by conducting a detailed physical examination. Although successful handrearing of owl monkeys is relatively simple, the infant must be strong with no insurmountable handicaps. For this reason the importance of early detection of infants with problems cannot be overemphasized.

Rejection of newborns does not always occur immediately after birth. An infant that has spent a few hours or days with its parents has had the opportunity to nurse on colostrum, which will enhance its ability to survive by providing passive immunity. The following guidelines should be used when assessing the prognosis of handrearing candidates:

- Hypothermic, dehydrated, and hypoglycemic animals are at a considerable disadvantage and require immediate medical attention.
- Severely injured animals are poor candidates, however, the ability of a strong infant to survive is remarkable. Examination of the oral cavity should always be included, as the tongue can frequently be missing

with no other evidence of abuse. If this is the case the infant should be euthanized appropriately as proper nursing is all but impossible.

- Animals with birth defects or other abnormalities can be handreared, provided the defect does not interfere with the animal's ability to survive and does not result in pain or distress to the animal. The convenience of doing so has to be assessed in light of the objectives of the laboratory (i.e., genetic defects have no place in a breeding colony, and a cleft palate will severely interfere with the animal's ability to survive).

V. HANDLING PROCEDURES

After a decision to handrear has been reached, effective action must be taken to ensure the survival of the infant. Hypothermia usually results from hypoglycemia and exposure to the cold surface of the cage. Metal cages and/or nest boxes are particularly efficient at dissipating body heat by acting as "heat sinks." A hypothermic animal should be immediately provided with an energy source, usually lukewarm (85°–90°F) 5% dextrose given *per os* or intraperitoneally (IP). The body temperature of a hypothermic animal must be raised as soon as possible; a bottle of warm water or a heating pad (105°–110°F) is usually effective. A clean towel should be placed between the heat source and the infant, and a second clean towel should then be placed over the infant. The newborn's temperature should be monitored every 15 min. A second dose of 5% dextrose should be provided 30 min after the first one, using the procedure described in the following. Only circulating-water heating pads should be used, as electrical pads can overheat with detrimental effects. The use of cardiac or respiratory stimulants has not proven beneficial at this stage.

To assess the infant's swallowing reflex, use a tuberculin syringe. Place a drop of lukewarm 5% dextrose (85°–90°F) at the lateral commissure of the mouth and allow the liquid to be drawn into the oral cavity by capillary action. When the liquid is in the oral cavity the newborn will swallow, if capable. If this is the case continue to provide lukewarm 5% dextrose up to a maximum of 0.5 ml. If the infant is incapable of swallowing, administer 1 ml of 5% dextrose IP.

If after an hour the infant is still lethargic, the likelihood of success is very limited, as success at handrearing requires the cooperation of an active infant. Nevertheless, continue treatment (i.e., provide 5% dextrose, maintain the heat source temperature between 105° and 110°F, and monitor body temperature). When body temperature approaches 100°F, decrease the temperature of the heat source to 100°F. If after 2 hr the animal is still lethargic and has no swallowing reflex, it is probably better to euthanize it in view of the diminishing chance of success.

VI. SUPPLIES AND EQUIPMENT

In contrast with other Neotropical nonhuman primates, the owl monkey is relatively easy to handrear. However, nursery personnel must be well trained, patient, and caring. Equipment and supplies necessary to handrear owl monkeys are described in Tables I and II.

VII. FEEDING FORMULAS

Several species of nonhuman primates are particularly sensitive to milk substitutes. Milk incompatibility will result in bloating, diarrhea, or both. Fortunately owl monkeys are relatively insensitive to this problem and have been handreared successfully using milk substitutes (Cicmanec and Campbell, 1977; Cicmanec *et al.*, 1979).

However, problems do occasionally occur. The feeding procedure described here is somewhat more complicated than using milk substitutes directly, but has proven reliably successful. The procedure was first developed at the Iquitos Primate Center (IPC), in Perú, where it was used for handrearing marmosets, tamarins, and owl monkeys. In turn, it was refined and extensively used for handrearing owl monkeys at Battelle Northwest Laboratories (BNW) in Richland, Washington, with excellent results.

The procedure is based on a series of formulas, each varying slightly. The infant is started on the simplest formula and is gradually moved to a complex diet. At the first sign of bloating or diarrhea the animal is returned to the previous formula, where it remains until symptoms disappear. Once the animal is stabilized it is again returned to the "appropriate" diet. A viable alternative when bloating or diarrhea occurs is to alternate feedings between 5% dextrose and the current diet until the animal is stabilized.

Table III presents the different diet designations, their composition, and the recommended period of use. It is important, when changing diets, to

TABLE I Equipment for Handrearing Owl Monkeys

- Circulating water heat pad or water bottle for warming hypothermic animals[a]
- Thermocare unit, incubator, or other means of controlling ambient temperature and humidity in a restricted space[a]
- Weight scale allowing measurement by grams
- Refrigerator
- Blender to powder monkey chow and homogenize diets appropriately

[a] Owl monkeys have been successfully handreared using nothing more than a woven basket and a light bulb as a heat source. The equipment described here is convenient but not indispensable and can be substituted with a little imagination.

TABLE II Supplies for Handrearing Owl Monkeys

- White towels, disposable, or abundant with the ability to clean and sanitize them satisfactorily
- Syringes, 1 ml (tuberculin) and 3 ml capacity
- Disposable containers, plastic or glass, with lids and 10 ml capacity
- 5% dextrose
- SMA (milk substitute)
- Baby cereal (mixed cereal, oatmeal cereal, and banana cereal, manufactured by Gerber, are recommended)
- Sustagen (high-protein supplement)
- BVMO (vitamin and mineral supplement)
- Nursing bottles, very small, disposable, or the means to sanitize them appropriately
- Nipples for nursing bottles, with opening only large enough to allow very slow dripping when inverted; too large an opening will result in aspiration of the formula

alternate formulas during the 1-day overlap period. This minimizes diarrhea and bloating, allowing the newborn time to adapt to the new diet.

VIII. FEEDING SCHEDULE

There are no fixed guidelines regarding the frequency of feeding during handrearing. Small infants, or those rejected within the first 72 hr after birth, should be fed every 2 hr around the clock. During the second week, depending on strength and weight gain, infants can be fed every 4 hr. If the infant is gaining weight adequately and is strong, feeding can be stopped between 12:00 pm and 6:00 am. As a rule of thumb, infants under 100 g should be fed every 2 hr, whereas animals over that weight can be fed every 4 hr. Obviously, the availability of 24 hr a day care significantly increases the chance of survival during the first 2 weeks of life. Small facilities and zoos might consider allowing interested personnel to take animals home overnight, where a light bulb can provide the necessary warmth without undue hardship.

After the fourth week of age, feeding intervals can be increased to every 6 hr. This schedule should be maintained until the animal is capable of consuming solid food, when it should be fed *ad libitum*. Infants fed moist diets should be given access to food for a limited time only, not to exceed 1 hr, as moist diets remaining at incubator temperatures for longer periods will spoil and cause diarrhea or bloating.

IX. FIRST WEEK FEEDING

Feeding just-born or debilitated infants is the most delicate procedure in handrearing owl monkeys. An absent or feeble sucking reflex is potentially dangerous as inhalation pneumonia can result from the effort to provide nourishment.

During the first 2 days of life, or when feeding a debilitated infant, the following procedure should be used:

1. Draw the formula (usually 5% dextrose solution) into a 1-ml tuberculin syringe.
2. Depress the plunger while holding the syringe upright until a drop of solution forms at the tip.
3. Bring the tip of the syringe, still in a vertical position, to the commissure of the lips of the infant. The solution will be drawn into the mouth by capillary action. This usually triggers licking and swallowing reflexes.
4. Repeat this procedure slowly until the infant has received 0.5 ml of formula, or until it refuses to accept more.

Do not force feed the infant under any circumstances, as this generally results in aspiration of the formula. If the infant does not swallow by reflex, gentle rubbing of the throat will usually cause the reflex. Newborn animals generally accept 1–2 ml of formula every 2 hr for the first week of life. In general, infants should be offered as much as they are willing to consume at any given feeding.

X. PROCEDURES IN CASE OF ASPIRATION OF FORMULA

Aspiration is usually noticed by bubbling of liquid through the nostrils or, less frequently, by coughing or choking of the infant. Detection of respiratory bubbling sounds by placing the infant close to the ear or using a stethoscope is a sure sign of aspiration. Aspiration of formula is usually caused by one of two circumstances:

1. A reluctant infant is force fed at a rate faster than it can swallow or
2. An active infant becomes overzealous in consuming the formula, resulting in rapid administration by the caretaker and consequent aspiration.

This occurrence can be prevented by ensuring careful and patient administration of the formula. However, if aspiration does occur, the following procedures should be performed immediately:

1. Remove formula from the nostrils and oral cavity of the infant, drying them with tissue paper.
2. Pick infant up by the tail and hold its head down, making it cry out. If crying does not occur spontaneously, pinch the tail progressively harder until the infant cries.
3. Repeat steps 1 and 2 alternatively until no more liquid can be collected with the tissue paper.

TABLE III Formulas for Handrearing Owl Monkeys

Diet designation	Formula[a]	Period used[b,c]
1	5% dextrose	First 36 hr; when animal is dehydrated, suffering from diarrhea or bloat
2	1 part SMA 9 parts 5% dextrose	24 hr to 4 days
3	2 parts SMA 8 parts 5% dextrose	3 days to 7 days
4	1 part SMA 2 parts 5% dextrose	6 days to 10 days
5	1 part SMA 1 part 5% dextrose	9 days to 30 days
6	4 parts SMA 4 parts 5% dextrose 1 part cereal	29 days to 37 days
7	4 parts SMA 4 parts 5% dextrose 2 parts cereal	36 days to 40 days
8	30 ml SMA 30 ml 5% dextrose 10 ml cereal 8 ml Sustagen 1 ml BVMO	39 days to 45 days
9	35 ml SMA 35 ml 5% dextrose 20 ml cereal 12 ml Sustagen 1 ml BVMO	44 days to 50 days
10[d]	50 ml SMA 50 ml 5% dextrose 40 ml cereal 25 ml Sustagen 1.5 ml BVMO	50 days to 59 days
11[d]	50 ml SMA 50 ml 5% dextrose 50 ml cereal 30 ml Sustagen 25 ml monkey chow[e] 1.5 ml BVMO	59 days to 2 1/2 months
12[d]	50 ml SMA 50 ml potable water 50 ml cereal 50 ml monkey chow[e] 30 ml Sustagen 1.5 ml BVMO 1/4 avg. size banana	2 1/2 months to 3 months
13[d]	100 ml monkey chow[e] 70 ml potable water 50 ml SMA	3 months onward

(*continues*)

TABLE III *(continued)*

Diet designation	Formula[a]	Period used[b,c]
	50 ml cereal 40 ml Sustagen 1.5 ml BVMO 1/3 avg. size banana	

[a] All diets should be properly homogenized; usage of a blender for the thicker diets is highly recommended.
[b] Ranges indicated in the "Period used" overlap. During the one day overlap period, formulas should be alternated.
[c] Periods indicated are recommendations only. Changes for each infant should be based on independent evaluation of its performance.
[d] Diets 10 through 13 should be sprinkled with a few pieces of monkey chow. Consumption of these should be recorded.
[e] Monkey chow should be powdered using a blender. Measuring the volume of the powdered diet using a beaker or large syringe is more expedient than weighing.

Usually these procedures are enough to prevent the occurrence of inhalation pneumonia. If unsuccessful and inhalation pneumonia does occur, the condition should be treated medically. The chance of infant survival under these circumstances is poor.

XI. THERMOCARE UNIT SETTING

During the first week the incubator's temperature should be set at 95°F (35°C). During the second and third weeks, if the infant is stable, the temperature should gradually be reduced to 85°F (29°C), where it should remain until the animal is self-feeding. Relative humidity should be kept close to 75%.

XII. SURROGATE MOTHERS

The use of furry objects or other apparatus functioning as surrogate mothers is encouraged in handrearing (Hennessy, 1985; Anderson, 1986; Dronzec *et al.*, 1986). These surrogates will enrich the environment of the newborn. However, they get soiled quickly and should be sanitized accordingly.

XIII. INTRODUCTION TO A FAMILY GROUP

The consequences of handrearing nonhuman primates are well known: the absence of adequate socialization can result in abnormal social behavior, the appearance of "stereotypies," the inability to form proper bonds with a mate, and, eventually, the inability to reproduce or to raise offspring appropriately (Hennessy, 1985; Anderson, 1986; Ruppenthal and Sacket, 1979).

It has been recognized that a period of socialization is imperative for the successful breeding of colony-born owl monkeys (Dixson and Fleming, 1981). At some facilities, such as IPC and BNW, socialization continues after the infant is old enough for weaning. This "social upbringing" includes, if possible, retaining the infant within the family group through the birth of another infant, which allows it to carry or play with the younger infant and gain parenting experience.

By definition, handreared nonhuman primates do not benefit from the parent's social group. In many institutions it is common practice, with other species, to form groups of handreared infants that are raised together. However, in the case of owl monkeys, this leads to clinging and other stereotypies, thus increasing the animal's inability to socialize properly.

Owl monkeys occasionally have trouble in forming stable breeding pairs (Cicmanec and Campbell, 1977; Elliot *et al.*, 1976; Rieckmann *et al.*, 1980). However, once a stable family unit has been formed, they are a suitable animal in which to investigate experimental techniques for reintroducing handreared infants into their own or other family units. Owl monkeys are relatively gentle compared to other species of nonhuman primates and usually accept the introduction of an infant into their group, if the procedure is handled with care.

Reintroduction of handreared infants into family groups has been successfully achieved in callitrichids (Dronzec *et al.*, 1986; Tardif *et al.*, 1985). It has also been performed repeatedly with monkeys at both IPC and BNW. Some situations under which introductions have been successful are described in the following sections.

A. Natural Parents

A breeding pair, known to be good parents, has had their infant removed because of impaired nursing (agalactia or the inability of the infant to suck properly). After the infant has regained its strength (within 24–48 hr), the infant is taken to "visit" its parents. This can be done one of two ways.

1. The infant is placed on a clean towel within the cage: the parents are usually suspicious of the towel and avoid it for a few minutes. Eventually one of the animals will overcome its suspicions and retrieve the infant by allowing it to crawl onto the parent's back.

The advantages of this procedure are twofold: the infant is kept warm by the body heat of the parents, which greatly increases the time it can stay "out," while at the same time it is actively imprinting on its own species. However, care must be exercised with the initial introduction of the infant to ensure that it is accepted adequately without aggression. It is important to retrieve the infant before it gets weak. Care must be exercised in this procedure as the possibility exists of harming the infant during retrieval.

2. The infant is placed in a small cage or nest box, which is positioned in contact with, or inside, the cage of the family unit. This procedure allows for the safe placement and retrieval of the infant. An additional advantage is that the animal can be examined more easily, allowing feeding and checking for hypothermia. If an appropriate heat source is used, the infant can remain for a considerable amount of time with its parents.

This method has the disadvantage of slower and incomplete imprinting. In addition, the openings in the cage or nest box must be smaller than 1/4 inch × 1/4 inch. Larger openings will allow the parents to introduce their hand into the cage, and they might grab the infant and try to retrieve it through too small an opening, with dire consequences.

B. Foster Parents

A newly rejected, strong infant, or one being handreared, can be introduced to a breeding pair that has recently lost a newborn infant owing to problems unrelated to poor parental care. In our experience there is little risk to the infant if fostering is performed within 72 hr after the loss of the original offspring. The introduction can be performed on foster parents that have been anesthetized with ketamine hydrochloride (10 mg/kg). Both the anesthetized foster parents and the infant are enclosed in a nest box, which is only opened once the parents have fully recovered from anesthesia. Release of the foster parents to the larger cage prior to recovery may lead to injury of the infant as the parents move around the cage in an uncoordinated fashion.

If a breeding pair that has recently lost an infant is not available as foster parents, then a pair of tame, easy to handle owl monkeys can be used. Initially the newborn is placed inside a small cage or nest box, which is then placed in contact with or inside the cage of the foster parents. If the infant's cage or nest box is provided with a suitable heat source, the infant can be removed for feeding with little disturbance to the foster parents until the infant becomes self-feeding, when it can be introduced permanently to its new family.

Alternatively, the infant can be introduced into the foster family after an initial introductory period during which they have limited contact. Foster parents can be trained to allow removal of the infant with a minimum of commotion by teaching them to accept small treats (raisins, pieces of banana, etc.) as a reward for this procedure. Animals so trained can be used again as foster parents once the original infant is approximately 6 months old.

During the introduction of a self-locomoting infant into a foster parent group, one must ensure that the infant has access to its feeding formula—which at this time has the consistency of a gruel—as quite often the foster parents develop a liking for the gruel and will consume it readily, preventing the infant from feeding properly. A strategy that has proven effective is to provide the gruel inside a nest box with an opening large enough to allow the infant to enter, but small enough to prevent the foster parents from doing so.

Handrearing owl monkeys can be both a challenging and a rewarding activity. The temperament and relative hardiness of the species guarantee that, with reasonable care, satisfactory results can be obtained. Handreared owl monkeys introduced into foster families develop into well-socialized animals that produce offspring of their own. The time and effort invested in developing a nursery for handrearing and in identifying suitable foster parents will be well rewarded. It will become a source of pride to all individuals involved in the effort.

REFERENCES

Anderson, J. H. (1986). Rearing and care of nonhuman primates. *In* "Primates: The Road to Self-Sustaining Populations" (K. Benirschke, ed.), 1st ed., pp. 747–762. Springer-Verlag, Berlin and New York.

Cicmanec, J. V., and Campbell, A. K. (1977). Breeding the owl monkey (*Aotus trivirgatus*) in a laboratory environment. *Lab. Anim. Sci.* **27,** 512–517.

Cicmanec, J. V., Hernández, D. M., Jenkins, S. R., Campbell, S. R., and Smith, J. A. (1979). Hand rearing infant Callitrichida (*Saguinus* spp. and *Callithrix jacchus*), owl monkeys (*Aotus trivirgatus*) and capuchins (*Cebus albifrons*). *In* "Nursery Care of Non-human Primates" (G. C. Ruppenthal, ed.), 1st ed., pp. 307–312. Plenum, New York and London.

Dixson, A. F., and Fleming, D. (1981). Parental behavior and infant development in owl monkeys (*Aotus trivirgatus griseimembra*). *J. Zool.* **194,** 25–39.

Dixson, A. F., Martin, R. D., Bonney, R. C., and Fleming, D. (1980). Reproductive biology of the owl monkey, *Aotus trivirgatus griseimembra. Non-Hum. Primate Models Study Hum. Reprod., Satell. Symp.*, Bangalore, *1979*, pp. 61–68.

Drozek, L. A., Savage, A., Snowdon, C. T., Whaling, C. S., and Ziegler, T. E. (1986). Techniques of handrearing and reintroducing rejected cotton-top tamarin infants. *Lab. Anim. Sci.* **36,** 243–247.

Elliot, M. W., Sehgal, P. K., and Chalifoux, L. V. (1976). Management and breeding of *Aotus trivirgatus. Lab. Anim. Sci.* **26,** 1037–1040.

Hennessy, M. B. (1985). Effect of surrogate-rearing of the infant squirrel monkey. In "Handbook of Squirrel Monkey Research" (L. A. Rosenblum and C. L. Coe, eds.), 1st ed., pp. 149–166. Plenum, New York and London.

Málaga, C. A., Weller, R. E., and Buschbom, R. L. (1991). Twinning in the karyotype I night monkey (*Aotus nancymae*). *J. Med. Primatol.* **20,** 370–372.

Rieckmann, K. H., Mrema, J. E. K., Marshall, P. H., and Hafner, D. M. (1980). Breeding of *Aotus* monkeys for human malaria research. *Pan Am. Health Organ. Bull.* **14,** 251–257.

Ruppenthal, G. C., and Sacket, G. P. (1979). Experimental and husbandry procedures: Their impact on development. *In* "Nursery Care of Nonhuman Primates" (G. C. Ruppenthal, ed.), 1st ed., pp. 269–284. Plenum, New York and London.

Tardif, S., Lenhard, A., Carson, R., and McArthur, A. (1985). Hand rearing of infant *Saguinus oedipus* with subsequent introduction into social groups. *Am. J. Primatol.* **8,** 368–369.

7

Infectious and Noninfectious Diseases of Owl Monkeys

Richard E. Weller
Battelle Pacific Northwest Laboratories
Richland, Washington 99352

I. INTRODUCTION

Owl monkeys are important among Neotropical primates because of their widespread use in biomedical research. They are among the important primate species kept in zoological gardens and primate centers. Diseases in owl monkeys are not as extensively documented as those in Old World monkeys and squirrel monkeys. Also, the information is scattered in the literature and is not readily available to the veterinarian confronted with disease diagnosis. This chapter is intended to focus some of that information and increase awareness of diseases that affect owl monkeys.

Recognition of a sign in an animal by a technician or veterinarian that might signify the presence of disease introduces the patient into the clinical realm of differential diagnosis (Madewell, 1987). The goal is to identify and localize the disease. A thorough general physical examination is the first step in the diagnostic process (Hornbuckle, 1988). The examination consists of several parts. First is a "hands-off" observation of the animal to assess its condition, attitude, conformation, nutritional status, and movements. Depending on the size of the monkey, this can be done while the animal is in its cage or in the examination room. Unless the monkey is well socialized, it is advisable to conduct the systemic "hands-on" portion of the examination under light sedation to minimize stress and the possibility of physical injury to the animal as well as animal care personnel. The examiner should establish a routine procedure for physical examinations to assure that all accessible systems, organs, structures, and orifices of the animal are examined. This is often facilitated by development of a physical examination form to ensure completeness and uniformity of the examination procedure. The emphasis and time spent on certain areas of the examination are determined from the history and initial visual observations. Routinely recorded observations should include body weight, rectal temperature, respiratory rate, heart rate, mucous membrane color, capillary refill time, and hydration status. Examination should always include careful palpation and examination of those regional lymph nodes that drain the anatomical area in which the suspected disease is located. Results of the physical examination are used to plan a problem-oriented diagnostic approach for eliminating other differential considerations and obtaining a specific diagnosis. The accuracy of the diagnostic approach is often enhanced by the availability of technologies such as clinicopathologic determinations, cytology, endoscopy, and imaging methodologies.

II. INFECTIOUS DISEASES

A. Bacterial Diseases

Bacterial infections in owl monkeys generally result in cellulitis, pneumonia, enteritis, or septicemias. Owl monkeys may be predisposed to some bacterial diseases owing to the cage system or husbandry program being used. The best approach to the management of bacterial diseases in a colony of owl monkeys is prevention rather than treatment. This is dealt with by a carefully controlled quarantine program, maintenance of a high level of environmental sanitation, adequate nutrition, and a responsive husbandry program. In a stable colony, bacterial diseases usually occur as sporadic isolated cases.

1. Streptococcal Infections Alpha-hemolytic and nonhemolytic streptococci were isolated from a variety of infectious conditions diagnosed in owl monkeys maintained at the Battelle Primate Facility (BPF) (Weller *et al.*, 1991a). Positive cultures were obtained from swabs taken from nasal discharges, heart blood, pericardial fluid, trachea, oropharynx, abscesses, areas of cellulitis, fistulous tracts, and a hepatic cyst (Fig. 1). Clinical signs depend on the organ system involved and can range from fever to septicemia. Definitive diagnosis requires isolation and identification of chains of gram-positive spherical bacteria that form small translucent to opaque colonies on blood agar. The hemolytic characteristics of various streptococci can be used to clinically classify different isolates (Coles, 1974; Goldstein and Joye, 1981). Immunologic grouping of streptococci is probably unnecessary in most cases.

Successful treatment requires the use of specific antibiotics selected on the basis of *in vitro* sensitivity testing. Penicillin, erythromycin, chloramphenicol, and cephalexin have been shown to be effective against streptococci.

2. Staphylococcal Infections *Staphylococcus aureus* is the prime culprit in infections caused by staphylococci. Coagulase-positive staphylococci were isolated from the oropharynx of 45% of *Aotus nancymae* received at the

FIGURE 1 Hepatic cyst localized near the wall of the gallbladder. An alpha-hemolytic streptococcus was isolated from swabs of the mucopurulent content of the cyst.

BPF over a one-year period. Coagulase-positive and coagulase-negative staphylococci were often isolated from the heart, trachea, nasal cavity, abscesses, ascitic fluid, ocular discharges, fistulous tracts, and liver abscesses of owl monkeys exhibiting signs of clinical disease. Clinical signs are similar to those of streptococcal infections. Diagnosis requires demonstration of gram-positive coccoid bacteria commonly arranged in grapelike clusters. Pathogenic staphylococci usually hemolyze blood and produce the enzyme coagulase (Coles, 1974).

Antibiotics are the primary therapy for staphylococci infections. Trimethoprim/sulfadiazine, ampicillin, penicillin, chloramphenicol, the aminoglycosides, and fluoroquinolones have all demonstrated clinical efficacy against common staphylococci.

3. Pseudomonas Infections *Pseudomonas aeruginosa* is a ubiquitous organism that can cause significant clinical problems in nonhuman primates. It occurs normally in the intestinal flora and on the skin of nonhuman primates (Wallach and Boever, 1983). In our facility, *P. aeruginosa* was occasionally cultured from oropharyngeal swabs routinely obtained from owl monkeys in quarantine. The organism can cause only local disease, or septicemia with widespread systemic involvement of multiple organs. Clinical signs vary with regard to the organ system involved (Wallach and Boever, 1983). A characteristic green or blue-green exudate is often observed in association with some lesions. Diagnosis is dependent on a positive culture of blood or exudate that reveals motile, slender, gram-negative, bacilli (Fossum, 1988). *Pseudomonas aeruginosa* forms beta-hemolytic, flat, spreading colonies with a fruitlike odor and a "metallic sheen" (Coles, 1974; Goldstein and Joye, 1981).

Treatment should include aggressive parenteral antibiotic therapy with carbenicillin, gentamicin, tobramycin, amikacin, ticarcillin, or cefotaxime. Animals should be maintained on a selected antibiotic for a minimum of 2 weeks after all clinical changes have resolved.

4. Colibacillosis Colibacillosis generally refers to diseases caused by hemolytic *Escherichia coli*. The organism normally inhabits the flora of the ileum and large bowel, but both invasive and enterotoxigenic isolates can cause diarrhea or systemic illness in compromised animals. An outbreak of colibacillosis occurred in our owl monkey colony that was characterized clinically by dehydration, depression, and profuse, foul-smelling, blood-tinged diarrhea. The causative agent was subsequently identified as a hemolytic *E. coli*. Identification of a pathogenic or nonpathogenic *E. coli* requires specific typing and/or studies of toxin production.

Treatment of colibacillosis should include orally administered intestinal protectants, parenteral fluid therapy, and parenteral antibiotics such as ampicillin, gentamicin, trimethoprim/sulfadiazine, and enrofloxacin. Motility modifiers are contraindicated in bacterial diarrhea.

5. Klebsiella Infections *Klebsiella*, a gram-negative, nonmotile rod, primarily affecting the respiratory tract, is highly pathogenic in owl monkeys. Several authors have described culturing *Klebsiella pneumoniae* from infected lungs and air sacs of owl monkeys (Good and May, 1971; Giles *et al.*, 1974; Gozalo and Montoya, 1990), and it has also been associated with meningitis and enteritis (Wallach and Boever, 1983). Clinical signs include depression, anorexia, sneezing, mucopurulent nasal discharge, coughing, dyspnea, and facial edema. A diagnosis of klebsiella infection is made through culture of pharyngeal or rectal swabs.

Treatment of acute klebsiella infections should include aggressive parenteral administration of antibiotics, whereas chronic or recurrent infections can often be managed by oral or parenteral antibiotic administration. Cephalothin, trimethoprim/sulfadiazine, gentamicin, amikacin, and kanamycin have been shown to be effective in the treatment of *K. pneumoniae*. Antibiotic therapy should be continued for 3–4 weeks.

6. Yersiniosis Pseudotuberculosis caused by *Yersinia enterocolitica* has been reported as an enzootic disease of the owl monkey (Baggs *et al.*, 1976). The organism has a worldwide distribution and causes disease in several species of nonhuman primates (Wallach and Boever, 1983). Wild rodents and birds are common reservoirs of the organism. Clinical signs can vary significantly, but commonly include weakness, depression, diarrhea, abdominal distension, and splenomegaly (Baggs *et al.*, 1976). Neutrophilia with left shift was a common hematologic finding in owl monkeys experimentally infected with *Y. enterocolitica* (Baggs *et al.*, 1976). The primary lesions observed following necropsy are purulent, necrotizing enteritis, hepatitis, and splenitis. Diagnosis of yersiniosis is made by culture of the feces.

Treatment for dehydration and electrolyte imbalances associated with diarrhea should include oral intestinal protectants, parenteral fluid therapy, and parenteral broad-spectrum antibiotics. Trimethoprim/sulfadiazine and gentamicin appear to be most effective against *Yersinia.*

7. Pasteurellosis Pasteurella infections occur in all species of nonhuman primates, and *Pasteurella multocida* is common to the flora of the upper respiratory tract. An outbreak of pasteurellosis in the owl monkey has been described (Benjamin and Lang, 1971). *Pasteurella multocida* is considered the principal respiratory pathogen of the owl monkey (Good and May, 1971). Pathologic conditions associated with pasteurella infections include abscesses, pneumonia, septicemia, otitis media, and encephalitis. Clinical signs can include localized swelling and exudation associated with abscesses, fever, anorexia, depression, coughing, dyspnea, seizures, and weight loss. Diagnosis depends on demonstrating small, gram-negative bipolar, ellipsoidal rods in positive cultures of blood or exudate (Wallach and Boever, 1983; Fossum, 1988).

Antibiotics are the primary therapy for pasteurellosis. The antibiotic of choice is dependent on culture and sensitivity testing. *Pasteurella multocida* is generally susceptible to oral or parenteral administration of chloramphenicol, penicillin, and ampicillin. Animals should be maintained on a selected antibiotic for a minimum of 2 weeks after all clinical changes have resolved.

8. Bordetella *Bordetella bronchiseptica* is considered a common pathogen of marmosets, patas monkeys, macaques, titi monkeys, and guenons (Wallach and Boever, 1983). However, it is an uncommon respiratory pathogen of owl monkeys (Good and May, 1971), although it was reported as one of the more common pathogens isolated from owl monkeys dying of pneumonia at the Center for Reproduction and Conservation of Nonhuman Primates (CRCP) in Iquitos, Peru (Gozalo and Montoya, 1990). The organism may persist for as long as 14 weeks in the respiratory tract. It attaches to the cilia of the bronchial epithelium and produces ciliary stasis. The bordetella organism causes tracheitis, bronchitis, and pneumonia. Uncomplicated bordetella infections are usually mild and self-limiting. Systemic signs of disease can include serous to mucopurulent nasal and ocular discharges, dyspnea, depression, anorexia, and weight loss.

Bordetella bronchiseptica can be diagnosed by demonstrating the presence of small colonies of gram-negative rods on blood agar plates and other metabolic characteristics following culture of nasal exudates or throat swabs. In addition to supportive care, treatment should include administration of chloramphenicol, tetracycline, or amoxicillin.

9. Tuberculosis All nonhuman primates are considered to be susceptible to tuberculosis (TB), although they vary in their level of susceptibility. Owl monkeys are relatively resistant to tuberculosis in comparison to Old World species (Snyder *et al.*, 1970), and a few cases of TB in owl monkeys have been described in the literature (Bone and Soave, 1970; Snyder *et al.*, 1970). Nonetheless, because of the zoonotic potential of the disease and the susceptibility of other nonhuman primates, owl monkeys should be routinely tested for TB. Most cases of tuberculosis in nonhuman primates are caused by three species of *Mycobacterium—M. tuberculosis, M. Bovis,* and *M. avium*—although atypical mycobacterial strains have been implicated in some cases (Whitney, 1979).

Infected animals may be asymptomatic until the disease is quite advanced. Clinical signs reported in owl monkeys include weight loss, dehydration, serous ocular and nasal discharge, and depression. Less common signs reported in other species of nonhuman primates include diarrhea, skin ulceration, suppurative lymphadenitis, and hepatosplenomegaly (Whitney, 1979).

Tuberculin testing is the preferred method of diagnosing the disease in owl monkeys and other nonhuman primates. The specifics of TB testing in owl monkeys are discussed in Chapter 5. Any reaction should be considered positive and the animal should be quarantined until it can be retested in 1 week. Other diagnostic methods, such as thoracic radiography and culturing blood, sputum, or feces, are of no diagnostic value and are not recommended. Because of their extensive use in biomedical research, many owl monkeys are given Freund's complete adjuvant (FCA) at some time in their life. These animals are likely to have false positive reactions to TB testing that wax and wane over their life span. This can be quite confusing and disconcerting to animal care personnel, particularly if they have no knowledge of the animal's prior research history. In addition, embolization of FCA can cause lesions that resemble those of tuberculosis in some animals, as shown in Fig. 2 (Broderson, 1989). This points out the importance of documenting the medical history of each animal and making that record available to each institution that holds that animal for even a brief period of time. Animals known to have received FCA should not be TB tested.

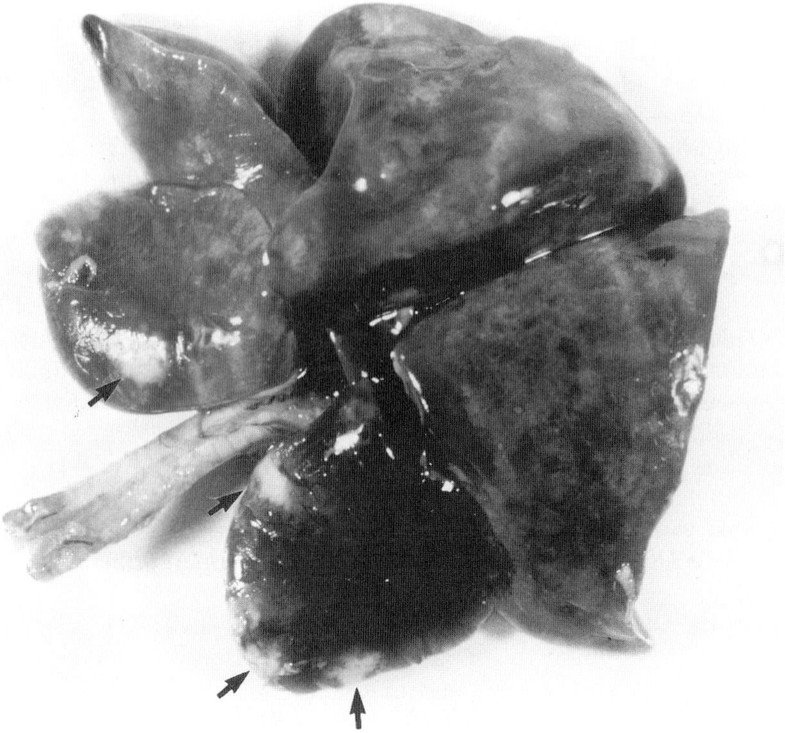

FIGURE 2 Photograph of lungs of an owl monkey that received multiple intramuscular injections of Freund's complete adjuvant showing multiple necrotic foci (arrows) containing caseouslike material.

Treatment is contraindicated, and positive animals should be removed and euthanatized. A complete necropsy should be done and appropriate laboratory tests performed to confirm the diagnosis. Any other animals in the room that housed the positive animal must be immediately quarantined and tested accordingly.

10. Campylobacter *Campylobacter jejuni* has frequently been reported as a cause of acute bacterial gastroenteritis in a number of species, including humans. No published reports on this problem exist for the owl monkey. However, at the CRCP in Iquitos, Peru, *C. jejuni* was recovered from several species of nonhuman primates, including the owl monkey. When evaluating gastrointestinal conditions in owl monkeys, one should always consider the possibility of campylobacteriosis.

B. Mycotic Infections

1. Histoplasmosis Histoplasmosis, the systemic fungal disease caused by *Histoplasma capsulatum*, has been described in a squirrel monkey (Bergeland *et al.*, 1970) and in an owl monkey (Weller *et al.*, 1990). The disease has a worldwide distribution, but is usually associated with humid, temperate climates. The endemic distribution of the disease is related to factors favoring the persistence and growth of the organism in soil frequently contaminated with droppings of fowl, bats, starlings, and other birds. *Histoplasma capslatum* var. *capsulatum* infection in Amazonian wildlife is of a low-order subclinical nature (Naiff *et al.*, 1985). Histoplasmosis is primarily a pulmonary disease, with marked affinity for the reticuloendothelial (RE) system; it is characterized by systemic granulomatous formations or chronic cutaneous ulcerations. In addition to the lung, affected organs often include the liver, spleen, regional lymph nodes, adrenal glands, intestine, bone, and bone marrow.

Clinical signs can include weight loss, diarrhea, fever, hepatomegaly, splenomegaly, lymphadenopathy, anorexia, dyspnea, and unhealed granulating ulcers. Diagnosis is made by finding the typical 2- to 4-μm-diameter, round, intracellular yeast forms in a biopsy of a regional lymph node, the spleen, or a postmortem sample.

Most clinically symptomatic cases of histoplasmosis are progressive and fatal if untreated. The preferred treatment for histoplasmosis is systemic ketaconazole. Amphotericin B may be used but is less desirable because of its toxicity. A promising drug is itraconazole, which has fewer gastrointestinal side effects than ketaconazole.

2. Moniliasis Moniliasis, the overgrowth of *Candida albicans*, is the most common systemic fungal infection reported in owl monkeys (Daniel *et al.*, 1976; Gozalo and Montoya, 1990). *Candida* spp. are normal flora of the oral, gastrointestinal, and genital mucosa. Disease usually develops only in animals that are severely debilitated or immunosuppressed, or those that have been on chronic antibiotic therapy. Clinical signs in owl monkeys include anorexia, erosive or ulcerative lesions at mucocutaneous sites, whitish pseudomembranes on mucous membranes, diarrhea, and dehydration. Diagnosis is made with a combination of identification by culture and histopathologic or cytologic demonstration of epithelial invasion and colonization.

Treatment begins with the identification and correction of predisposing causes if possible. Mucosal and cutaneous lesions can be treated with topical nystatin or ketaconazole. In widespread or chronic disease, systemic ketaconazole is the treatment of choice.

C. Viral Infections

1. Measles Natural outbreaks of measles (rubeola) have been reported in owl monkeys. Symptomatology often includes fever, diarrhea, and giant cell pneumonia. Diagnosis is usually dependent on demonstrating increased antibody titers to measles, the finding of inclusion bodies in biopsy material, and on virus culture. Treatment is symptomatic and includes broad-spectrum antibiotics, intravenous fluid and electrolyte replacement, and antidiarrheals.

2. Herpes T *Herpesvirus tamarinus* causes severe disease and death in owl monkeys (Barahona *et al.*, 1976; Wallach and Boever, 1983). The virus is carried asymptomatically by squirrel and spider monkeys (Wallach and Boever, 1983; Abee, 1985), therefore, it is very important that these two species be maintained well separated from facilities housing owl monkeys. The clinical signs of Herpes T infection are oral and labial ulcers, ulcerative dermatitis, conjunctivitis, edema of the eyelids and the periorbital area, conjunctivitis, anorexia, weakness, and marked swelling of the parotid salivary gland (Barahona *et al.*, 1976; Wallach and Boever, 1983). Most affected animals die within 2 to 6 days following the onset of clinical signs, although some may spontaneously recover within 10 days (Wallach and Boever, 1983).

Treatment is symptomatic, plus a broad-spectrum antibiotic as prophylaxis against opportunistic bacterial infections.

3. Herpes Simplex Herpes simplex, caused by *Herpesvirus hominis*, produces a severe disease in owl monkeys characterized by clinical signs indistinguishable from those of Herpes T. Therapy has little effect on the course of the disease. Diagnosis is made by serological techniques and demonstration of inclusion bodies at necropsy (Wallach and Boever, 1983). Personnel with "cold sores" should be excluded from facilities housing owl monkeys.

4. Infectious Hepatitis Virus Infectious hepatitis has been experimentally induced in owl monkeys with hepatitis A virus (LeDuc *et al.*, 1983) and hepatitis E virus (Ticehurst *et al.*, 1992). This suggests the possibility that hepatitis could be passed to owl monkeys by personnel with the disease, or that the species could serve as a source of infection for human subjects. A newer clinical entity in Neotropical primates termed viral callitrichid hepatitis has been described; the viral agent has been classified as an *Arenavirus* resembling lymphocytic choriomeningitis virus (LCMV) carried by mice (Montali, 1992).

5. Encephalomyocarditis Virus The encephalomyocarditis virus (EMCV) is an enterovirus in the family *Picornaviridae*. The virus occurs worldwide and has been associated with fatal epizootics in New World monkeys as well as other mammalian species (Kessler *et al.*, 1982). Rodent populations are suspected of being the reservoir for EMCV in nature (Kessler *et al.*, 1982). EMCV has been isolated from two wild-caught owl monkeys that died acutely with few clinical signs (Roca-Garcia and Sanmartin-Barber, 1957). The onset of disease appears to be peracute and fulminate in nature.

Reported clinical signs included acute dyspnea, marked depression, and pulmonary hemorrhage and edema characterized by large quantities of blood-tinged froth about the mouth and nares. Diagnosis is based on isolation of the virus from the brain, heart, spleen, lung, and other tissues at necropsy.

D. Protozoal Infections

1. Toxoplasmosis Spontaneous toxoplasmosis, caused by *Toxoplasma gondii,* has been reported in the owl monkey (Seibold and Wolf, 1971). Neotropical primates are considered highly susceptible to infection with toxoplasma (Wallach and Boever, 1983). Those clinical signs reported in nonhuman primates are fever, diarrhea, weight loss, depression, abdominal pain, and death. Antemortem diagnosis is made by serological techniques. Fecal examination for oocysts can also yield diagnostic information, however, oocyst shedding generally occurs for only 1–6 weeks after primary exposure. Therapy for polysystemic disease should include antibiotics such as clindamycin, trimethoprim/sulfadiazine, and sulfonamides. It is important to remember that this is a disease with zoonotic potential, and transmission to human subjects and other animals must be avoided.

III. Disorders of the Skin

A. Trauma

The most common causes of cutaneous disease in the owl monkey are trauma and poor husbandry conditions. Trauma is usually the result of aggression between adult animals or between adult parents and juvenile or subadult offspring. Aggression between siblings in family groups is uncommon, however, older siblings can often be observed denying younger siblings access to food. The time course to aggression is quite unpredictable in either case and can be observed in male:female pairs or owl monkey families that have been together for some time, as well as in animals that have recently been placed together to form new breeding pairs. Owl monkeys lack prominent canine teeth, which are characteristic of other Neotropical species, therefore, the wounds resulting from fighting are typically puncture wounds complicated by crushing injury to surrounding tissue. Lacerations are uncommon, although tearing of the skin can occur as the victim attempts to pull away and escape from the aggressor. Bleeding is often minimal, making the wounds difficult to detect in most instances. Undetected and/or untreated wounds, even minuscule wounds, quickly become secondarily infected with bacteria from the oral cavity of the aggressor or the normal cutaneous flora of the victim that are driven into the subcutis during wounding. Organisms frequently isolated from septic wounds include coagulase-positive beta-hemolytic staphylococcus, alpha-hemolytic streptococcus, *Escherichia coli*, *Proteus mirabilis*, and *Pasteurella multocida*.

Early recognition and treatment are critical to the successful management of wounds affecting the skin and subcutis of the owl monkey. The longer the wound remains undetected, the more likely it will become infected, resulting in cellulitis or septicemia. Formation of well-circumscribed abscesses is uncommon. This characteristic of owl monkeys often delays diagnosis and can make treatment of cutaneous wounds a long and tedious process. If fighting is observed, the animals should be separated immediately, removed from the cage, and the entire body thoroughly examined for evidence of trauma to the skin or other organ systems. Even minor appearing lesions should be treated.

Treatment usually consists of sedating the animals with ketamine hydrochloride given intramuscularly at a dosage of 10 mg/kg body weight. The traumatized area(s) is then clipped to remove the hair and provide better visualization of the wound(s). Prior to initiating any further treatment, it is important to obtain multiple swabs from several locations and submit these for bacterial culture and antibiotic sensitivity testing. This can be crucial to the management of contaminated wounds, since appropriate antibiotic therapy is often necessary to control or prevent cellulitis or septicemia. After obtaining swabs for culture, each lesion should be probed to determine its depth and extent. Clinicians often have the urge to pour things into wounds to kill bacteria. All too often, the urge to "kill germs" results in toxic substances being applied to wounds. In most cases, the critical question is what concentration is bactericidal yet does not retard healing.

Solutions commonly used for lavage of wounds include povidine–iodine solution, chlorhexidine, hydrogen peroxide, acetic acid, and Dakin's solution. Hydrogen peroxide, a commonly used wound irrigant, has little value as an antiseptic, but, if carefully used, it provides good flushing action. However, it can be damaging to tissues if not used carefully. Povidine–iodine (PI) solutions are commonly used for lavage of wounds. When using PI for wound irrigation, it is important to remember that it is absorbed systemically. Excess systemic iodine concentrations can result from iodine absorption through the skin and mucous membranes and cause transient thyroid dysfunction. Inasmuch as the owl monkey has been shown to be hypometabolic (Whittow *et al.*, 1979), the use of PI is probably contraindicated for the species. Chlorhexidine was our choice as a wound irrigation solution in owl monkeys because of its wide spectrum of antimicrobial activity. In fact, studies have shown it to be superior to PI in preventing infection in wounds contaminated with *Staphylococcus aureus* (Swaim and Lee, 1987). We routinely used 0.05% chlorhexidine diacetate solution to irrigate wounds. Older lesions and areas of devitalized skin must be debrided to promote acceptable first or second intention healing. For the most part, wounds should not be closed by suturing, although with large wounds it is often helpful to reduce the defect by bringing the edges into closer approximation. It has been our observation and experience that owl monkeys tend to have an "exuberant" granulation response and have fewer complications if large wounds are allowed to heal through second intention healing. This approach requires daily treatment and a great deal of patience on the part of the animal care and veterinary staff.

Numerous formulations of antibiotics and antibacterials are available for topical application. We have used Triple Antibiotic Ointment (Rugby) and gentamicin sulfate ointment. Studies of skin wounds in pigs treated with Triple Antibiotic Ointment demonstrated that epithelialization occurred 25% faster in treated wounds than in untreated wounds (Swaim and Lee, 1987). Wounds that did not respond to treatment with Triple Antibiotic Ointment usually responded to treatment with gentamicin sulfate ointment. Although this ointment is effective in controlling bacteria in wounds, clinicians need to be aware that certain formulations of gentamicin can delay wound contraction and epithelialization (Swaim and Lee, 1987).

Wounds to the extremities are particularly difficult to treat even if medical intervention takes place shortly after wounding occurs. Trauma to the digits and toes quickly results in local swelling with impaired circulation distal to the wound (Figs. 3A and 3B). The hemorrhage from digital wounds contributes to the formation of eschars or scabs that can extend the area of injury and form a ring around the injured extremity. When this scab retracts, it causes impaired circulation that can result in necrosis. Therapy is directed toward preventing sepsis, reducing swelling, and maintaining perfusion distal to the site of the injury. This requires complete removal of constricting scabs and wound examination. Acute injuries should be treated with broad-spectrum antibiotics both systemically and locally, cold packs to reduce local swelling and inflammation, and subsequently hydrotherapy or warm compresses to stimulate vasodilation and circulation in the injured extremity. Chronic injuries often present with significant avascular necrosis distal to the injury. In most cases, surgical amputation of the affected extremity at the most proximal unaffected articulation is necessary to cure the animal. In some cases, recovery is more prompt if the entire extremity is removed, eliminating repeated trauma and infection of the stump. Where possible, it is preferable to culture these wounds and place the patient on appropriate antibiotic therapy for 2 to 3 days prior to surgery to reduce the likelihood of postoperative complications. Antibiotic therapy should be continued for an additional 5 to 7 days following surgery.

Poor husbandry conditions can result in the formation of callosities or decubital ulcers, "pressure sores," on the feet, hocks, and tails of owl monkeys. This usually results from the use of improperly sized materials for perches. We found that 2-inch-diameter polyvinylchloride (PVC) pipe

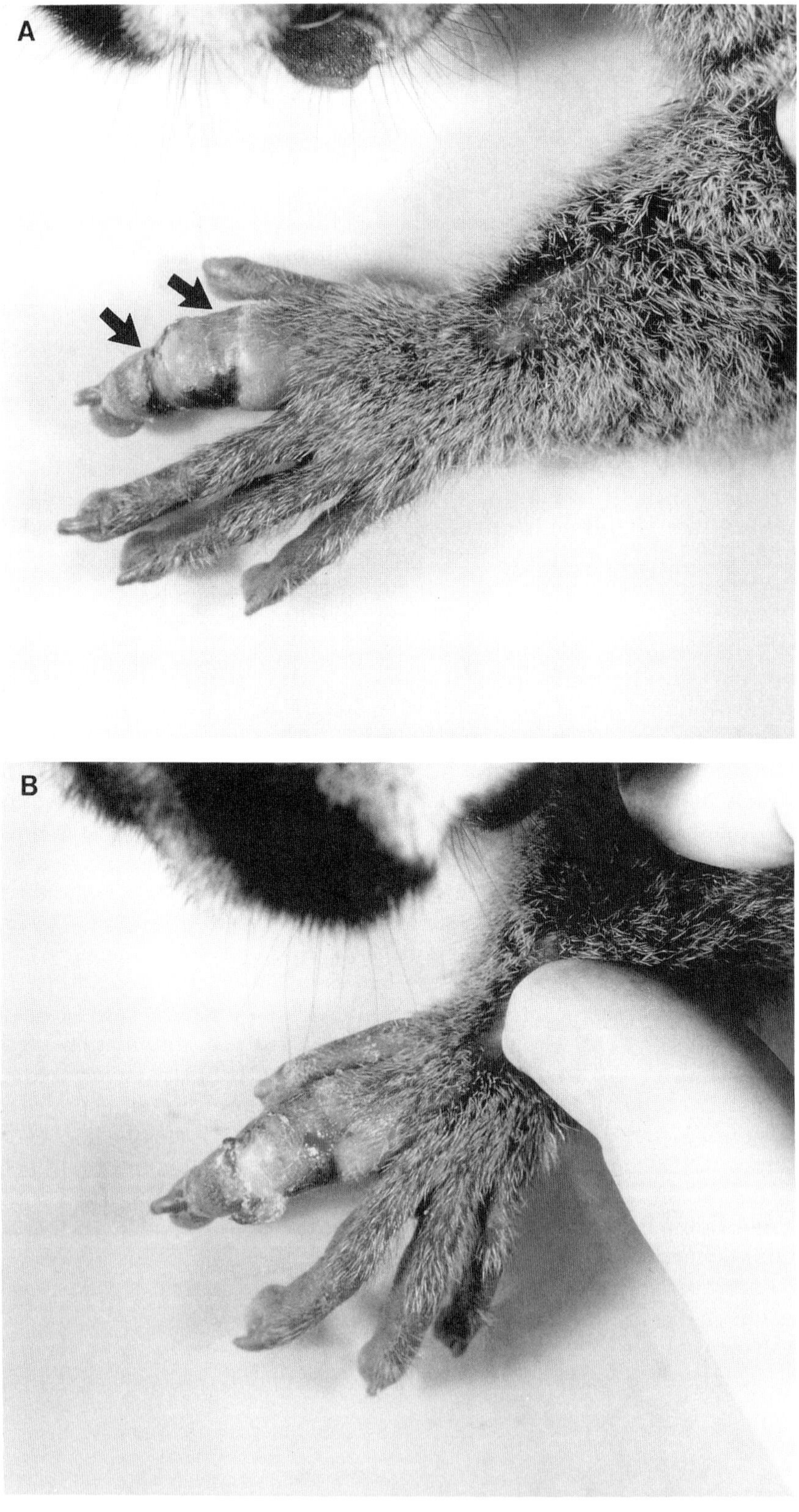
A
B

was the best material to use for perches since it presented sufficient surface area over which the animal could distribute its weight, and could be easily cleaned and sanitized. Changing the composition and size of the perch material is usually sufficient to resolve these problems. In some cases, however, it is becomes necessary to intervene medically, particularly if the lesion is ulcerated or has become secondarily infected. In those cases, the wound should be irrigated and debrided if necessary, and the monkey placed on systemic antibiotics and topical antibiotic ointment therapy until healing occurs.

B. Dermatophilosis

Cutaneous streptothricosis is a chronic infection of the cutaneous layers of the epidermis caused by *Dermatophilus congolensis* (King *et al.*, 1971). The disease has been occasionally described in owl monkeys. It is characterized by the development of raised, alopecic, nonpruritic, or papillomatous lesions covered by thick, keratinaceous incrustations (King *et al.*, 1971). The lesions may affect the skin or any portion of the body and may be well circumscribed or confluent (King *et al.*, 1971; Fox *et al.*, 1973).

Clinical signs can include the typical cutaneous lesions, weight loss, and dehydration. Diagnosis is made by demonstrating characteristic mycelia with transverse banding in skin scrapings and biopsy specimens, by culturing the organism on blood agar, and by performing biochemical tests (King *et al.*, 1971; Fox *et al.*, 1973).

Treatment with penicillin and streptomycin in combination, or penicillin alone, has been shown to be effective in treating cutaneous streptothricosis in owl monkeys. Ampicillin and tetracycline are also effective against *D. congolensis*. It is advisable to remove and destroy the crusts by incineration or chemical disinfectants.

Our own experience with the disease consists of one animal received from the CRCP in Iquitos, Peru, and housed at the BPF. This animal developed proliferative, exudative lesions on its extremities that became confluent and eventually involved large areas of the skin (Fig. 4). The animal failed to respond to topical medication and was subsequently euthanatized and necropsied. Microscopic examination of slides prepared from several lesions revealed a "palisading" crust with hyperkeratosis, neutrophils, and branching parallel chains of gram-positive cocci. The source of the infection in this animal is unknown, although it is likely that it was contracted in

FIGURE 3 (A) Wounded extremity (finger) showing formation of two "scab bands" (arrows), swelling, and loss of circulation distal to lesion. (B) Wound has been debrided revealing extensive tissue damage found underneath and between the scabs. (Photographs courtesy of Dr. C. A. Málaga.)

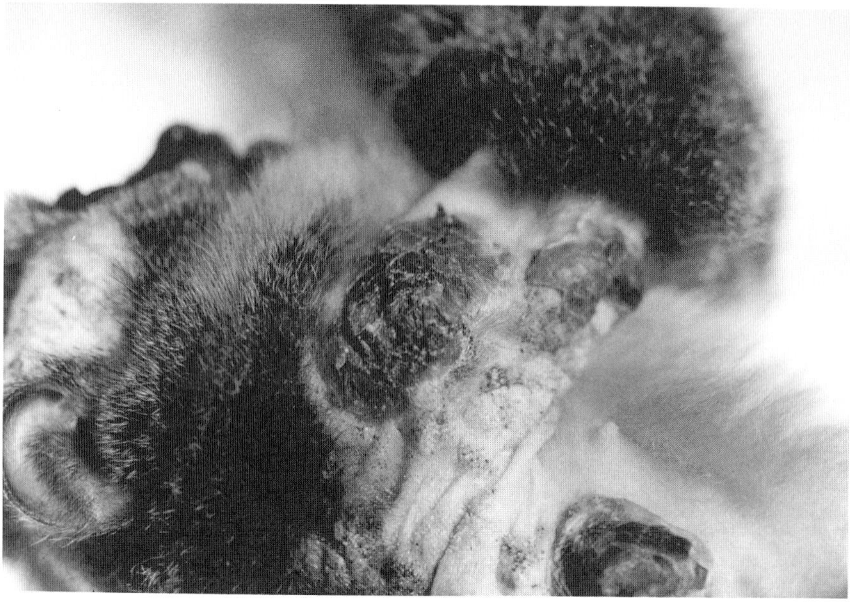

FIGURE 4 Severe case of cutaneous streptothricosis in an *Aotus nancymae*. Note confluent cutaneous lesions characterized by proliferation and areas of thick, keratinaceous incrustations.

the wild or during quarantine and conditioning at the CRCP when it was brought into close approximation to other owl monkeys captured in different locales.

C. Foreign Bodies

Penetrating thorns have been described in subcutaneous tissues and the thoracic cavity of owl monkeys (Shimp *et al.*, 1973). Those in the thoracic cavity were up to 2.7 cm in length and penetrated the lungs and heart, and were thought to be the cause of death in affected animals. The thorns were incidental findings at necropsy and resembled those found on a number of South American clump palms that produce a favorite fruit of the owl monkey (Fig. 5). It is postulated that the animals acquire these foreign bodies as they attempt to reach the fruit on these particular palm trees. Similar thorns were identified in owl monkeys imported from Peru that were housed at the BPF. In a few cases, the foreign bodies were identified in the subcutis or peritoneal cavity in living animals that were undergoing surgical procedures for medical problems. It is worth noting that the thorns appeared to cause no gross evidence of inflammation.

FIGURE 5 Typical South American clump palm tree growing on Padre Isla in the Peruvian Amazonia.

IV. DISEASES OF THE CIRCULATORY SYSTEM

The cardiovascular system of the owl monkey is similar in anatomy and physiology to that of humans, the domestic dog, and other nonhuman primates.

A. Cardiomyopathic Diseases

Diseases affecting the heart have only been briefly described in owl monkeys, although cardiopathic disease is one of the major causes of morbidity and mortality in the species (Rodger *et al.*, 1986; Gozalo *et al.*, 1992b). The prevalence of the disease is somewhat difficult to ascertain, however, approximately 40% of *Aotus* dying from all causes at the BPF had gross evidence of myocardial hypertrophy or ventricular dilatation at necropsy. This hypertrophy was noted to almost obliterate the left ventricular chamber as well as cause marked thickening of the left ventricular free wall and the interventricular septum. Many animals evidenced no signs of symptoms of disease prior to death and displayed no gross or microscopic etiology for the death. A prominent feature of the spontaneous deaths in this colony of *Aotus* was "sudden death," particularly during periods of high physical activity or psychological stress. Other clinical signs are mouth breathing, fatigue, reduced exercise tolerance, dyspnea, ascites, and subcutaneous edema.

The etiopathogenesis of cardiac disease in owl monkeys is unknown, however, gross hypertrophy of the left ventricle has many causes in humans and animals. In the absence of gross evidence of resistive or shunting lesions, hypertrophic cardiomyopathy and hypertension are the two most likely causes of or stimuli for observed left ventricular hypertrophy. Spontaneous hypertension has been reported as causing dilative cardiomyopathy in wooly monkeys (Giddens *et al.*, 1987), and vitamin E deficiency has been reported as a cause of cardiomyopathy (primarily dilative) in gelada baboons as well as other primates (Liu *et al.*, 1984, 1985). Cardiac changes have also been reported secondary to experimental trypanosomiasis in vervet and squirrel monkeys (Poltera and Sayer, 1983; Pung *et al.*, 1988), as well as subsequent to group A streptococci infection in rhesus monkeys (Mohan *et al.*, 1987). However, to date there are only three reports detailing spontaneous myocardial hypertrophic changes in a nonhuman primate (Rodger *et al.*, 1986; Gozalo *et al.*, 1992a, 1994). It has been speculated that the etiology of left ventricular hypertrophy in owl monkeys is the result of renal-induced hypertension given the prevalence of renal disease in the species. Presently, however, there has been no relationship established in *Aotus* between antemortem blood pressure and the presence or absence of renal lesions and myocardial disease. Work conducted in our laboratory indicated that blood pressure differences existed among normal owl monkeys and those with either hypertrophic or dilative cardiac disease, and those changes appeared to be unrelated to the presence or absence of concurrent renal disease. The clinical presentation of individual animals may progress over time from primarily hypertrophic to dilative, suggesting a continuum of disease. These monkeys usually present with signs associated with congestive heart failure.

Diagnosis of cardiac disease in owl monkeys is best accomplished by echocardiography, although thoracic radiographs, electrocardiography, and auscultation are useful ancillary tools. Examination may reveal cardiac hypertrophy, ventricular dilatation, arrhythmias, pericardial effusion, pleural effusion, pulmonary edema, and ascites. Neither creatine kinase (CK) nor its isozymes are reliable indicators of myocardial disease in the owl monkey (Weller *et al.*, 1991b). Figure 6 is a thoracic radiograph of an owl monkey with dilative cardiac disease, whereas Fig. 7 shows the M-mode echocardiogram obtained from the same animal.

The only drug that we have employed in treating either form of cardiac disease has been furosemide given either orally or parenterally at a dosage of 2 mg/kg of body weight once or twice daily depending on the severity of clinical signs. Low-sodium diets have also been formulated for individual animals. This approach has been relatively successful in the long-term

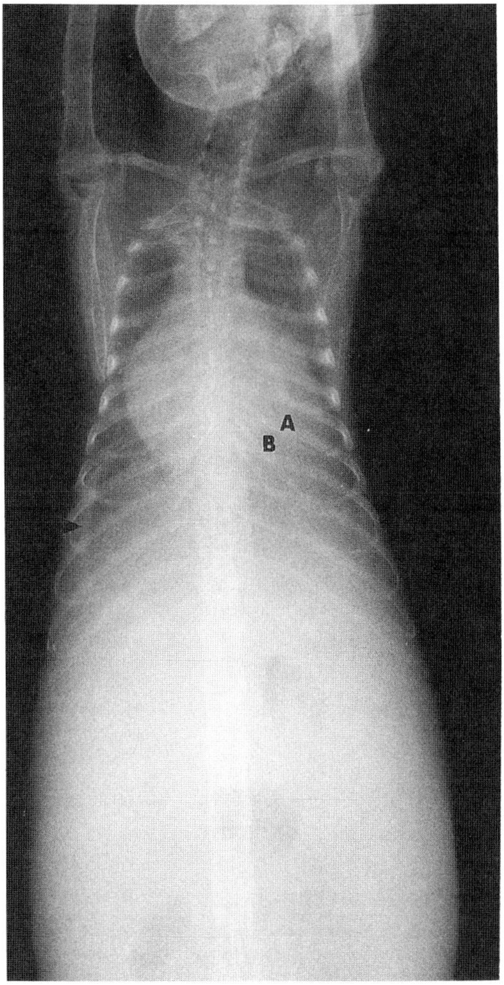

FIGURE 6 Dorsoventral thoracic radiograph showing marked cardiomegaly with enlarged arteries (A) and veins (B). The lung lobes in the right hemithorax are separated by pleural fluid (arrowhead). (Photograph courtesy of Dr. R. L. Perry.)

clinical management of several monkeys with marked cardiomegaly, pulmonary edema, pleural effusion, or ascites. Animals with marked pleural effusion or ascites might benefit from thoracentesis or abdominocentesis, respectively, to remove fluid and relieve pressure on surrounding organs. Although it is likely that other classes of drugs would be beneficial in the medical management of these animals, to date, none have been evaluated in owl monkeys.

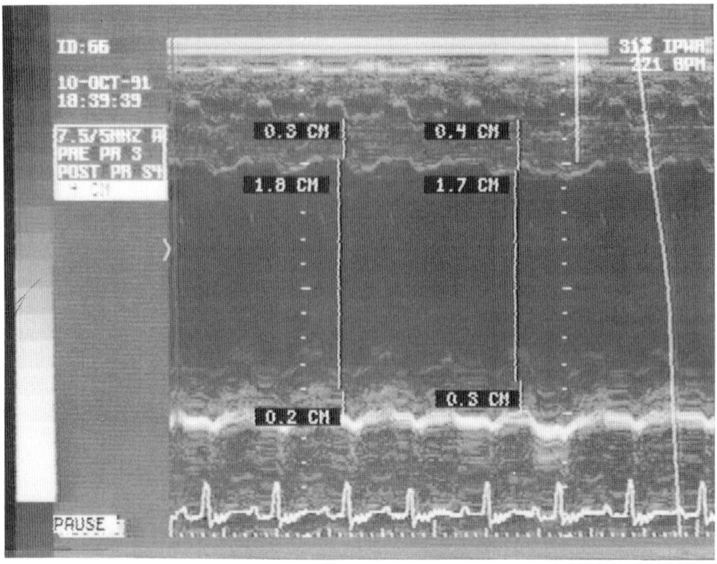

FIGURE 7 M-mode echocardiogram of same monkey shown in Fig. 6. The monkey was judged to have dilative cardiac disease. Echocardiogram was performed using a two-dimensional guided M-mode technique with a 7.5-mHz mechanical sector scanner. Image was obtained from a right or left parasternal position with the monkey in dorsal recumbency.

B. Dissecting Aneurysm of the Aorta

The reported prevalence of aortic aneurysms in nonhuman primates is very low. In addition to the owl monkey, spontaneously occurring aortic aneurysms have been described in the gorilla and squirrel, howler, capuchin, patas, African green, and spider monkeys (Baer *et al.*, 1992). Clinical signs of aortic aneurysm in the owl monkey include weight loss, lethargy, depression, facial and/or subcutaneous edema, tachypnea, bradycardia, and CNS disturbances. The etiology is unknown.

Diagnosis of aortic aneurysm can be made on the basis of a radiographic examination, ultrasonographic examination, abdominal palpation, and auscultation of bruits; definitive diagnosis may require angiography or computer-assisted tomography. Antemortem clinical pathology findings are quite variable and nondiagnostic. When examined grossly, aortic lesions can range in size from 2 mm to 3 cm in diameter (Fig. 8). The majority of aneurysms in owl monkeys have been histologically classified as dissecting (Baer *et al.*, 1992).

Treatment of dissecting aortic aneurysm has not been attempted in the owl monkey. In human subjects, treatment of this disease is best accomplished through implantation of an aortic prosthesis.

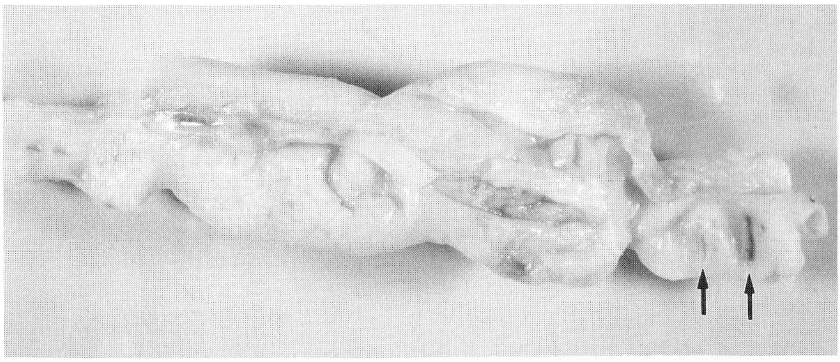

FIGURE 8 Dissecting aortic aneurysm from an owl monkey that died acutely. Arrows point to defects in the wall of the aorta leading to two smaller saccular aneurysms proximal to the larger lesion.

C. Hematopoietic Disorders

Primary diseases affecting the hematopoietic system are rare in owl monkeys. Malignant lymphoma has been reported in several Neotropical primate species, either as spontaneous disease or related to oncogenic herpesvirus infection. Spontaneous myeloproliferative disorders have been described in three owl monkeys (Chalifoux and King, 1983; Matherne *et al.*, 1986).

D. Idiopathic Eosinophilia

Although hardly a disease, idiopathic eosinophilia in owl monkeys is worth noting since it has not been associated with any diseases that normally stimulate eosinophil production in other species. Susceptibility to idiopathic eosinophilia is related to karyotype, occurring more frequently in karyotype 2, 3, 4, and 6 monkeys (Beland *et al.*, 1981). Parasitism or other subclinical diseases can probably be excluded as causes of the eosinophilia, especially since colony-born offspring of susceptible karyotypes also exhibit this hematologic trait.

V. DISEASES OF THE RESPIRATORY SYSTEM

A. Air Sacculitis

The owl monkey possesses very prominent air sacs that extend from the larynx into the perilaryngeal tissue. The degree of perilaryngeal development is very similar to that observed in Old World species of nonhuman primates (Giles *et al.*, 1974). Infection of the air sacs has been frequently noted in the owl monkey, although the pathogenesis of the disease is unknown. *Klebsiella pneumoniae* is the most common microorganism isolated from affected animals (Giles *et al.*, 1974). *Proteus mirabilis, Escherichia coli,* beta-hemolytic *Streptococcus,* and *Streptococcus fecalis* have also been isolated from animals with mixed infections (Giles *et al.*, 1974). We observed one case of iatrogenic air sacculitis in our laboratory that occurred in an infant *Aotus nancymae* that was being hand fed formula following rejection by its parents. The presumed cause was aspiration of the liquid formula mixture. A mixed infection of *Pseudomonas aeruginosa* and *Proteus mirabilis* was isolated from the air sacs at necropsy. In addition, we have isolated *K. pneumoniae* from the oropharynx of several asymptomatic adult owl monkeys during quarantine.

The disease usually follows an acute course making antemortem diagnosis difficult. Reported clinical signs include mucopurulent nasal discharge, swelling of the cervical and intermandibular areas, listlessness, depression, and anorexia. Diagnosis depends on isolating pathogenic microorganisms from swabs obtained from exudates or the oropharynx, and from blood since a hematogenous origin is suspected. Radiographs of the laryngeal area might reveal evidence of thickening of the mucosa of the air sac and the presence of exudate.

Treatment is surgical drainage of a dependent area of the infected air sac and maintenance of a patent drain (Wallach and Boever, 1983). Care must be taken to prevent aspiration of the exudate. The infected sac should be flushed and treated daily with a bactericidal solution. Parenteral cephalothin, trimethoprim/sulfadiazine, gentamicin, amikacin, and kanamycin are the antibiotics of choice.

B. Pneumonia

Pneumonia is frequently observed in owl monkeys in captivity; stress, chilling, crowding, shipping, and other untoward conditions can contribute to the development of this disease. Infection of the respiratory tract accounted for 24–27% of all owl monkey deaths at the CRCP in Iquitos, Peru, over an 8-year period (Gozalo and Montoya, 1990) *Escherichia coli, K. pneumoniae, Streptococcus pneumoniae, P. multocida,* and *B. bronchiseptica* are the principal microorganisms isolated in cultures taken from own monkeys with pneumonia (Good and May, 1971; Gozalo and Montoya, 1990). In addition, we have isolated *Haemophilus influenzae, Neisseria* spp., and *Corynebacterium pyogenes* from owl monkeys with clinical evidence of upper respiratory disease. Signs of upper respiratory infection may precede pulmonary involvement. Once pneumonia is established, fever, depression, anorexia, cough, dyspnea, exercise intolerance, cyanosis, and nasal discharge are common clinical signs.

Diagnosis may be suggested by clinical signs and auscultation of the chest. Thoracic radiography is required to characterize the pulmonary changes. Respiratory or nasal secretions should be obtained for cytology and culture. Laboratory tests usually reflect inflammation, and hematologic tests may provide clues of the etiology of the pneumonia. For example, leukopenia is common with viral infections whereas leukocytosis with a left shift is more characteristic of bacterial or mycotic infections.

Antibiotics are the primary therapy. Monkeys should be maintained on a selected antibiotic for a minimum of 2 weeks after all clinical and radiographic changes have resolved. Nebulization/humidification may increase expectoration of secretions. This can be accomplished by placing the animal in an incubator or enclosed area. Oxygen is indicated if cyanosis is present. Maintaining hydration with subcutaneous or intravenous fluids is also critically important.

VI. DISEASES OF THE DIGESTIVE SYSTEM

The normal anatomy and physiology of the owl monkey digestive system are similar to those of humans. Diseases of the alimentary canal are the number one disease of caged owl monkeys presented to the clinician.

A. Diarrhea

The most common manifestation of diseases involving the alimentary canal is diarrhea. Causes of diarrhea in owl monkeys can include dietary factors, bacteria, fungi, viruses, and poor environmental conditions. Dietary factors would appear to be the primary cause of diarrhea in owl monkeys. These can include changes in foodstuffs, changes in dietary formulation, outdated foodstuffs, and contamination of feed with toxins of mycotic or bacterial origin (Gozalo and Montoya, 1990). Environmental factors such as stress, crowding, poor sanitation, poor ventilation, and suboptimal temperature and humidity can be contributing factors. *Escherichia coli, Proteus mirabilis, P. vulgaris, Klebsiella ozaenae, K. oxytoca, Campylobacter jejuni,* and *Shigella* spp. have all been cultured from owl monkeys with either acute catarrhal or acute hemorrhagic enteritis (Gozalo and Montoya, 1990), although their presence was considered to be secondary to other factors. *Salmonella* spp. have also been isolated from the gastrointestinal tract of owl monkeys (Daniel *et al.*, 1976).

Clinical signs obviously include diarrhea of varied consistency, frequency, and severity. Other clinical signs include dehydration, weakness, weight loss, anorexia, vomiting, and circulatory collapse. Diagnosis is based on clinical signs, laboratory findings, and culturing the stool and rectal swabs obtained from affected animals. Acid–base disturbances, electrolyte imbalances, and evidence of prerenal azotemia are common findings in owl monkeys with protracted disease and significant fluid loss.

Treatment is primarily symptomatic unless a pathogenic microorganism is isolated. Sick monkeys should be removed to a clinic or hospital area and placed in an intensive care chamber (incubator) where temperature, humidity, and sanitation can be more closely regulated and monitored. The intestinal tract should be rested by withholding food for 24–48 hr. Small amounts of bland food can be offered only after resolution of vomiting and diarrhea. Oral fluid therapy may be attempted, such as Enterolyte (Beecham), at a rate of 40–80 ml/kg/day. Rehydrating and maintaining hydration are critical to the survival of severely affected animals. This can be accomplished by the administration of sterile Ringer's lactate or 0.9% sodium chloride solutions. The preferred route of administration is intravenous, however, either the subcutaneous, intraperitoneal, or intramedullary route is acceptable if one cannot locate and successfully catheterize a peripheral vein. Ringer's lactate solution is also preferred over 0.9% saline solutions since most of these animals are hyponatremic, hypokalemic, hypochloremic, and acidotic. It may be necessary to add potassium chloride to the fluids or supplement by mouth if hypokalemia is severe. Young animals may become hypoglycemic and benefit from oral or intravenous glucose solutions. Motility modifiers are contraindicated, but antiemetics are indicated when vomiting is severe. Local intestinal protectants, such as bismuth subsalicylate or Kaopectate (UpJohn), may be used. If pathogenic microorganisms are identified, then aggressive parenteral antibiotic therapy is the treatment of choice. Ampicillin, gentamicin, trimethoprim/sulfadiazine, erythromycin, and enrofloxacin are antibiotics of choice. Patient monitoring should include the daily determination of the state of hydration by skin turgor or the determination of packed cell volume (PCV). Serum electrolytes should also be evaluated as clinical signs warrant.

B. Acute Gastric Dilatation

Although acute gastric dilatation has been observed with some frequency in a variety of nonhuman primates (Wallach and Boever, 1983), it is an uncommon finding in owl monkeys. It is characterized by sudden onset of abdominal distention, discomfort, depression, and subnormal temperature. Animals are usually found on the floor of the cage or trying to assume a position of relief. Diagnosis is made during physical examination and may be confirmed by abdominal radiography. Treatment involves passage of a stomach tube to relieve the gas and administration of corticosteroids. In the rare event that a stomach tube cannot be successfully passed, emergency decompression can be accomplished by passing an 18-gauge needle through the abdominal wall into the lumen of the stomach.

C. Cholelithiasis

Spontaneous diseases affecting the biliary system are rare in nonhuman primates. Only two published reports have described cholelithiasis in the owl monkey (Anver *et al.*, 1972; Baer *et al.*, 1990), and in all cases the diagnosis was made at necropsy (Figs. 9A and 9B). Although gallstones have been induced in the cebus monkey and squirrel monkey by feeding diets high in cholesterol, the pathogenesis of cholelithiasis in the owl monkey remains unknown. Possible etiologies include chronic hemolytic anemia, systemic bacterial infections, parasitic infections, and diet. In one of the published reports, gallstones from five animals were analyzed to determine their chemical composition (Baer *et al.*, 1990). Results showed the gallstones to be composed primarily of cholesterol (89% by dry weight). The composition of gallstones in owl monkeys is very similar to that of choleliths found in human subjects, but very different from the composition of gallstones incuded in other animal models.

Clinical signs and antemortem diagnosis of gallstones are rare in nonhuman primates, although weight loss, emesis, and abdominal pain are seen occasionally, and icterus may be evident if obstruction of the common bile duct has occurred. Laboratory findings are nonspecific for cholelithiasis per se, however, increases in alkaline phosphatase, or gamma glutamyl transpeptidase levels, followed by hyperbilirubinemia, may indicate cholestasis. Ultrasonographic imaging of gallstones within the gallbladder is the best technique available for non-invasive clinical diagnosis of cholelithiasis in the owl monkey. Choleliths as small a 1 mm in diameter can be visualized *in vivo*. Cholecystectomy has been recommended as the treatment of choice for gallstones in nonhuman primates with clinical disease. Alternative therapies could include cholecystotomy (Moscovici *et al.*, 1988) or dehydrocholic acid (Decholin) given orally every 8 hr at a dosage of 10 mg/kg body weight (Cornelius and Bjorling, 1988).

D. Hepatopathy

Hepatic lipidosis, acute toxic hepatitis, and cholangitis have been diagnosed as causes of death in owl monkeys (Gozalo and Montoya, 1990). The etiology of these diseases is unknown. Idiopathic hepatic lipidosis was diagnosed in several owl monkeys that died while housed at the BPF. Hepatic lipidosis is a disorder characterized by severe accumulation of fat in the liver that may result in hepatic failure and death. In cats, abnormalities in lipoprotein metabolism or protein-calorie malnutrition may be involved in the pathogenesis of the disease (Cornelius and Bjorling, 1988). Owl monkeys of karyotypes susceptible to vitamin-E-responsive hemolytic anemia have been shown to have altered lipoprotein metabolism, which could

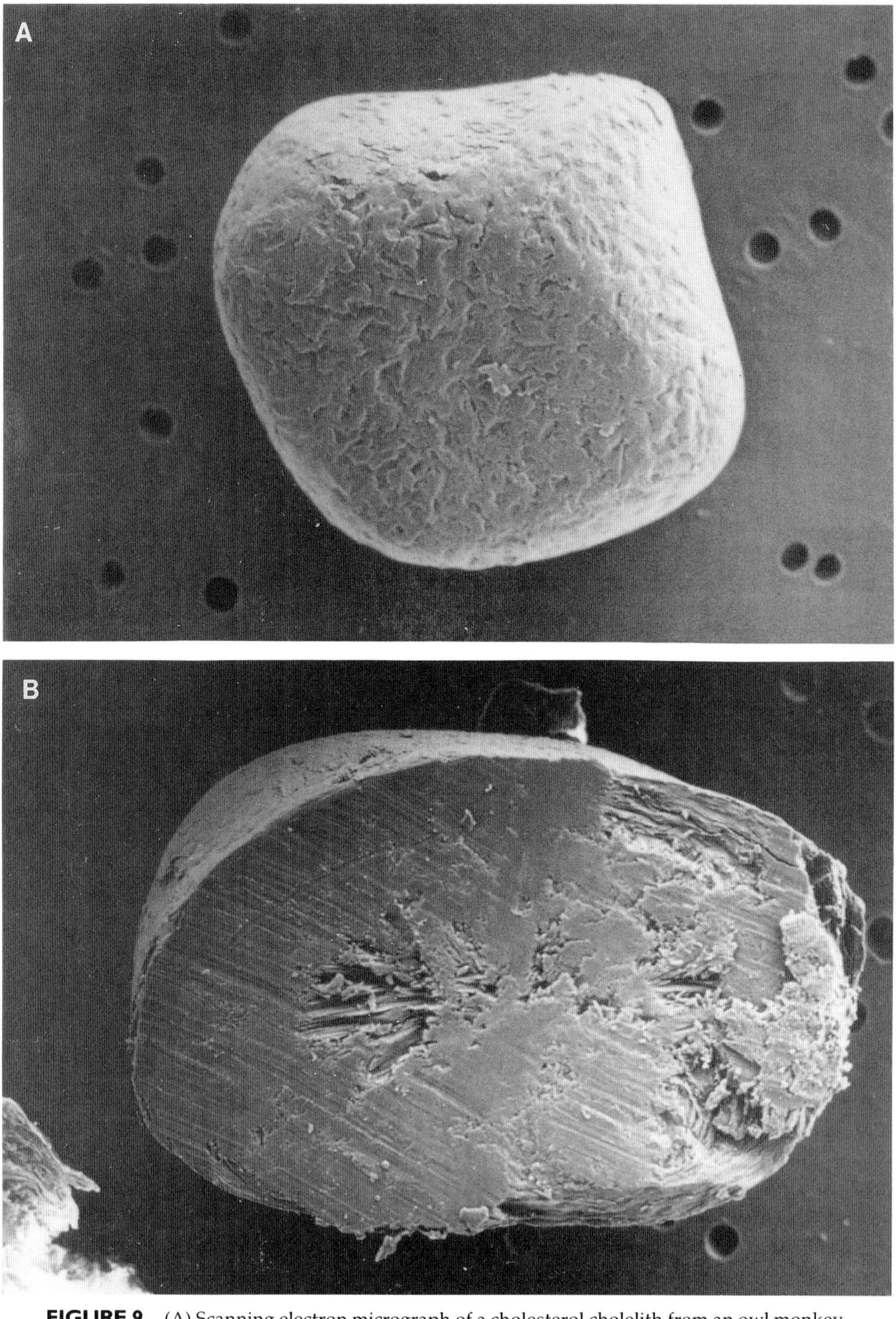

FIGURE 9 (A) Scanning electron micrograph of a cholesterol cholelith from an owl monkey (45×). (B) Cut surface of another gallstone from the same animal (45×). Note the laminar appearance of the cut surface. (Photographs courtesy of K. E. Luahala and R. P. Schumacher.)

contribute to the pathogenesis of hepatic lipidosis in the species (Meydani *et al.*, 1983). Clinical signs have included anorexia, depression, lethargy, incoordination, vomiting, diarrhea, and icterus. Diagnosis is accomplished by demonstrating markedly increased serum alkaline phosphatase, alanine aminotransferase, and bilirubin. Definitive diagnosis requires a liver biopsy. A percutaneous or laparoscopic needle biopsy using a Vim Tru-cut needle is preferred (Cornelius and Bjorling, 1988). Treatment in cats includes appetite stimulants, enteral feeding, and nutritional and fluid support (Cornelius and Bjorling, 1988).

VII. DISEASES OF THE NERVOUS SYSTEM

A. Hydrocephalus

Spontaneous hydrocephalus is a rare condition in nonhuman primates and has been described only once in the owl monkey (Málaga *et al.*, 1991). Clinical signs in that case were indistinguishable from those associated with any seizure disorder. Diagnosis was determined by the presence of cortical thinning and an open posterior fontanelle on survey radiograph of the skull, and gross pathological examination of the brain, which revealed marked dilatation of the lateral, third and fourth ventricles (Fig. 10). There is no documented treatment for this disorder in nonhuman primates.

B. Cerebrovascular Accidents

Cerebrovascular accidents or strokes were noted in a few older (age > 6 years) owl monkeys housed in the BPF. The most likely pathogenesis was occlusion of cerebral circulation. There did not appear to be a sex or karyotype predilection. Clinically, these events were characterized by the sudden onset of hemiparesis, depressed consciousness, facial paralysis, impaired mastication and swallowing, and impaired direct and consensual pupillary light response. Diagnosis is usually the result of ruling out other possible causes of paralysis, especially thromboembolism due to cardiomyopathy.

Treatment is symptomatic and supportive. Initially this can include intravenous heparin, glucocorticoids, and fluids. Chronic management is directed toward preventing recurrence of the occlusive event. This can require platelet suppressant agents or anticoagulants. Meeting the daily nutritional and fluid requirements of the animal by assisted alimentation (i.e., hand feeding or tube feeding) and physical therapy are important aids to recovery in monkeys that survive the acute event.

C. Seizure Disorders

In our owl monkey colony, these were characterized by infrequent tonic-clonic seizures that ranged in severity from petit mal to grand mal seizures.

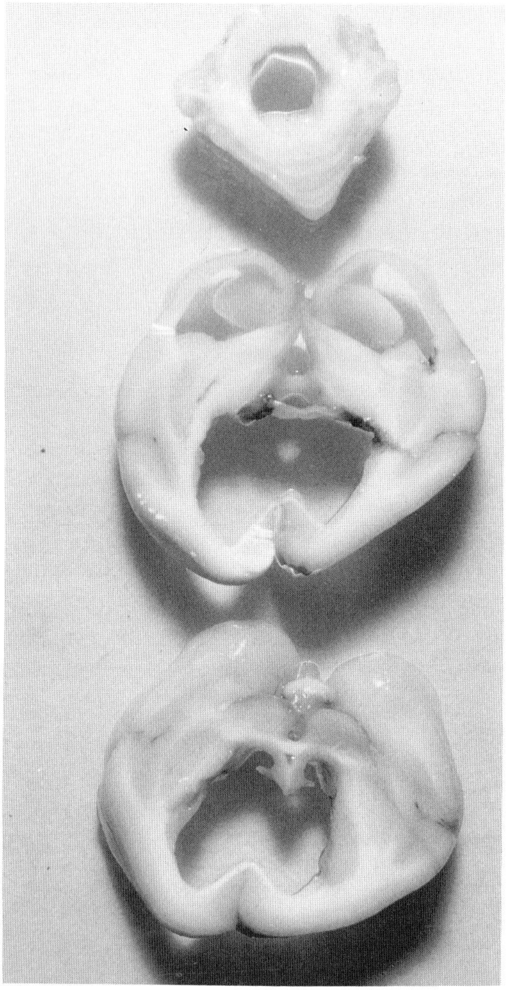

FIGURE 10 Sagittal sections of the cerebrum demonstrating dilatation of the lateral, third and fourth ventricles, and the cerebral aqueduct in an adult owl monkey (*Aotus nancymae*) with spontaneous hydrocephalus.

Observant animal care personnel occasionally described an aura that immediately preceded the seizure itself. Clinical signs often included extensor rigidity of all four limbs with opisthotonos, jerky or paddling limb movements, chewing movements, and visceral motor activity such as salivation, defecation, and urination. The duration of the seizure was usually short, lasting less than 5 min. During the postictal phase, affected monkeys sometimes appeared disoriented and confused. Diagnosis in this instance is also usually made by excluding all other causes of seizuring, such as hypoglycemia, hydrocephalus, syncope, and metabolic disorders.

Anticonvulsant therapy is indicated if the seizures increase in frequency, duration, or severity. The therapy must be individualized and adjusted according to the monkey's response, and relapses ("breakthroughs") should be expected. Therapy was not required for any of the monkeys diagnosed with seizure disorders in our facility.

VIII. DISEASES OF THE GENITOURINARY SYSTEM

Spontaneous renal disease is one of the more common causes of morbidity and mortality reported in the owl monkey, although the etiopathogenesis, in most cases, remains unknown. Nephrotic syndrome, glomerulonephropathy, and interstitial nephritis have been observed and described (Hunt *et al.*, 1976; Chalifoux *et al.*, 1981; Aikawa *et al.*, 1988). Infectious diseases, hemoparasites, endoparasites, environment, and dietary factors have all been suggested as possible causes of nephropathy in wild-caught and colony-born owl monkeys. However, the lack of reference data related to the urinary system for the species makes difficult the antemortem diagnosis and determination of the pathogenesis of these diseases and their effect on the renal function. Determination of the prevalence of spontaneous renal disease is often confounded by the lack of information regarding the research history, if any, of individual or groups of owl monkeys.

A. Glomerulonephropathy

Spontaneous and malaria-associated glomerulonephropathies have been described in the owl monkey; the cause of spontaneous renal lesions in the species is unknown (Hunt *et al.*, 1976; Aikawa *et al.*, 1988). Several authors have described the association between infection with various species of human and nonhuman primate malaria parasites and induced glomerulonephropathy in owl monkeys (Hutt *et al.*, 1975; Aikawa *et al.*, 1988). The pathogenesis of glomerular disease in those cases is immune-complex disease or type III hypersensitivity. Histologically, glomeruli of affected monkeys characteristically show pathological changes consistent with either membranoproliferative glomerulonephropathy or mesangioproliferative glomerulonephropathy (Aikawa *et al.*, 1988). Regardless of etiology, a hallmark of glomerular disease is proteinuria. Abnormal proteinuria can result from hemorrhage into the urinary tract or inflammatory disease at any point along the urinary tract; however, in the absence of pathologic urine sediment, it indicates renal disease (Cowgill, 1983).

The detection of proteinuria is important to the diagnosis of glomerular disease, because abnormally high protein loss may occur without detectable impairment of glomerular filtration or tubular concentration (Cowgill, 1983). Animals not exhibiting nephrotic syndrome (see Section VIII.C.) may be asymptomatic, although weight loss, ascites, and peripheral edema can occur. Laboratory findings can include anemia, azotemia, increased serum creatinine, and increased serum phosphate concentrations if renal failure is present. Hypoproteinemia and proteinuria are always present. Urine protein quantitation is necessary to document proteinuria. The method most commonly used to measure urinary protein relies on timed urine collections, which are time-consuming, cumbersome, and occasionally inaccurate. Work conducted in our laboratory suggested that determination of urine protein/creatinine ratio was a better predictor of 24-hr urine protein loss than measurement of urine protein concentration alone. Its predictive value is based on the fact that protein and creatinine are excreted proportionately through the glomerular basement membrane. Our study suggested that, in general, a urine protein/creatinine ratio (UP/Cr) in excess of 2.5 in a voided urine sample in an owl monkey with stable renal function could be taken to represent impaired glomerular function (Weller *et al.*, 1991c). The amount of protein excreted could be calculated from the equation

$$\text{Urine protein loss (mg/kg/day)} = 29.8 \times \text{UP/Cr}$$

Renal biopsy with examination of tissue by light and electron microscopy, as well as immunofluorescent antibody staining, is necessary to confirm and determine the type of glomerular lesion. In owl monkeys, a blind percutaneous technique, the keyhole technique, laparoscopy, and ultrasound-guided biopsy can be employed to obtain a sample of renal tissue (Osborne *et al.*, 1974). When available, ultrasound-guided biopsy is the technique of choice.

The best treatment for glomerulonephropathy would be to identify and eliminate the source of antigenic stimulation leading to the formation of antibodies and immune complexes. This is unlikely in most circumstances. The use of immunosuppressive drugs to treat glomerulonephritis in owl monkeys appears logical, but clinical studies are lacking. A new class of drugs that holds promise for treating immune-complex disease are the thromboxane synthetase inhibitors. Thromboxane synthetase inhibitors have been used successfully to treat experimentally induced glomerulonephritis in several species and may be important in the treatment of spontaneous glomerular disease (Grauer *et al.*, 1992). We have used prednisolone given intramuscularly once daily at a dosage of 2 mg/kg of body weight to treat a few proteinuric monkeys with mixed results. Furosemide, given orally or intramuscularly at a dosage of 2–4 mg/kg body weight every 12 to 24 hr, can be used as needed to manage edema. Affected animals should also be fed restricted protein diets with 12–15% dietary calories as protein.

B. Interstitial Nephropathy

Interstitial or tubulo-interstitial renal disease has not received as much attention in the owl monkey as glomerular disease. Although unlikely to be a common cause of chronic renal injury, it is recognized as an end stage of a variety of renal injuries. In human nephrology, a broad spectrum of discrete interstitial diseases have been recognized (Mujais and Quintanilla, 1988). Tubulo-interstitial nephritis is a clinico-pathologic diagnosis that includes a variety of diseases. Common to these diseases, however, is a pathologic picture characterized by injury localized primarily to the renal tubules and interstitium, commonly, but not exclusively, of inflammatory origin, and associated with the chronic sequelae of interstitial fibrosis and parenchymal atrophy (Hostetter *et al.*, 1988). It is also important to remember that many of these diseases can affect both the glomerular and tubular portion of the nephron (Cowgill, 1983). Tubulo-interstitial lesions can range in severity from minor foci of inflammatory cell infiltration to chronic pyelonephritis. Interstitial infiltrates and glomerulonephritis have been described in owl monkeys with hemolytic anemia (Chalifoux *et al.*, 1981), along with chronic pyelonephritis (Roberts *et al.*, 1972). The relationship between malaria and interstitial disease in owl monkeys is well documented (Hutt *et al.*, 1975; Aikawa *et al.*, 1988).

Regardless of etiology, the clinical signs will vary with the severity and extent of renal pathology and/or functional impairment. Renal failure (or renal insufficiency) implies that 66 to 75% or more of the functional capacity of the nephrons of both kidneys has been impaired (Cowgill, 1983) (Figure 11). Clinical signs of chronic renal disease in owl monkeys can include polyuria, polydipsia, anorexia, dehydration, depression, vomiting, diarrhea, and weight loss. Diagnosis is based on laboratory findings that show significant alterations in those indices of renal function for the owl monkey (Weller *et al.*, 1992a, 1993). Common among those are reduced glomerular filtration rate (GFR), azotemia, increased serum creatinine, hyperphosphatemia, hypocalcemia, impaired ability to dilute or concentrate urine, anemia, hypoproteinemia, and proteinuria. In monkeys with subclinical disease, the measurement of GFR, fractional clearance of electrolytes, and urinary enzyme activities can detect the presence of reduced renal function before clinical signs appear (Weller *et al.*, 1992a, 1993). Survey radiographs or ultrasonography can provide evidence of kidney size, shape, and density (Fig. 11).

Conservative medical management is the only therapy available, and treatment must be formulated in relation to the degree of renal dysfunction and the value of individual monkeys. Fresh water should be provided *ad libitum* and may be supplemented with water-soluble vitamins to compensate for urinary losses. As with glomerulonephropathy, owl monkeys with chronic renal disease also need to be fed sodium- and protein-restricted

FIGURE 11 Small, pale, irregularly shaped kidneys removed from an owl monkey that died of chronic renal failure. Note the thin cortices, uneven corticomedullary junctions, and possible hemorrhagic infarct of right kidney.

diets with 12–15% dietary calories as protein. Therapy for hyperphosphatemia can include oral phosphate-binding gels in addition to dietary protein restriction. Anabolic steroids may be used to stimulate erythropoiesis in monkeys with hematocrits lower than 20%. Stanozolol (Winstrol-V; Sterling) and nandrolone decanoate (Deca-Durabolin; Organon) were both used at the BPF. Although a prospective study was not done to determine the relative efficacy of either drug, nandrolone appeared to be most effective clinically. Owl monkeys with peripheral edema or those thought to be hypertensive were also treated with furosemide. Intravenous fluids were occasionally employed to correct volume deficits and improve renal perfusion. Patient monitoring is important and should include weekly determination of body weight and fluid intake, and quarterly determinations of BUN, serum creatinine, hematocrit, serum electrolytes, and urinalysis.

C. Nephrotic Syndrome

Nephrotic syndrome describes a complex of clinical and laboratory signs that include proteinuria, hypoalbuminemia, hyperlipoproteinemia (hypercholesterolemia), and edema. It can occur in conjunction with any glomerular disorder and can appear in owl monkeys with and without clinical evidence of renal disease.

Marked facial edema is the most common clinical sign observed in owl monkeys with nephrotic syndrome. Weight loss, oliguria, ascites, and subcutaneous edema have also been described. Diagnosis may be confirmed by demonstration of proteinuria, hypoalbuminemia, and renal biopsy.

Treatment is similar to that described for glomerulonephropathy. In addition, one should avoid draining ascitic fluid unless the monkey is dyspneic or uncomfortable, as it potentiates albumin loss. Edema or ascites should be assessed daily, and the animal should be weighed twice weekly to monitor body water. Serum protein and albumin, and urine protein:creatinine ratio should be determined every 2 to 4 weeks to monitor response to therapy.

IX. NUTRITIONAL AND METABOLIC DISORDERS

A. Vitamin-E-Responsive Hemolytic Anemia

The owl monkey spontaneously develops a vitamin-E-responsive hemolytic anemia that is a function of phenotype and karyotype (Beland *et al.*, 1981; Meydani *et al.*, 1983). Those owl monkeys of the gray-necked or "agouti" phenotype and karyotypes 2, 3, 4, 5, and 7 are susceptible to developing the disease, whereas those monkeys of the "red-necked" phenotype and karyotypes 1 and 6 are resistant. Neonatal, juvenile, and subadult animals are at greater risk than adult animals for experiencing hemolytic episodes. Prior to the recognition that vitamin E and selenium were effective therapy for the disease, annual mortality of owl monkeys with anemia at the New England Regional Primate Center was reported to be between 10 and 20% (Sehgal *et al.*, 1980). The disease is associated with red blood cell lipid peroxidation and altered red blood cell membrane lipid composition (Brady *et al.*, 1982; Meydani *et al.*, 1983). Susceptible animals have depressed plasma levels of vitamins E and A and an altered lipoprotein metabolism characterized by elevated ratios of low/high-density lipoprotein cholesterol and free to esterified cholesterol in these lipoproteins (Meydani *et al.*, 1983).

Affected animals exhibit weakness, mucosal pallor, prolonged capillary refill, hypothermia, dehydration, heart murmur, and anorexia. Severely affected animals can be icteric and exhibit convulsions and opisthotonus (Sehgal *et al.*, 1980). Diagnosis is often based on clinical signs, karyotypic susceptibility, and demonstration of reduced volume of packed red blood cells (<15%).

Anemic monkeys should be placed in oxygen therapy units maintained at 90°C and 50% humidity to reduce stress and stabilize body temperature. Treatment consists of Seletoc (Schering Corp.) given intramuscularly once daily and vitamin E drops given orally twice daily until the VPRC stabilizes and begins to increase. Affected animals rarely require transfusion, however, if this becomes necessary, we recommend the use of fresh heparinized blood obtained from an owl monkey of the same karyotype. It is preferable to administer the transfusion intravenously, however, if one cannot locate a vein to catheterize, the intramedullary or intraperitoneal routes can also be used. Those karyotypes at risk and monkeys that have had anemic episodes should be treated with prophylactic injections of *dl*-α-tocopherol acetate in oil (Anthony Products) given subcutaneously once a month.

B. Hypothermia and Hyperthermia

The two extremes of body temperature are usually the result of environmental extremes related to weather or improper functioning of heating and cooling systems in animal care facilities. Hypothermia is most commonly observed in infant or young owl monkeys whose ability to thermoregulate is compromised by trauma, poor husbandry conditions, stress, or denial of access to food sources. Besides being cold, these animals are also likely to be hypoglycemic and predisposed to opportunistic infectious agents. They should be warmed rapidly, which can best be accomplished by whole-body immersion, with the exception of the head, in warm water (105–110°F) until the rectal temperature approximates the normal range for owl monkeys (100–102°F) and then rubbing the monkey briskly with a warm towel to dry it. Rewarming can also be accomplished by warm enemas with 5% dextrose–saline solutions, or by placing the animal in an incubator and maintaining the humidity and temperature at 50% and 90°F, respectively. This is a slower process, but fraught with less potential for complications. However, in our experience, it carries a higher failure rate than immersion in warm water. Parenteral glucocorticoids (Solu-Delta-Cortef; UpJohn) and warm 5% dextrose solutions are indicated for monkeys showing signs of shock or suspected of being hypoglycemic. Patient monitoring includes determination of the rectal temperature at 15- to 20-min intervals, capillary refill time, and auscultation of the chest for signs of fluid buildup. Oral or parenteral antibiotics are indicated during recovery to preclude the development of secondary bacterial infections.

Hyperthermia, other than fever associated with infectious disease, was not a significant cause of morbidity and mortality among owl monkeys maintained in our laboratory. The most likely cause of clinically significant hyperthermia in most animal care facilities housing owl monkeys would be failure of cooling or ventilation systems resulting in extreme temperature elevations. Nonetheless, if hyperthermia is diagnosed in an animal, the first order of business is to lower the body temperature as quickly as possible. Immersing the monkey in ice or cold-water baths is very effective at lowering body temperature. Avoid immersing the head, and monitor the rectal temperature every 5 min since it is important not to "overshoot" the normal temperature range and make the patient acutely hypothermic. Once an acceptable body temperature is achieved, the monkey should be dried briskly with a towel and moved to a stable environment where temperature and humidity can be carefully controlled. Parenteral glucocorticoids and dextrose solutions may be necessary to preclude circulatory collapse and provide energy. Once the animal has stabilized, additional nutrition can be provided through the oral administration of supplements with high caloric content such as Nutri-Cal (Evsco Pharmaceuticals). As with hypothermic animals, oral or parenteral antibiotics are indicated during recovery to preclude the onset of opportunistic infections.

C. Hypoglycemia

Hypoglycemia occurs more commonly in younger owl monkeys, but can be observed in older, debilitated animals as well. Predisposing factors include stress, higher metabolic rates, a lower percentage of body adipose stores, limited glycogen reserves, limited gluconeogenic enzymes, limited ability to utilize ketones or fatty acids, inanition, or reduced food intake. Clinical signs are lethargy, weakness, disorientation, convulsions, and coma. The severity of clinical signs is directly related to the blood glucose concentration and how rapidly that concentration changes over time. Rapid changes in concentration can result in severe neurological signs. Diagnosis is usually based on clinical signs and demonstration of a low resting blood glucose concentration (<50 mg/dl). We have found glucose meters developed for human diabetics to be useful in this regard since the procedure only requires 1 or 2 drops of blood to obtain a reasonable determination of blood glucose concentration.

Treatment consists of oral or parenteral administration of warm 5% dextrose solutions (Abee, 1985) or fruit juices given orally. Identification and elimination of the underlying cause of hypoglycemic episodes, if possible, are important in preventing recurrence in monkeys at risk.

X. NEOPLASTIC DISEASE

Spontaneous neoplasms have rarely been described in owl monkeys, particularly if one considers tumors other than those arising from the hematopoietic and lymphoreticular systems. Reviews and case reports of spontaneous neoplasms in nonhuman primates have revealed only a seminoma (Gozalo *et al.*, 1992a) and an oligodendroglioma (Weller, 1994, unpublished data) in the owl monkey.

REFERENCES

Abee, C. R. (1985). Medical care and management of the squirrel monkey. *In* "Handbook of Squirrel Monkey Research" (L. A. Rosenblum and C. L. Coe, eds.), 2nd ed., pp. 447–488. Plenum, New York.

Aikawa, M., Broderson, J. R., Igarashi, I., Jacobs, G., Pappaioanou, M., Collins, W. E., and Campbell, C. C. (1988). "An Atlas of Renal Disease in *Aotus* Monkeys With Experimental Plasmodial Infection." Am. Inst. Biol. Sci., Washington, DC.

Anver, M. R., Hunt, R. D., and Chalifoux, L. V. (1972). Cholesterol gallstones in, *Aotus trivirgatus*. *J. Med. Primatol.* **1,** 241–246.

Baer, J. F., Weller, R. E., Dagle, G. E., Málaga, C. A., and Lee, S. P. (1990). Cholelithiasis in the owl monkey: Seven cases. *Lab. Anim. Sci.* **40,** 629–633.

Baer, J. F., Gibson, S. V., Weller, R. E., Buschbom, R. L., Málaga, C. A., and Leathers, C. W. (1992). Naturally occurring aortic aneurysms in owl monkeys (*Aotus* spp.). *Lab. Anim. Sci.* **42,** 463–466.

Baggs, R. B., Hunt, R. D., Garcia, F. G., Hajema, H. H., Blake, B. J., and Fraser, C. E. O. (1976). Pseudotuberculosis (*Yersinia enterocolitica*) in the owl monkey (*Aotus trivirgatus*). *Lab. Anim. Sci.* **26,** 1084–1087.

Barahona, H., Melendez, L. V., Hunt, R. D., and Daniel, M. D. (1976). The owl monkey (*Aotus trivirgatus*) as an animal model for viral diseases and oncologic studies. *Lab. Anim. Sci.* **26,** 1104–1112.

Beland, M. F., Bronson, R. T., Ma, N. S. F., McIntyre, K. W., Sehgal, P. K., and Keadle, T. L. (1981). Karyotypic variation in susceptibility to hemolytic anemia and idiopathic eosinophilia of owl monkeys, *Aotus trivirgatus*. *Primates* **22,** 551–556.

Benjamin, S. A., and Lang, C. M. (1971). Acute pasteurellosis in owl monkeys (*Aotus trivirgatus*). *Lab. Anim. Sci.* **21,** 258–262.

Bergeland, M. E., Barnes, D. M., and Kaplan, M. (1970). "Spontaneous Histoplasmosis in a Squirrel Monkey," Primate Zoonosis Surveillance, Rep. No. 1. U.S. Department of Health, Education and Welfare, Centers for Disease Control, Atlanta, GA.

Bone, J. F., and Soave, O. A. (1970). Experimental tuberculosis in owl monkeys. *Lab. Anim. Care* **20,** 946–948.

Brady, P. S., Sehgal, P. K., and Hayes, K. C. (1982). Erythrocyte characteristics in vitamin E-responsive anemia of the owl monkey (*Aotus trivirgatus*). *Am. J. Vet. Res.* **43,** 1489–1491.

Broderson, J. R. (1989). A retrospective review of lesions associated with the use of Freund's adjuvant. *Lab. Anim. Sci.* **39,** 400–405.

Chalifoux, L. V., and King, N. V. (1983). Eosinophilic myelocytoma in an owl monkey (*Aotus trivirgatus*). *Lab. Anim. Sci.* **33,** 189–191.

Chalifoux, L. V., Bronson, R. T., Sehgal, P., Blake, B. J., and King, N. W. (1981). Nephritis and hemolytic anemia in owl monkeys (*Aotus trivirgatus*). *Vet. Pathol.* **18,** Suppl. 6, 23–37.

Coles, E. H. (1974). Microbiology. *In* "Veterinary Clinical Pathology" (E. H. Coles, ed.), pp. 400–462. Saunders, Philadelphia.

Cornelius, L. M., and Bjorling, D. E. (1988). Diseases of the liver and biliary system. *In* "Handbook of Small Animal Practice" (R. V. Morgan, ed.), pp. 437–458. Churchill-Livingstone, New York.

Cowgill, L. D. (1983). Diseases of the kidney. *In* "Textbook of Veterinary Internal Medicine: Diseases of the Dog and Cat" (S. J. Ettinger, ed.), pp. 1793–1879. Saunders, Philadelphia.

Daniel, M. D., Fraser, C. E. O., Barahona, H. H., Hajema, E. M., and Melendez, L. V. (1976). Microbial agents of the owl monkey. *Lab. Anim. Sci.* **26,** 1073–1078.

Fossum, T. W. (1988). Pleural cavity diseases. *In* "Handbook of Small Animal Practice" (R. V. Morgan, ed.), pp. 215–228. Churchill-Livingstone, New York.

Fox, J. G., Campbell, L. H., Reed, C., Snyder, S. B., and Soave, O. A. (1973). Dermatophilosis (cutaneous streptothricosis) in owl monkeys. *J. Am. Vet. Med. Assoc.* **163,** 642–644.

Giddens, W. E., Combs, C. A., Smith, O. A., and Klein, E. C. (1987). Spontaneous hypertension and its sequelae in wooly monkeys (*Lagothrix lagotricha*). *Lab. Anim. Sci.* **37,** 750–756.

Giles, R. C., Hildebrandt, P. K., and Tate, C. (1974). Klebsiella air sacculitis in the owl monkey (*Aotus trivirgatus*). *Lab. Anim. Sci.* **24,** 610–616.

Goldstein, E., and Joye, N. (1981). Laboratory diagnosis of specific bacterial and fungal diseases. *In* "The Laboratory in Clinical Medicine: Interpretation and Application" (C. H. Halsted and J. A. Halsted, eds.), pp. 829–870. Saunders, Philadelphia.

Good, R. C., and May, B. D. (1971). Respiratory pathogens in monkeys. *Infect. Immun.* **3,** 87–93.

Gozalo, A., and Montoya, E. (1990). Mortality causes of owl monkeys (*Aotus nancymae* and *Aotus vociferans*) in captivity. *J. Med. Primatol.* **19,** 69–72.

Gozalo, A., Nolan, T., and Montoya, E. (1992a). Spontaneous seminoma in an owl monkey in captivity. *J. Med. Primatol.* **21,** 39–41.

Gozalo, A., Dagle, G. E., Montoya, E., Weller, R. E., and Málaga, C. A. (1992b). Spontaneous cardiomyopathy and nephropathy in captive owl monkeys. *J. Med. Primatol.* **21,** 279–283.

Gozalo, A., Dagle, G. E., Montoya, E., and Weller, R. E. (1994). Spontaneous cardiomyopathy resembling acute rheumatic heart disease in an owl monkey. *J. Med. Primatol.* **21,** 279–283.

Grauer, G. F., Frisbie, D. D., Snyder, P. S., Dubielzig, R. R., and Panciera, D. L. (1992). Treatment of membranoproliferative glomerulonephritis and nephrotic syndrome in a dog with a thromboxane synthetase inhibitor. *J. Vet. Intern. Med.* **6,** 77–81.

Hornbuckle, W. E. (1988). General physical examination of the cat and dog. *In* "Handbook of Small Animal practice" (R. V. Morgan, ed.), pp. 3–11. Churchill-Livingstone, New York.

Hostetter, T. H., Nath, K. A., and Hostetter, M. K. (1988). Infection-related chronic interstitial nephropathy. *Semin. Nephrol.* **8,** 11–16.

Hunt, R. D., Van Zwieten, M. J., Baggs, R. B., Sehgal, P. K., King, N. W., Roach, S. M., and Blake, B. J. (1976). Glomerulonephritis in the owl monkey (*Aotus trivirgatus*). *Lab. Anim. Sci.* **26,** 1089–1092.

Hutt, M. S., Davies, D. R., and Voller, A. (1975). Malarial infections in *Aotus trivirgatus* with special reference to renal pathology. *Br. J. Exp. Pathol.* **56,** 429–438.

Kessler, M. J., Brown, R. J., Kalter, S. S., and Altman, N. H. (1982). A fatal epizootic of undetermined etiology in New World monkeys. *Am. J. Primatol.* **3,** 253–261.

King, N. W., Fraser, C. E. O., Garcia, F. G., Wolf, L. A., and Williamson, M. E. (1971). Cutaneous streptothricosis (dermatophiliasis) in owl monkeys. *Lab. Anim. Sci.* **21,** 67–74.

LeDuc, J. W., Lemon, S. M., Keenan, C. M., Graham, R. R., Marchwicki, R. H., and Binn, L. N. (1983). Experimental infection of the New World owl monkey (*Aotus trivirgatus*) with hepatitis A virus. *Infect. Immun.* **40,** 766–772.

Liu, S. K., Dolensek, E. P., Tappe, J. P., Stover, J., and Adams, C. R. (1984). Cardiomyopathy associated with vitamin E deficiency in seven gelada baboons. *J. Am. Vet. Med. Assoc.* **185,** 1347–1350.

Liu, S. K., Dolensek, E. P., and Tappe, J. P. (1985). Cardiomyopathy and vitamin E deficiency in zoo animals and birds. *Heart Vessels, Suppl.* **1,** 288–293.

Madewell, B. R. (1987). Cancer diagnosis. *In* "Veterinary Cancer Medicine" (G. H. Theilen and B. R. Madewell, eds.), 2nd ed., pp. 3–12. Lea & Febiger, Philadelphia.

Málaga, C. A., Weller, R. E., Dagle, G. E., Perry, R. L., and Baer, J. F. (1991). Spontaneous hydrocephalus in an adult owl monkey (*Aotus nancymae*). *Lab. Anim. Sci.* **41,** 182–184.

Matherne, C. M., Gibson, S. V., Kelley, S. T., and Wagner, J. E. (1986). Myeloproliferative syndrome in two owl monkeys. *Lab. Anim. Sci.* **36,** 551 (abstr.).

Meydani, S. N., Nicolosi, R. J., Sehgal, P. K., and Hayes, K. C. (1983). Altered lipoprotein metabolism in spontaneous vitamin E deficiency of owl monkeys. *Am. J. Clin. Nutr.* **38,** 888–894.

Mohan, C., Ganguly, N. K., and Chakravarti, R. N. (1987). Experimental production of cardiac injury in rhesus monkeys by L-forms of group-A streptococci. *Indian J. Med. Res.* **86,** 361–371.

Montali, R. J. (1992). Marmoset and tamarin disease awareness. *AAZPA Commun.* **8,** 11.

Moscovici, W. G., Marks, J. O., and McSherry, C. K. (1988). An experimental model for combined medical and surgical treatment of gallstones. *Surg. Endosc.* **2,** 66–70.

Mujais, S. K., and Quintanilla, A. (1988). Chronic tubulo-interstitial nephritis: Saga of the ubiquitous. *Semin. Nephrol.* **8,** 4–10.

Naiff, R. D., Mok, W. Y., and Naiff, M. F. (1985). Distribution of *Histoplasma capsulatum* in Amazonian wildlife. *Mycopathologia* **89,** 165–168.

Osborne, C. A., Stevens, J. B., and Perman, V. (1974). Kidney biopsy. *Vet. Clin. North Am.* **4,** 351–365.

Poltera, A. A., and Sayer, P. D. (1983). Cardiac lymph drainage in experimental African trypanosomiasis in vervet monkeys. *Bull. Soc. Pathol. Exot. Ses Fil.* **76,** 614–621.

Pung, O. J., Hulsebos, L. H., and Kuhn, R. E. (1988). Experimental Chagas' disease (*Trypanosoma cruzi*) in the Brazilian squirrel monkey (*Saimiri sciureus*): Hematology, cardiology, cellular and humoral immune responses. *Int. J. Parasitol.* **18,** 115–120.

Roberts, J. A., Clayton, J. D., and Seibold, H. R. (1972). The natural incidence of pyelonephritis in the nonhuman primate. *Invest. Urol.* **9,** 276–281.

Roca-Garcia, M., and Sanmartin-Barber, C. (1957). The isolation of encephalomyocarditis virus from *Aotus* monkeys. *Am. J. Trop. Med.* **6,** 840–852.

Rodger, R. F., Hartley, L. H., Ringler, D. J., and Nicolosi, R. J. (1986). Hypertrophic cardiomyopathy in owl monkeys (*Aotus trivirgatus*): Clinical diagnosis and clinicopathologic correlations. *Lab. Anim. Sci.* **36,** 561 (abstr.).

Sehgal, P. K., Bronson, R. T., Brady, P. S., McIntyre, K. W., and Elliott, M. W. (1980). Therapeutic efficacy of vitamin E and selenium in treating hemolytic anemia of owl monkeys (*Aotus trivirgatus*). *Lab. Anim. Sci.* **30,** 92–98.

Seibold, H. R., and Wolf, R. H. (1971). Toxoplasmosis in *Aotus trivirgatus* and *Callicebus moloch*. *Lab. Anim. Sci.* **21,** 118–120.

Shimp, R. G., Fiske, R. A., Woodard, J. C., and Moreland, A. F. (1973). Thorns of South American clump palm trees in the thoracic organs of owl monkeys (*Aotus trivirgatus*). *Lab. Anim. Sci.* **23,** 285–286.

Snyder, S., Peace, T., Soave, O., and Lund, J. (1970). Tuberculosis in an owl monkey (*Aotus trivirgatus*). *J. Am. Vet. Med. Assoc.* **157,** 712–713.

Swaim, S. F., and Lee, A. H. (1987). Topical wound medications: A review. *J. Am. Vet. Med. Assoc.* **190,** 1588–1593.

Ticehurst, J., Rhodes, L. L., Jr., Krawczynski, K., Asher, L. V. S., Engler, W. F., Mensing, T. L., Caudill, J. D., Sjogren, M. H., Hope, C. H., LeDuc, J. W., Bradley, D. W., and Binn, L. N. (1992). Infection of owl monkeys (*Aotus trivirgatus*) and cynomolgus monkeys (*Macaca fasicularis*) with hepatitis E from Mexico. *J. Infect. Dis.* **165,** 835–845.

Wallach, J. D., and Boever, W. J. (1983). Primates. *In* "Diseases of Exotic Animals: Medical and Surgical Management" (J. D. Wallach and W. J. Boever, eds.), pp. 3–133. Saunders, Philadelphia.

Weller, R. E., Dagle, G. E., Málaga, C. A., and Baer, J. F. (1990). Hypercalcemia and disseminated histoplasmosis in an owl monkey. *J. Med. Primatol.* **19,** 675–680.

Weller, R. E., Wierman, E. L., Málaga, C. A., and LeMieux, T. P. (1991a). Battelle primate facility. *J. Med. Primatol.* **20,** 133–137.

Weller, R. E., Buschbom, R. L., Martell, S. L., Baer, J. F., Málaga, C. A., and Ragan, H. A. (1991b). Total serum creatine kinase and isozyme concentrations in the owl monkey. *J. Med. Primatol.* **20,** 290–294.

Weller, R. E., Málaga, C. A., Buschbom, R. L., Baer, J. F., and Ragan, H. A. (1991c). Protein concentration in urine of normal owl monkeys. *J. Med. Primatol.* **20,** 365–369.

Weller, R. E., Baer, J. F., Málaga, C. A., Buschbom, R. L., and Ragan, H. A. (1992a). Renal clearance and excretion of endogenous substances in the owl monkey. *Am. J. Primatol.* **28,** 115–123.

Weller, R. E., Collins, W. E., Buschbom, R. L., Málaga, C. A., and Ragan, H. A. (1992b). Impaired renal function in owl monkeys (*Aotus nancymae*) infected with *Plasmodium falciparum*. *Mem. Inst. Oswaldo Cruz* **87,** 435–442.

Weller, R. E., Málaga, C. A., Buschbom, R. L., and Ragan, H. A. (1993). Urinary enzyme concentrations in the owl monkey (*Aotus nancymae*). *J. Med. Primatol.* **22,** 340–347.

Whitney, R. A. (1979). Primate medicine and husbandry. *Vet. Clin. North Am: Small Anim. Pract.* **9,** 429–445.

Whittow, G. C., Guernsey, D. L., and Morishige, W. K. (1979). Thyroid activity in a hypometabolic primate, the owl monkey (*Aotus trivirgatus*). *Arch. Int. Physiol. Biochim.* **87,** 963–967.

8

The Owl Monkey as a Model for Malaria

William E. Collins
Centers for Disease Control
Atlanta, Georgia 30333

I. INTRODUCTION

In 1966, Young *et al.* described the experimental infection of owl monkeys in Panama with the human malaria parasite *Plasmodium vivax*. Reports of the experimental infection of *Aotus lemurinus griseimembra* monkeys with *Plasmodium falciparum* by Geiman and Meagher (1967) and *Plasmodium malariae* by Geiman and Siddiqui (1969) soon followed. The availability of animal models for the development and testing of antimalarial drugs was an advance of major importance because resistance of *P. falciparum* and *P. vivax* to standard antimalarial therapy was becoming widespread. Within a few years, methods for the management of malarial infections and the testing of drugs in owl monkeys were instituted (Schmidt, 1973, 1978a,b,c) and led to the development of techniques for biologic and immunologic studies in these animals.

The different species of *Aotus* vary in their ability to serve as hosts for the human malarias. However, once an isolate of a human malaria parasite (*P. falciparum, P. vivax,* or *P. malariae*) or a monkey malaria parasite (*P. brasilianum, P. simium, P. cynomolgi, P. fragile, P. knowlesi,* or *P. inui*) is established in one species of *Aotus,* it can be readily passaged to other species. Splenectomy is vital for the maintenance of long-term infections and the production of large numbers of infective gametocytes. Thus, almost all studies with newer isolates are conducted in splenectomized animals.

Reported here are the results of studies that have been conducted at the U.S. Public Health Service laboratories in Chamblee, Georgia, since 1966 on the infection of various *Aotus* species with a variety of malaria parasites.

II. NATURAL INFECTIONS

Extensive studies of owl monkeys from Panama and northern Colombia (= *Aotus lemurinus lemurinus* and *Aotus lemurinus griseimembra,* respectively) in the 1960s and 1970s found none naturally infected with malaria parasites. Later, owl monkeys from other areas of South America were examined, but infections were found in only two monkeys; both were *Aotus vociferans* monkeys from Iquitos, Peru, infected with *Plasmodium brasilianum,* a parasite common to many species of monkey from South America (Collins *et al.,* 1985b).

III. EXPERIMENTAL INFECTIONS

A. Methods

Animals are housed doubly or singly to avoid injuries due to fighting with cagemates. All monkeys are fed a diet that has been shown to provide adequate nutrition and calories in captive owl monkeys used in malaria-related research. Animals are handled as little as possible. They are anesthetized with 0.1 cc ketamine hydrochloride intramuscularly when they are to be given a physical exam, have blood drawn, and be weighed. On days when they are to have blood drawn or to be weighed only, anesthesia is not given.

1. Infection with Trophozoites Blood from a donor animal is collected into a heparinized tube. Based on the parasite count, the blood is then aseptically diluted in phosphate-buffered saline (pH 7.0) to give the required number of parasites to be inoculated in 1 ml or less. The recipient animal is restrained, and the blood is slowly injected into the femoral vein using a 25-gauge needle.

2. Infection with Sporozoites For infection with mosquito bites, the monkeys are anesthetized and immobilized on a restraining board. Infected mosquitoes are caged in an ice-cream carton, the top of which is covered with nylon netting to allow feeding on the monkey to engorgement (usually 10 to 15 min). Feeding mosquitoes are then dissected and their salivary glands examined to determine the number of infected mosquitoes that fed upon the animal. The number of sporozoites remaining in the salivary glands is graded 1+ (1–10 sporozoites), 2+ (11–100 sporozoites), 3+ (101–1000 sporozoites), or 4+ (greater than 1000 sporozoites). The total number of grades is recorded for all mosquitoes feed on the animal to determine the exposure index for the monkey.

For infection with sporozoites dissected from the salivary glands, infected mosquitoes are chloroformed and then decapitated. Legs and wings are removed and the bodies transferred to a clean slide. Salivary glands are teased from the neck opening with a minutin needle into a drop of 20% heat-treated monkey serum/saline or fetal bovine serum/saline. After the salivary glands are crushed under a coverslip, the coverslip is gently removed and the sporozoites are "washed" from the coverslip onto the slide with several drops of serum/saline. The fluid is then aspirated into a syringe using a 25-gauge needle. The suspension is transferred to a chilled NUNC vial and the volume increased to approximately 1 ml. After being further dissociated from salivary gland tissue by repeated expiration and expression from the syringe, sporozoites are counted in a Neubauer cell-counting chamber. The number of sporozoites per milliliter is calculated and the sporozoite suspension diluted to the desired number of sporozoites in 1 ml of serum/saline. The sporozoite suspension is then injected into the monkey's femoral vein with a 25-gauge needle.

To prepare blood films, the animal is removed from the cage and immobilized. The tail or back of the leg is thoroughly cleaned and then pricked with a lancet; care must be taken to avoid the vein. From the drop of blood that is expressed, blood films are made by the method of Earle and Perez (1932). Blood is collected in a 5-μl pipette and then expressed onto a prescribed area (15 mm $\times$ 5 mm) on a precleaned, prelabeled microscope slide. A second drop of blood is added to the same slide, and a thin film is prepared. After being labeled with the animal identification number and the date, the slides are stained with Giemsa and the number of parasites per microliter is recorded. Gametocytes are counted on the thin film and recorded per 100 WBC. Whenever possible, separate counts are recorded for the micro- and macrogametocytes.

3. Surgery Most monkeys used in malarial studies are splenectomized either before or during infection. All surgical procedures are conducted under general anesthesia by a qualified veterinary surgeon in an approved surgical suite appropriate for aseptic surgery. All animals are fasted for 12 hr before surgery. Ketamine HCl, the basic anesthetic agent employed for initial induction, is used at a dosage of 15–25 mg/kg BW. Atropine sulfate is also used adjunctively. For prolonged procedures, or when more prolonged muscle relaxation is required, halothane gas is administered at a concentration of 1–2% by using a standard closed oxygen delivery apparatus with CO_2 absorption. For shorter procedures, thiamyl sodium is given intravenously to effect. Anesthesia is monitored by respiratory rate and depth, skeletal muscle tone, and, with noncateleptic agents, cranial nerve reflexes.

The animal is initially prepared for splenectomy by clipping the ventral body region from the xyphoid to the pubis with a #40 surgical blade. The area is scrubbed three times with surgical soap, swabbed three times with 70% alcohol, and allowed to dry. The final surgical field is draped with an autoclaved muslin or paper drape fenestrated for the appropriate surgical approach. Instruments are sterilized by autoclaving. Splenectomy is performed through a midline incision above the umbilicus approximately one-third of the distance toward the xyphoid cartilage. The incision, usually 1 to 2 cm, is extended through the skin and linea alba. The spleen is exteriorized by using an ovariohysterectomy hook or allis forceps. The vasculature is fenestrated at intervals to allow for the most effective hemostasis and minimal tissue trauma and then ligated with 3-0 or 4-0 chromic gut sutures. The spleen is excised at the lesser curvature using curved Mayo or Metzenbaum scissors. The ligatures are checked for hemostasis and replaced into the peritoneal cavity. The abdominal laparotomy incision in the linea is closed with 3-0 adsorbable sutures. The skin is closed with nonabsorbable sutures using simple interrupted placement at 0.5-cm intervals.

4. Infection of Mosquitoes with Plasmodia Monkeys are anesthetized and immobilized on restraining boards with ketamine or leather thongs. Mosquitoes are allowed to feed to engorgement (usually 8 to 10 min). Unfed mosquitoes are removed, and the fed mosquitoes are placed in a large holding incubator (25°C; 60% relative humidity). Mosquitoes are fed a sugar solution on a cellulose sponge daily. Aliquots of mosquitoes are dissected 7 to 10 days after feeding (depending on the species of *Plasmodium*), and the midguts are examined for the presence of oocysts, which are counted and recorded. Infected lots of mosquitoes are returned to the incubator and maintained until sporozoites are present in the salivary glands (10 to 18 days postfeeding, depending on the species of *Plasmodium*).

B. New World Monkey Malarias

1. Plasmodium brasilianum Two different strains of *P. brasilianum* were established from naturally acquired infections in *Aotus* monkeys. This parasite is characterized by a 72-hr asexual developmental cycle. Cellular enlargement is minimal and stippling of the infected erythrocyte (Ziemann's stippling) is faint. Some sequestration of the more mature forms may occur. There is no evidence of relapse from the fixed tissue.

An *Aotus vociferans* monkey received from Iquitos, Peru, was shown upon blood film examination to be infected with *P. brasilianum* (Peruvian I strain). The parasite was passaged to splenectomized *A. lemurinus griseimembra* monkey AO-117 (Fig. 1). Low-density parasitemia of $<1000/\mu l$ persisted until Day 48, after which time the parasite density increased to a peak of 10,400/μl on Day 58; parasitemia of greater than 1000/μl persisted beyond Day 100. Passage of parasites from monkey AO-117 to splenectomized *A. azarae boliviensis* money AO-200 resulted in a more rapid increase in parasite density to $>1000/\mu l$ by Day 17, with a maximum parasite count of 19,500/μl on Day 41 (Fig. 1). The parasite count declined to undetectable levels by Day 84, and the infection was treated with chloroquine (10 mg base daily for 3 days). A third linear passage was made to splenectomized *A. lemurinus griseimembra* monkey AO-179. A peak parasite density of 280,000/μl occurred on Day 44. The infection was treated on Day 89. Infection of *Anopheles freeborni* mosquitoes was attempted on 46 separate occasions. However, only 5 of 1389 mosquitoes dissected and examined had oocysts.

Another *A. vociferans* monkey from Iquitos, Peru, was naturally infected with *P. brasilianum* (Peruvian II strain). The parasite was passaged to splenectomized *A. vociferans* monkey AO-361 (Fig. 2). A maximum parasite count of 25,500/μl occurred on Day 20. Before the infection was treated with chloroquine on Day 23, passage was made to intact *A. nancymae* monkey AO-387. A maximum parasite count of 1675/μl occurred on Day 27. Low-density parasitemia persisted until the animal was splenectomized on Day 78, after which time the parasite count rose to a maximum of 27,300/μl on Day 92. The infection was treated on Day 105. The third linear passage was to monkey AO-334 (Fig. 2). A maximum parasite count of 21,390/μl occurred on Day 29, and the animal was treated with 15 mg of chloroquine on Days 30 and 31. After 59 days of undetectable parasitemia, parasites were again seen. The maximum parasite count of 17,100/μl occurred on Day 111; thereafter, the parasite count slowly declined. A second recrudenscence of the infection occurred on Day 188. No infection of *Anopheles freeborni* mosquitoes was obtained from 30 feedings on these animals.

Only one of five attempts to transmit *P. brasilianum* by sporozoites from *An. stephensi* was successful; the prepatent period was 28 days. However, the parasite is readily transmitted by the inoculation of parasitized blood.

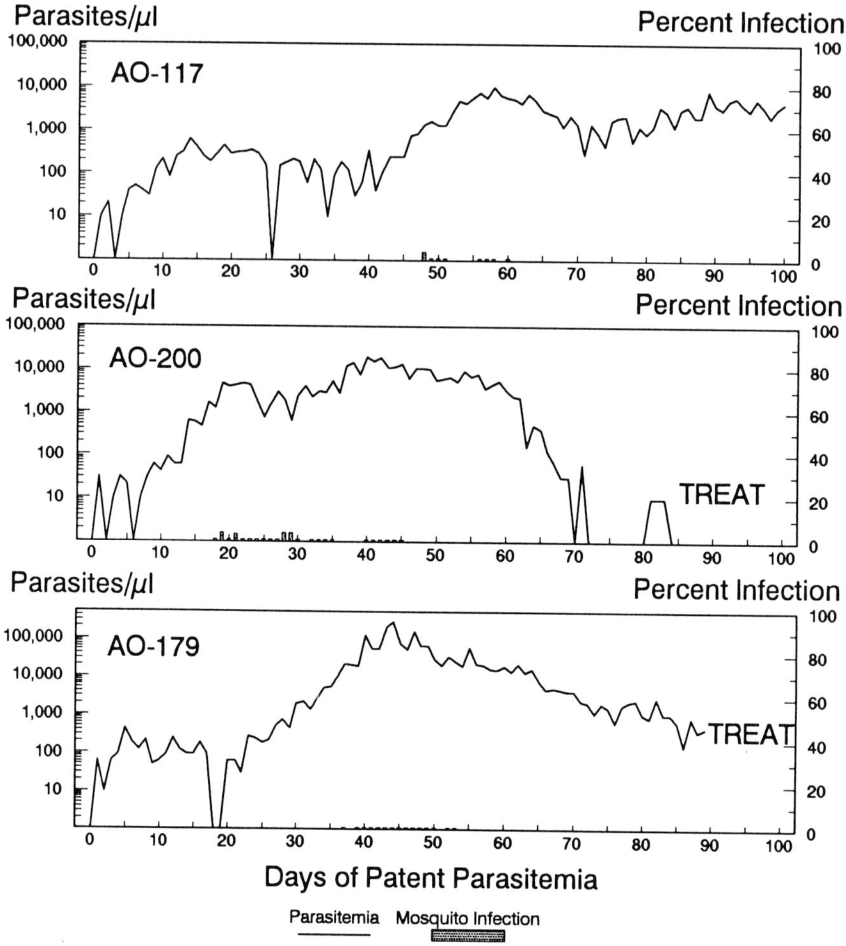

FIGURE 1 Course of parasitemia in three splenectomized *Aotus* monkeys infected with the Peruvian I strain of *Plasmodium brasilianum*.

The long periods of low- to moderate-density parasitemia observed in these animals are typical of the quartan malarias of primates. Thus, *Aotus* monkeys infected with *P. brasilianum* are suitable models for studies of chronic malarial infection.

2. Plasmodium simium *Plasmodium simium* and *P. brasilianum* are the only malaria parasites described from New World monkeys. *Plasmodium simium* has not been reported as naturally infective to *Aotus* monkeys. However, a strain originating from a howler monkey in Brazil has been passaged in different species of *Aotus* (Collins *et al.*, 1987b). The parasite, which is morphologically similar to *Plasmodium vivax*, has a 48-hr asexual develop-

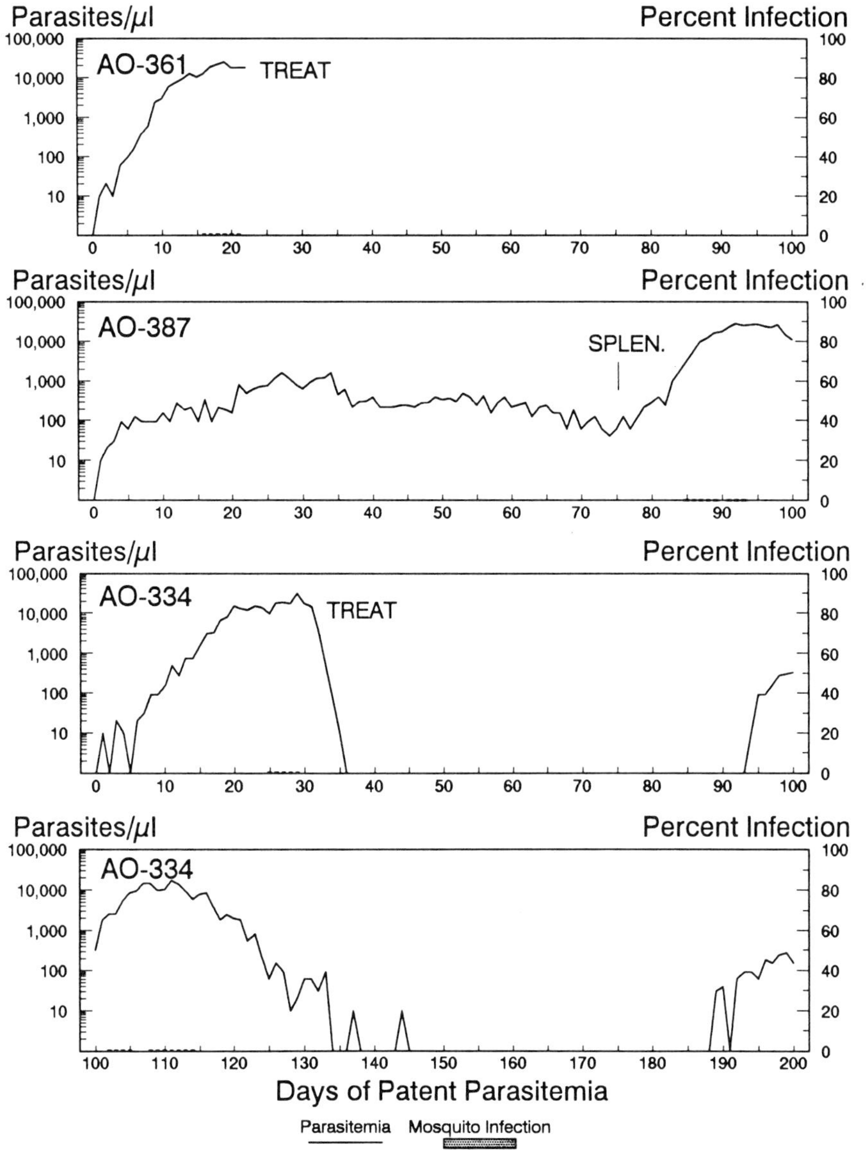

FIGURE 2 Course of parasitemia in three splenectomized *Aotus* monkeys infected with the Peruvian II strain of *Plasmodium brasilianum*.

ment cycle. Four splenectomized *A. azarae boliviensis* monkeys (AO-475, AO-476, AI-280, and AI-281) were infected by the bites of infected *Anopheles stephensi* mosquitoes (Fig. 3). Prepatent periods were 15, 17, 17, and 16 days, respectively. Maximum parasite counts were 6420, 32,300, 21,950, and 29,200/μl, respectively. Infection of *An. freeborni* mosquitoes was ob-

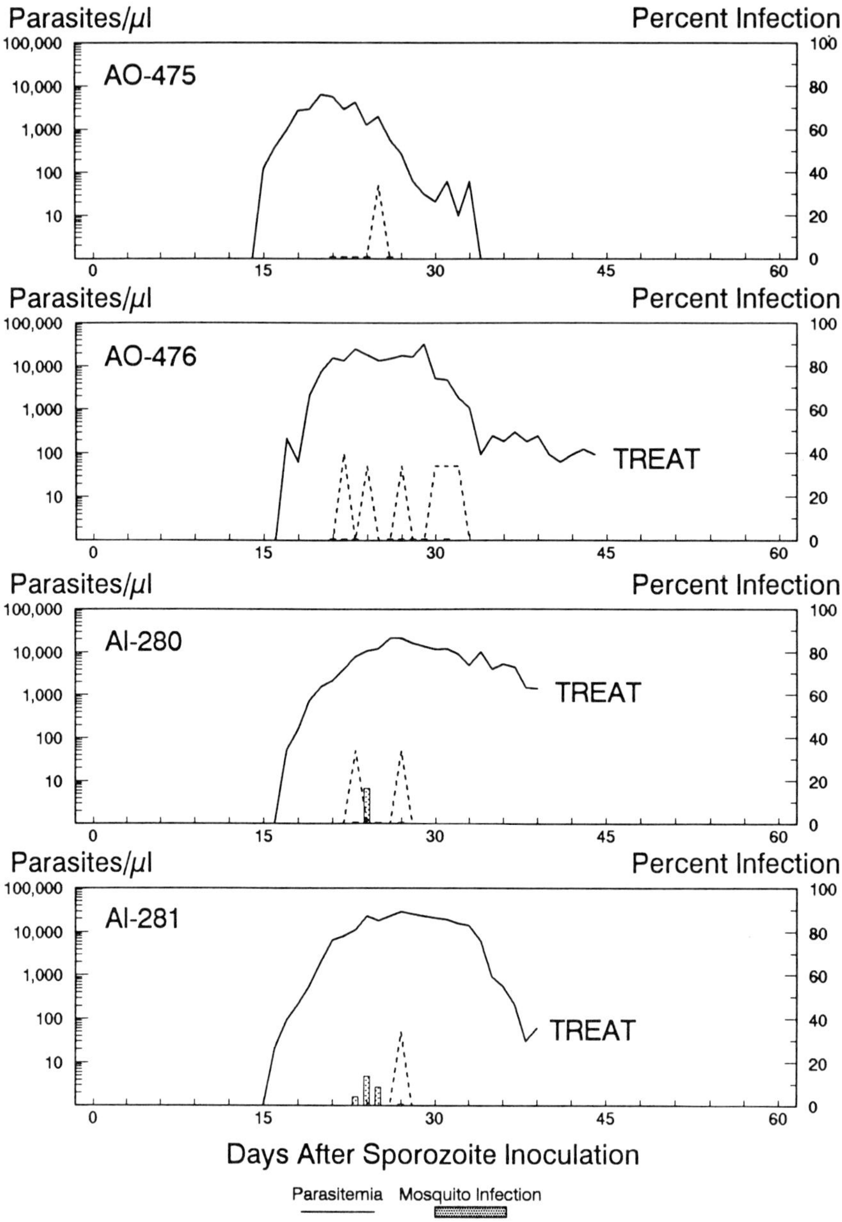

FIGURE 3 Prepatent periods and course of parasitemia in four splenectomized *Aotus azarae boliviensis* monkeys infected with *Plasmodium simium* via the bites of *Anopheles stephensi* mosquitoes.

tained on 4 occasions out of 21 feedings. Infections were cured by treatment with chloroquine (10 mg base daily for 3 days).

In a splenectomized *A. nancymae* monkey (AO-438) fed upon by infected *An. stephensi* mosquitoes, the prepatent period was 15 days, and the maximum parasite count was 16,200/μl (Fig. 4). No infections were obtained in *Anopheles freeborni* mosquitoes fed on 14 of 15 days between the 8th and 22nd days of patent parasitemia. The infection was successfully treated with chloroquine when the parasite count had dropped to 155/μl.

Two splenectomized *A. lemurinus griseimembra* monkeys (AI-299 and AI-300) were infected via the bites of infected *An. stephensi* (Fig. 5). After prepatent periods of 15 days, the parasite counts rose rapidly to 47,600 and 47,800/μl, respectively. These parasite counts were markedly higher than those seen in the other species of *Aotus*. Both animals were treated when the parasite counts were still relatively high. Of the seven lots fed on these animals, one mosquito was infected.

The similarity between this parasite and *P. vivax* suggests that this model could be used to study the immunologic and biologic consequences of infection with a benign tertian malaria parasite.

C. Old World Monkey Malarias

1. Plasmodium cynomolgi *Plasmodium cynomolgi* is one of the most-studied monkey malaria parasites because of its many similarities to the human-infecting parasite, *Plasmodium vivax*. This parasite has a 48-hr developmental cycle in the asexual stage. It is characterized by the presence of marked cellular enlargement of the more mature blood stage parasites, the presence of Schuffner's stippling, no evidence of sequestration of the ma-

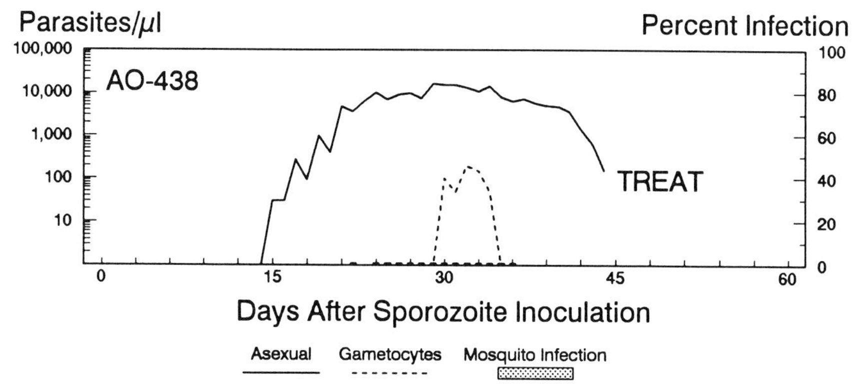

FIGURE 4 Prepatent period and course of parasitemia in a splenectomized *Aotus nancymae* monkey infected with *Plasmodium simium* via the bites of infected *Anopheles stephensi* mosquitoes.

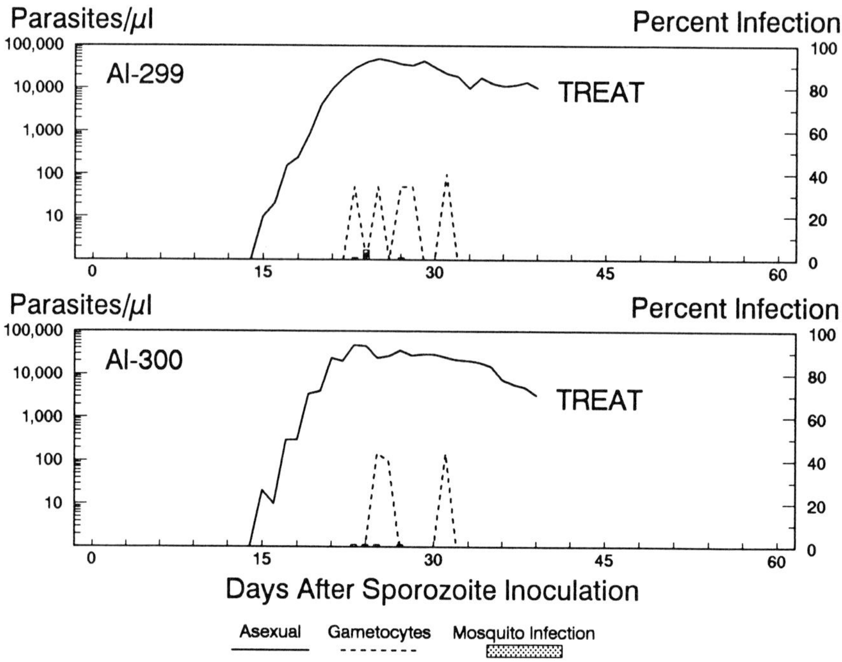

FIGURE 5 Prepatent periods and course of parasitemia in two splenectomized *Aotus lemurinus griseimembra* monkeys infected with *Plasmodium simium* via the bites of infected *Anopheles stephensi* mosquitoes.

ture forms in the deep circulation, and a proven mechanism for relapse from the fixed tissue. The subspecies *P. cynomolgi cynomolgi, P. cynomolgi bastianellii*, and *P. cynomolgi ceylonensis* are recognized (Coatney *et al.*, 1971); many strains with distinct characteristics also have been described. Four strains of *P. cynomolgi* (Berok, Cambodian, Gombak, and RO), in addition to *P. cynomolgi bastianelli* (B strain) have been studied in *Aotus* monkeys.

Transmission of the Cambodian, Berok, Ro, and B strains to splenectomized *A. lemurinus griseimembra, A. nancymae*, and *A. azarae boliviensis* via the bites of infected *Anopheles maculatus, An. freeborni*, and *An. dirus* mosquitoes has been reported (Collins *et al.*, 1975a; 1985c). Prepatent periods ranged from 9 to 30 days. In *A. azarae boliviensis* monkey AO-184 (Fig. 6), the parasite count rose rapidly to a peak of 115, 500/μl on the 13th day of patent parasitemia. The infection was cured on Day 16. Low-level infection was obtained in *An. freeborni* mosquitoes on Days 8 through 13 of patent parasitemia. In monkey AO-342, the peak parasite count of 25, 800/μl occurred on the 9th day of patent parasitemia. Despite the presence of gametocytes, no *An. freeborni* mosquitoes were infected when fed on Days 6 through 14. The monkey was treated on the 18th day of patent parasitemia when the parasite count had dropped to 186/μl. Monkey AO-200 had a

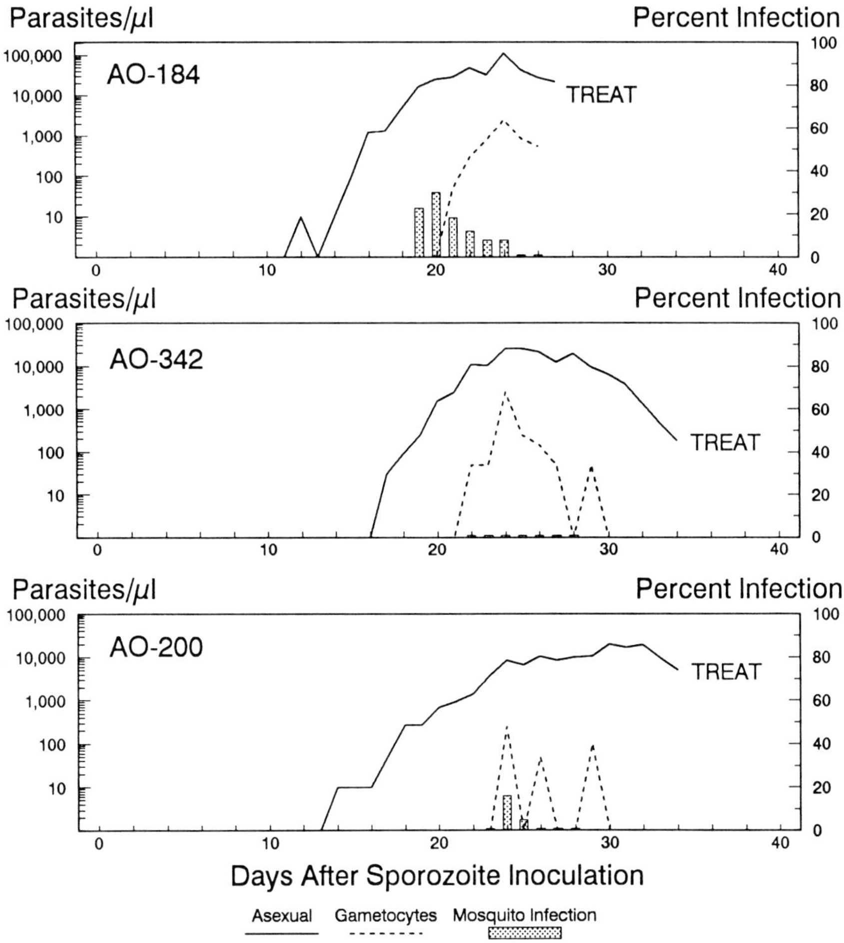

FIGURE 6 Prepatent periods and course of parasitemia in three splenectomized *Aotus azarae boliviensis* monkeys inoculated with the RO strain of *Plasmodium cynomolgi* via the bites of infected *Anopheles dirus* mosquitoes.

maximum parasite count of 20,460/μl on the 17th day of patent parasitemia. Mosquitoes were infected on Days 11 and 12. The animal was treated on the 21st day of patent parasitemia, when the parasite count had dropped to 5200/μl.

Aotus monkeys of different species are susceptible to both sporozoite and trophozoite infection with different strains of *P. cynomolgi*. Transmission via sporozoites from *An. freeborni, An. maculatus,* and *An. dirus* mosquitoes was obtained in 22 of 30 attempts; prepatent periods ranged from 11 to 33 days with a mean of 17.4 days. Parasitemia is usually transient. Although mosquito infections are usually of a low order, cyclic transmission studies from monkey to monkey via mosquitoes are possible. Thus different *Aotus* spp. may be useful as models for studies with this vivax-like malaria parasite.

2. Plasmodium fragile *Plasmodium fragile* is found in macaque monkeys in southern India and Sri Lanka. This parasite, which has a 48-hr asexual developmental cycle, is characterized by deep circulation schizogony, no marked enlargement of the red cell, no Schuffner's stippling, heavy pigmentation of the mature forms, schizonts that do not completely fill the red cell, and no true relapse mechanism from the fixed tissue stages (Coatney *et al.*, 1971).

Susceptibility of different *Aotus* monkeys to infection has been reported (Collins *et al.*, 1974b). Infections can be induced both by trophozoites passaged in parasitized blood and by sporozoites. Following the bites of infected *An. dirus* mosquitoes, prepatent periods in two splenectomized *A. lemurinus griseimembra* monkeys (AO-547 and AO-548) were 26 and 20 days, respectively (Fig. 7.). Parasite counts rose rapidly to over 1.6 million parasites/μl in AO-547 on the 13th day of patent parasitemia, and the animal

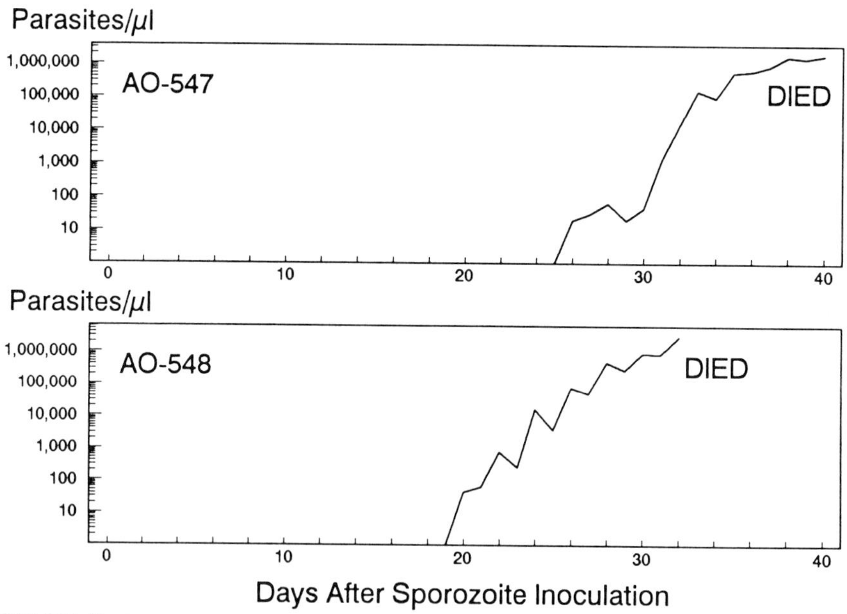

FIGURE 7 Prepatent periods and course of parasitemia in two splenectomized *Aotus lemurinus griseimembra* monkeys infected with *Plasmodium fragile* via the bites of infected *Anopheles dirus* mosquitoes.

died 2 days later. The parasite count in AO-548 rose to over 3 million parasites/μl on the 13th day, and the monkey died. Not all attempts to transmit *P. fragile* to this species via sporozoite inoculation were successful.

Similar studies were made with *A. nancymae* monkeys. Transmission via sporozoites to this species is rare. However, one splenectomized monkey, AI-552, developed a patent parasitemia following the intravenous inoculation of 3500 sporozoites dissected from the salivary glands of *A. dirus* mosquitoes (Fig. 8). After a prepatent period of 23 days, the parasite count rose to 218,000 /μl on the 10th day of patent parasitemia and then dropped rapidly to an undetectable level on Day 15.

When two intact *A. nancymae* monkeys (AO-602 and AO-596) were inoculated with parasitized blood, parasite counts rose rapidly to 1.2 million/μl on Day 15 and 1.24 million/μl on Day 13, respectively (Fig. 9). Monkey AO-602 died on Day 16; AO-596 was treated with chloroquine (10 mg base daily for 3 days) beginning on Day 22 and survived.

Transmission via sporozoites from *An. dirus* mosquitoes was successful in 5 of 26 attempts; prepatent periods ranged form 20 to 27 days, with a mean of 24.6 days. In all the monkeys, *P. fragile* exhibited a high–low periodicity pattern characteristic of a parasite in which the mature parasites are sequestered and not abundant in the peripheral blood. Because of this characteristic and the high level of mortality associated with *P. fragile* infection, this parasite is suitable as a biological model for the human malaria parasite *P. falciparum*. The high level of mortality was present in animals who had previously been infected with *P. falciparum*, *P. vivax*, or other species of monkey-infecting plasmodia.

3. Plasmodium knowlesi *Plasmodium knowlesi* is the only species of primate malaria parasite with a 24-hr asexual cycle. It is highly fatal to *Macaca*

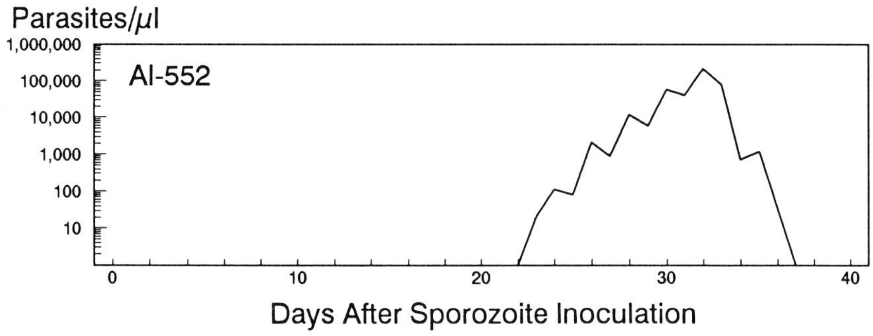

FIGURE 8 Prepatent period and course of parasitemia in a splenectomized *Aotus nancymae* monkey infected with *Plasmodium fragile* via the intravenous inoculation of sporozoites from *Anopheles dirus* mosquitoes.

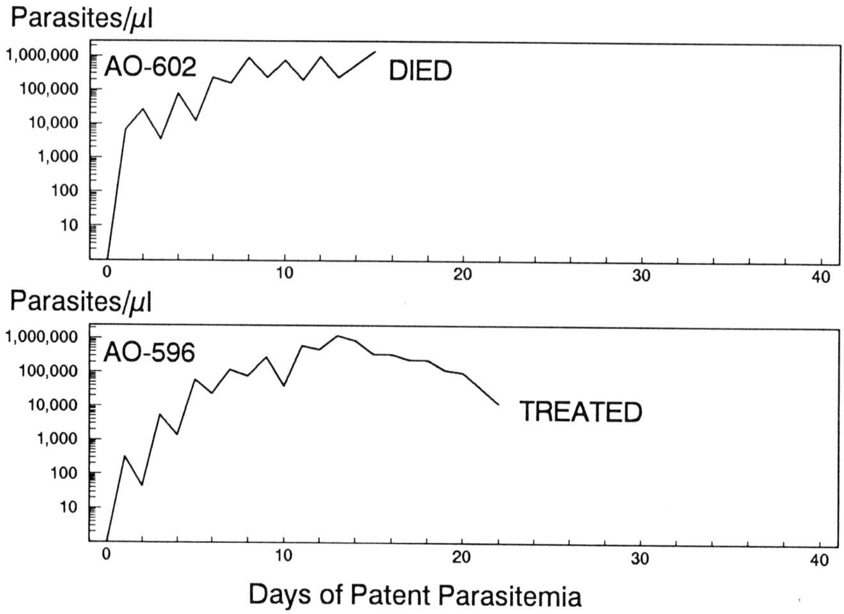

FIGURE 9 Parasitemia in two intact *Aotus nancymae* monkeys following inoculation with trophozoites of *Plasmodium fragile*.

mulatta monkeys, in which it has been studied extensively. The parasite is characterized by minimal cellular enlargement, the absence of Schuffner's stippling, no sequestration of the mature blood stage parasites, and the absence of a relapse mechanism from the fixed tissue. Development of this parasite in *A. lemurinus griseimembra* has been reported (Siddiqui *et al.*, 1974).

Transmission of the H strain of *P. knowlesi* to *Aotus* monkeys was obtained via the bites of *An. dirus* mosquitoes infectd by feeding on *M. mulatta* monkeys. In three *A. lemurinus griseimembra* monkeys (Fig. 10), the prepatent periods were 27, 16, and 12 days. Once parasites were detected, the parasite counts rose rapidly. In monkey AO-483, a maximum parasite count of 1,680,000/μl occurred on the 8th day of patent parasitemia. With AO-484, the maximum parasite count of 1,040,000/μl occurred on the 14th day. Both of these animals died. Monkey AO-698 had a maximum parasite count of 640,000/μl on Day 15; thereafter the parasite count declined, and the infection was eliminated with chloroquine (10 mg base daily for 3 days) 35 days after sporozoite inoculation.

Three *A. nancymae* monkeys (AO-637, AO-638, and AO-639) were also infected via the bites of infected *An. dirus* mosquitoes (Fig. 11). Prepatent periods were 17, 12, and 17 days and maximum parasite counts were 1,000,000, 1,640,000, and 2,040,000/μl, respectively. All the animals died.

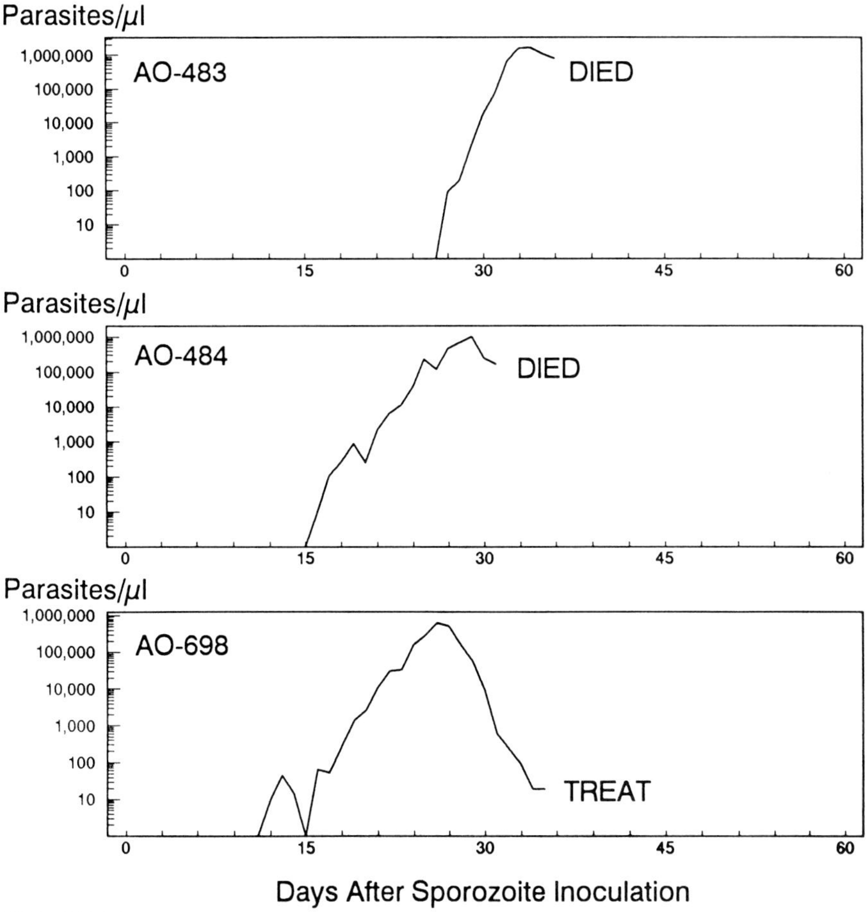

FIGURE 10 Prepatent periods and course of parasitemia in three splenectomized *Aotus lemurinus griseimembra* monkeys infected with the H strain of *Plasmodium knowlesi* via the bites of infected *Anopheles dirus* mosquitoes.

Transmission via sporozoites from *An. dirus* mosquitoes was successful in six of eight attempts. The prepatent periods ranged from 12 to 27 days, with a mean of 16.8 days. Infection with this parasite by sporozoite inoculation is often lethal to *Aotus* spp. Prepatent periods in *Aotus* monkeys were much longer than the 5 to 9 days usually seen in *M. mulatta* monkeys. However, once established, high-density parasitemia develops in these monkeys. Mosquito infection was attempted on very few occasions. However, all the *An. dirus* that were fed on monkey AO-483 when the parasite count was 680,000/µl (Day 6 of patent parasitemia) had oocysts on midgut examination.

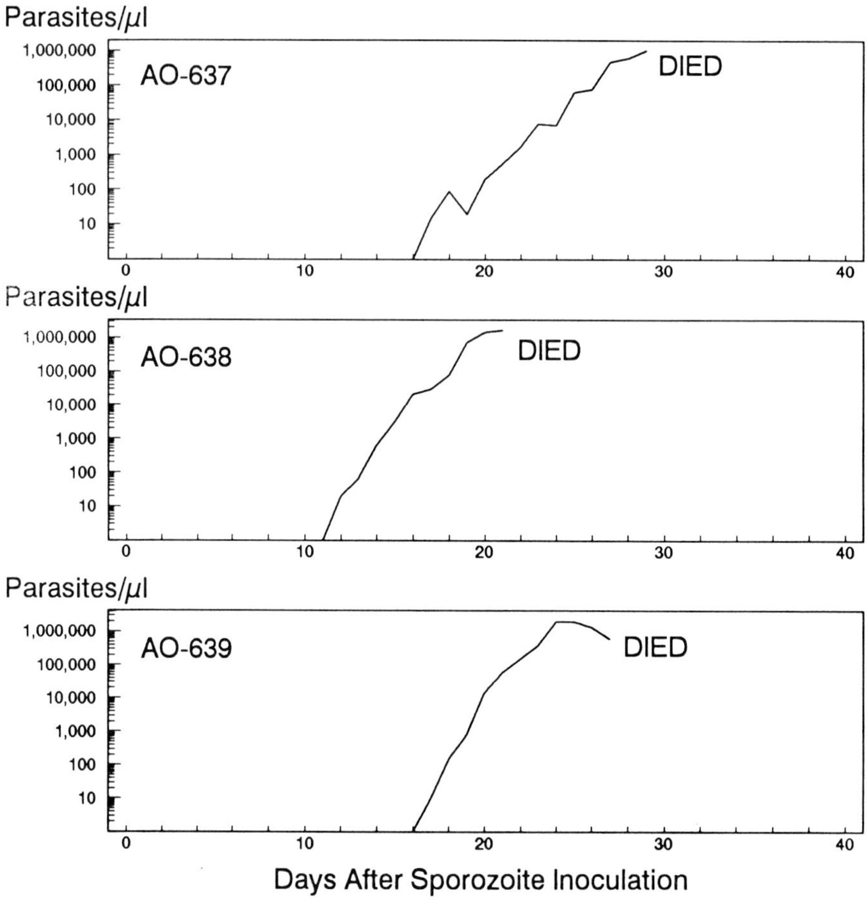

FIGURE 11 Prepatent periods and course of parasitemia in three splenectomized *Aotus nancymae* monkeys infected with the H strain of *Plasmodium knowlesi* via the bites of infected *Anopheles dirus* mosquitoes.

4. Plasmodium inui *Plasmodium inui*, a parasite of macaques and leaf monkeys, has a range extending from Sri Lanka to the Philippines. It is characterized by a 72-hr asexual developmental cycle, minimal cellular enlargement, the absence of Schuffner's stippling, lack of sequestration of the mature forms, and no evidence of relapse from the fixed tissue. This group of parasites is a large complex of closely related parasites that are usually referred to as strains of *P. inui* even though there are obvious differences in the sporogonic stages and the patterns of development in macaque monkeys. Several subspecies have been described (Eyles, 1963; Coatney *et al.*, 1971). The two most-studied parasites are the M strain (*P. inui inui*) and the OS strain (*P. inui shortti*).

Not all strains of *P. inui* inoculated into *Aotus* monkeys have developed. However, infection of *A. lemurinus griseimembra* and *A. azarea boliviensis* with the OS strain has been reported (Collins *et al.*, 1981, 1988b). This strain was adapted to develop in splenectomized *A. lemurinus griseimembra* monkeys following trophozoite inoculation (Fig. 12). Monkey AO-68 was inoculated with trophozoites from a *M. mulatta* monkey. Parasitemia of $>1000/\mu l$ was not evident until 30 days after inoculation. However, by Day 49, the parasite count had climbed to $101{,}600/\mu l$. *Anopheles dirus* mosquitoes were fed on 28 occasions between Days 27 and 55; only one mosquito was infected (Day 29). The animal died on Day 56. When parasites were then passaged to splenectomized monkey AO-20, the parasite count exceeded $1000/\mu l$ on Day 9. Following a maximum parasite count of $157{,}400/\mu l$ on Day 17, the animal died on Day 19. Low-level mosquito infection was obtained on Days 9, 10, 13, 14, and 16. The third linear passage to monkey AO-31 resulted in a maximum parasite count of $212{,}000/\mu l$ on Day 15. This animal was monitored for 185 days and then cured of its infection by treatment with chloroquine. Three distinct peaks in the parasite count occurred during recrudescence on Days 58, 113, and 167. Low-level mosquito infection was obtained on Days 11 and 12. As in monkeys AO-68 and AO-20, gametocytes were rarely observed.

One of two splenectomized *A. azarae boliviensis* monkeys developed a patent parasitemia following the bites of infected *An. dirus* mosquitoes (Collins *et al.*, 1988b). The prepatent period was 28 days; the maximum parasitemia of $380{,}000/\mu l$ occurred on the 13th day of patent parasitemia. Five *A. nancymae* monkeys similarly exposed to the bites of heavily infected mosquitoes failed to develop patent parasitemia. All of these monkeys had been previously infected with *P. falciparum* and *P. vivax*. Infection of *Aotus* monkeys with trophozoites is more predictable than with sporozoites.

5. Plasmodium coatneyi, Plasmodium fieldi, Plasmodium simiovale,* and *Plasmodium gonderi Attempts were made to passage these four additional malaria parasites of Old World monkeys to splenectomized *Aotus* monkeys, but none developed to detectable levels of parasitemia.

D. Susceptibility to Human Malarias

The report of Young *et al.* (1966) on the passage to *A. lemurinus lemurinus* of different indigenous strains of *P. vivax* by trophozoites and sporozoites provided the much sought-after model for studies on the chemotherapy of human malaria. Eventually, these monkeys became the model for immunologic and vaccine studies. Strains of *P. vivax*, *P. falciparum*, and *P. malariae* have been adapted to different species of *Aotus*. Major *P. vivax* strains that have been established in different species of *Aotus* monkeys are Chesson (Ward *et al.*, 1969; Schmidt, 1973, 1978a,b,c; Collins *et al.*, 1980), Achiote

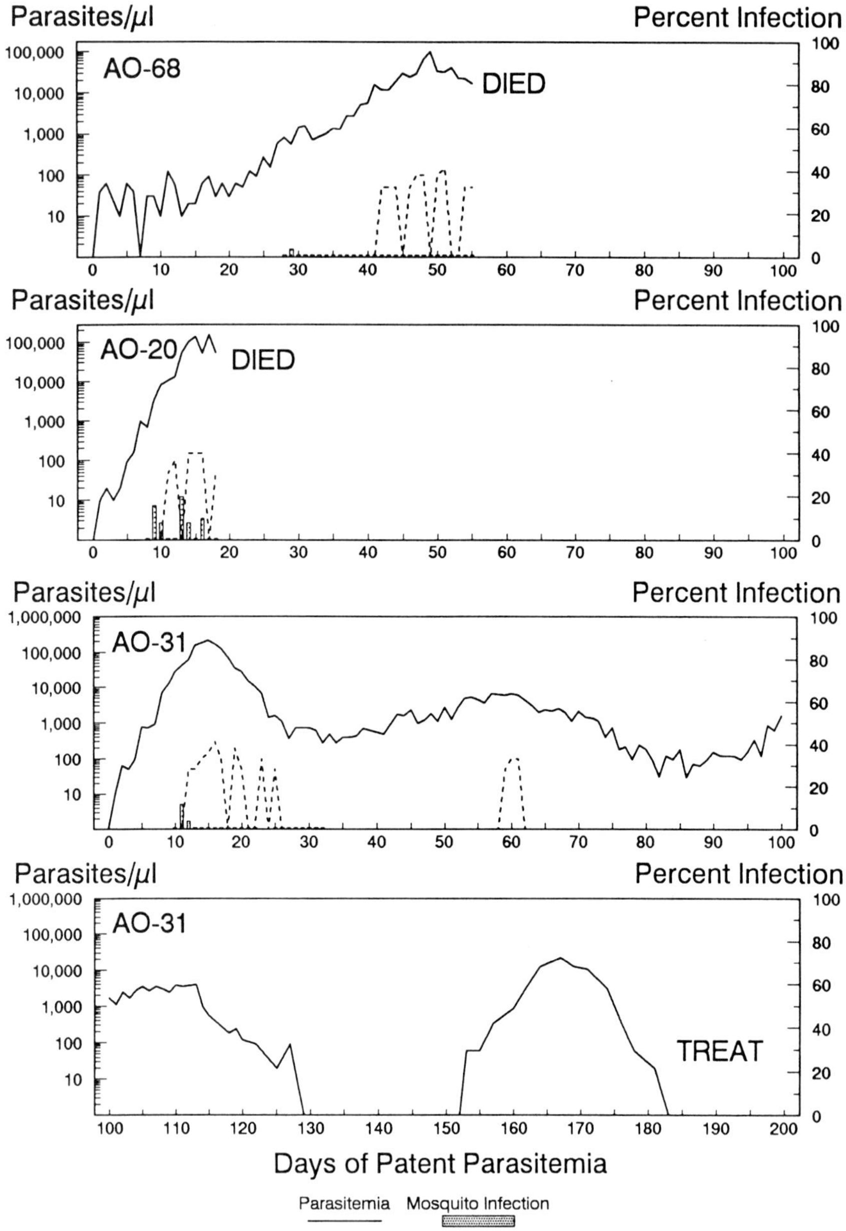

FIGURE 12 Parasitemia in three splenectomized *Aotus lemurinus griseimembra* monkeys inoculated with the OS strain of *Plasmodium inui*.

(Porter and Young, 1966; Porter, 1971), North Korean (Collins *et al.*, 1985c), Salvador II (Collins *et al.*, 1973a,b, 1974a, 1979, 1985a), and Salvador I (Collins *et al.*, 1985a, 1987a, 1988a). Major *P. falciparum* strains passaged to *Aotus* are Santa Lucia (Collins *et al.*, 1977, 1979, 1988a), Uganda Palo Alto (Schmidt, 1973, 1978a,b,c; Geiman and Meagher, 1967), Vietnam Oak Knoll (Schmidt, 1973, 1978a,b,c; Siddiqui *et al.*, 1972; Ruebush *et al.*, 1990), and Malayan Camp (DeGowin and Powell, 1965; Wellde *et al.*, 1979; Schmidt, 1973, 1978a,b,c). Only two strains of *P. malariae*, the Nigerian (Collins *et al.*, 1975b) and the Uganda I/CDC (Collins *et al.*, 1984, 1989), have been studied extensively.

Here, presentation of data on the course of parasitemia will be restricted to the Salvador II strain of *P. vivax*, the Santa Lucia strain fo *P. falciparum*, and the Uganda I/CDC strain of *P. malariae* because *An. freeborni* mosquito infection data are available for these strains.

I. Plasmodium vivax The Salvador II strain of *P. vivax* was isolated from a natural human infection in La Paz, El Salvador. Parasitized blood was inoculated into an *A. lemurinus griseimembra* monkey that was then shipped to the United States. The strain has been maintained by serial or sporozoite passage for many years.

When the splenectomized *A. lemurinus griseimembra* monkey (AO-21) was inoculated with sporozoites via the bites of infected *An. freeborni* mosquitoes, the prepatent period was 15 days (Fig. 13). The maximum parasite count was 21,200/μl on the 15th day of patent parasitemia. *Anopheles freeborni* mosquitoes were readily infected by feedings during the primary episode of parasitemia and on each of the three succeeding peaks in the parasite count. Parasites spontaneously disappeared 149 days after inoculation, and the monkey was eventually treated with 10 mg chloroquine (base) daily for 3 days, and 2.5 mg primaquine (base) daily for 7 days.

When an intact *A. lemurinus griseimembra* monkey (AO-83) was inoculated with trophozoites of the Salvador II strain, the peak parasite count during the primary episode was 5580/μl (Fig. 14). The monkey was splenectomized 20 days after inoculation. This resulted in only a minimal increase in parasite count; however, increased mosquito infection was obtained during subsequent recrudescences of the infection. The parasitemia declined to 62 parasites/μl on Day 100; the infection was treated with chloroquine on Day 104.

Transmission of the Salvador II strain of *P. vivax* to splenectomized *Aotus* monkeys via sporozoites was not predictable. Of 105 attempts with sporozoites produced by *An. freeborni, An. dirus, An. maculatus,* and *An. albimanus* mosquitoes, only 52 (49%) were successful. Prepatent periods ranged from 14 to 55 days, with a mean of 19.2 days. Mosquito infection was predictable only in splenectomized animals and more often occurred during the ascending or early rise in the parasitemia than during the descending part of a recrudescent episode. Recrudescence was common, but

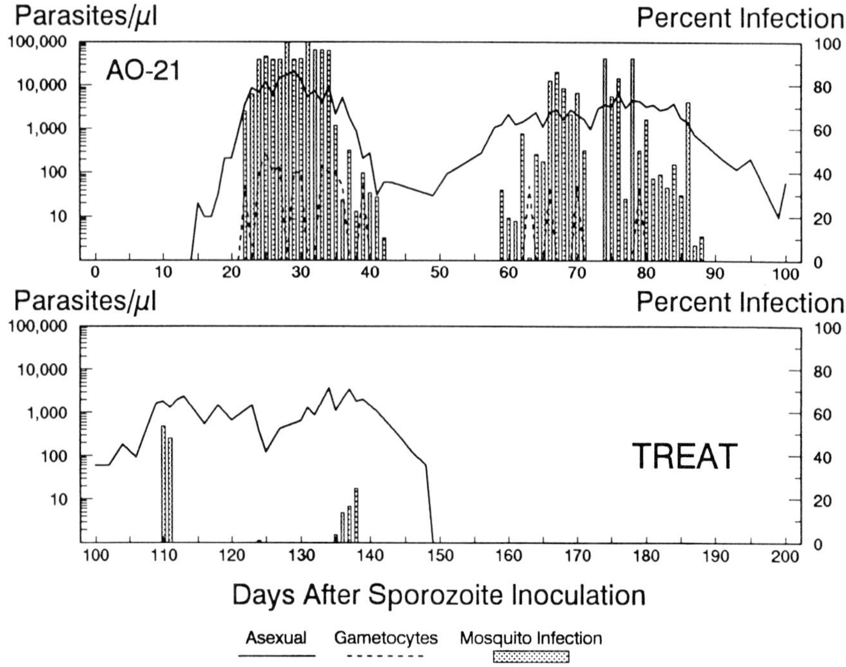

FIGURE 13 Prepatent period, course of parasitemia, and percentage mosquito infection of a splenectomized *Aotus lemurinus griseimembra* monkey infected with sporozoites of the Salvador II strain of *Plasmodium vivax* via the bites of infected *Anopheles freeborni* mosquitoes.

relapse from the fixed tissue of sporozoite-induced infections was infrequent following treatment with chloroquine.

2. Plasmodium falciparum Since infection of *Aotus* monkeys with *P. falciparum* was first reported by Geiman and Meagher in 1967, many strains have been reported to be infectious to different species of *Aotus*. These monkeys have become the standard animal used for developing and testing antimalarial drugs (Schmidt 1978a,b,c). Many strains have been adapted to develop in these monkeys by linear passage of the trophozoite stages. Production of infective gametocytes usually disappears after as few as 10 such passages. Thus, to maintain the capacity for mosquito infection, periodic transfer via sporozoites is essential.

Infection of splenectomized *A. lemurinus griseimembra* monkey AO-19 with the Santa Lucia strain of *P. falciparum* via the bites of infected *An. freeborni* mosquitoes resulted in a prepatent period much longer than would be expected in humans (Fig. 15). However, the parasite count rose rapidly to a peak of 1,920,000/μl on the 40th day of patent parasitemia. Infection

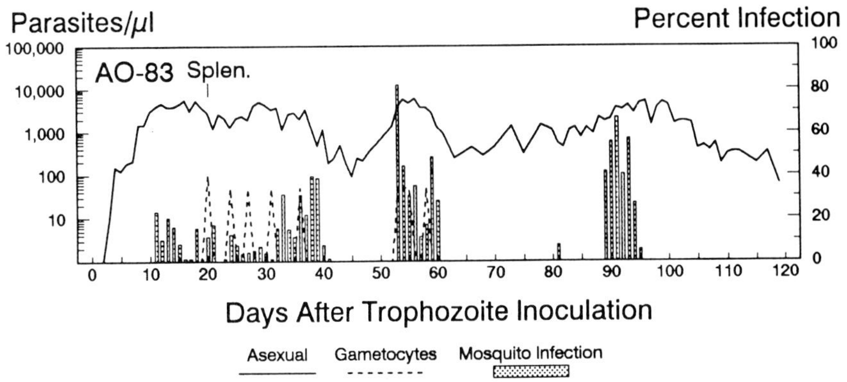

FIGURE 14 Course of parasitemia and percentage mosquito infection of a splenectomized *Aotus lemurinus griseimembra* monkey infected with trophozoites of the Salvador II strain of *Plasmodium vivax*. Splenectomy was performed 20 days after inoculation.

of *An. freeborni* mosquitoes was initially low; however, a high percentage of mosquito infection was later obtained for approximately 6 weeks. Following a recrudescence of the infection, high-density mosquito infection was again obtained between Days 107 and 127. Parasitemia gradually declined, and the infection was cured by treatment with chloroquine.

Transmission has also been successful to intact *A. lemurinus griseimembra* monkeys. In monkey AO-26 (Fig. 16), the prepatent period was 21 days. A maximum parasite count of 63,800/μl occurred on the ninth day of patent parasitemia. Secondary peak parasite counts of 276,000 and 113,800/μl occurred on Days 34 and 64, respectively, of patent parasitemia. Even though this monkey was not splenectomized, mosquito infection was obtained on 22 of the 34 days that mosquitoes were fed.

Transmission of the Santa Lucia strain of *P. falciparum* to splenectomized owl monkeys was not predictable. Of 95 attempts with sporozoites from *An. freeborni, An. maculatus, An. stephensi, An. dirus,* and *An. albimanus* mosquitoes, only 41 (43.2%) were successful. Prepatent periods ranged from 10 to 56 days, with a mean of 24.6 days. In general, sporozoite-induced infection resulted in longer periods of infectivity to mosquitoes than did trophozoite-induced infection. Although splenectomy was not a requirement for the production of infective gametocytes, splenectomized animals generally infected mosquitoes over a longer period. Linear trophozoite passage resulted in increased maximum parasite counts but reduced numbers of infective gametocytes. The Santa Lucia strain of *P. falciparum* is susceptible to treatment with chloroquine. With strains that are resistant, infections are cured with other drugs such as mefloquine (20-mg single dose) and quinine (50 mg daily for 5 or more days).

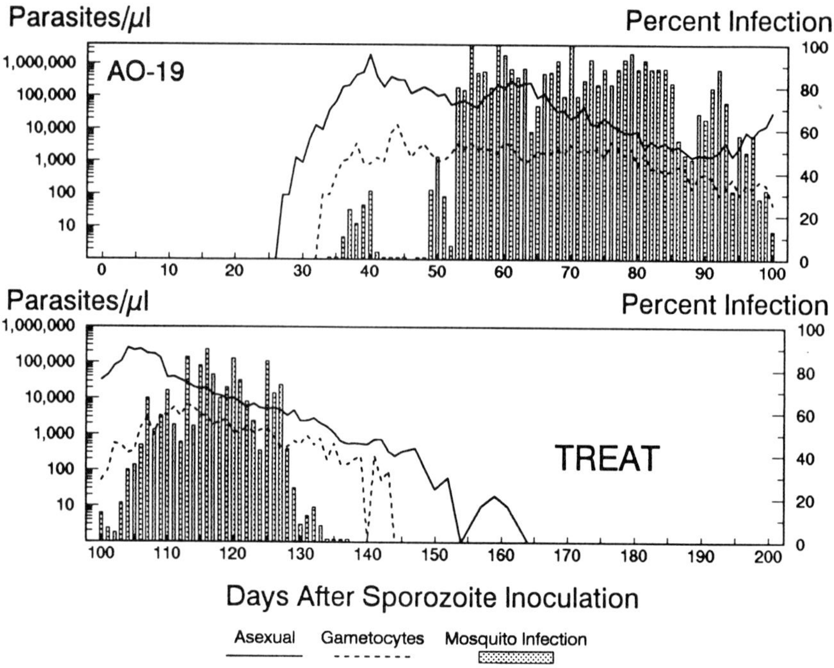

FIGURE 15 Prepatent period, course of parasitemia, and infectivity to mosquitoes of a splenectomized *Aotus lemurinus griseimembra* monkey infected with the Santa Lucia strain of *Plasmodium falciparum* via the bites of infected *Anopheles freeborni* mosquitoes.

3. Plasmodium malariae Only a few strains of *P. malariae* have been adapted to develop in *Aotus* monkeys. Following passage of infected human or chimpanzee blood to a monkey, the prepatent period can be from 2 to 4 months. However, subsequent linear passage results in very short prepatent periods, suggesting that this initial extended prepatent period is required for this minor population of parasites to adapt to these hosts. No successful transmissions have been obtained via sporozoite inoculation.

Blood from a patient infected with the Uganda I/CDC strain of *P. malariae* (Collins *et al.*, 1984) was passaged to a splenectomized chimpanzee. After the infection became apparent, parasitized blood was passaged to splenectomized *A. azarae boliviensis* monkey AO-206 (Fig. 17). After a prepatent period of 63 days, parasitemia developed to a maximum parasite count of 2110/µl on the 44th day of patent parasitemia; parasitemia persisted until Day 148. After a subpatent period of 45 days, parasites were again detectable. Low-density parasite counts were maintained for the next 100 days, after which the animal was treated with chloroquine (10 mg base daily for 3 days). Infection of *An. freeborni* mosquitoes was obtained on only 3 of the 20 days they were fed.

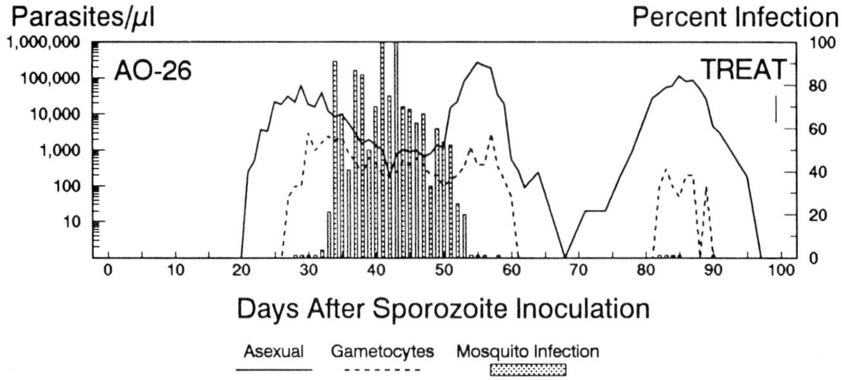

FIGURE 16 Prepatent period, course of parasitemia, and infectivity to mosquitoes of an intact *Aotus lemurinus griseimembra* monkey infected with the Santa Lucia strain of *Plasmodium falciparum* via the bites of infected *Anopheles freeborni* mosquitoes.

Blood from monkey AO-206 was then passaged to splenectomized *A. azarae boliviensis* monkey AO-207 (Fig. 18). The prepatent period was now only 4 days, and the parasite count rose rapidly to 19,300/μl at Day 26 after inoculation. The animal died unexpectedly on Day 53. Mosquito infection was obtained on 9 of 34 days; 8 of the infection days were prior to the peak parasite count. Infections with *P. malariae* have also been induced in *A. lemurinus griseimembra* monkeys (Collins *et al.*, 1989).

IV. *AOTUS* MONKEYS AS MODELS FOR MALARIA VACCINE TRIALS

Different species of *Aotus* have been used in vaccine trials to test the immunogenicity and efficacy of various antigens. Different combinations of species and strains of *Plasmodium* combined with different species and subspecies of *Aotus* have been studied to select appropriate models for vaccine trials. Almost all the effort has been directed to the testing of vaccines for *P. falciparum*, although similar studies could be done with *P. vivax* and *P. malariae*. To date, *Plasmodium ovale* has not been adapted to development in *Aotus* monkeys.

A. Sporozoite Vaccines

Aotus lemurinus griseimembra and *A. vociferans* are susceptible to sporozoite infection at a level that may be only marginally suitable for vaccine trials. Since only 50 to 80% of splenectomized monkeys develop infection

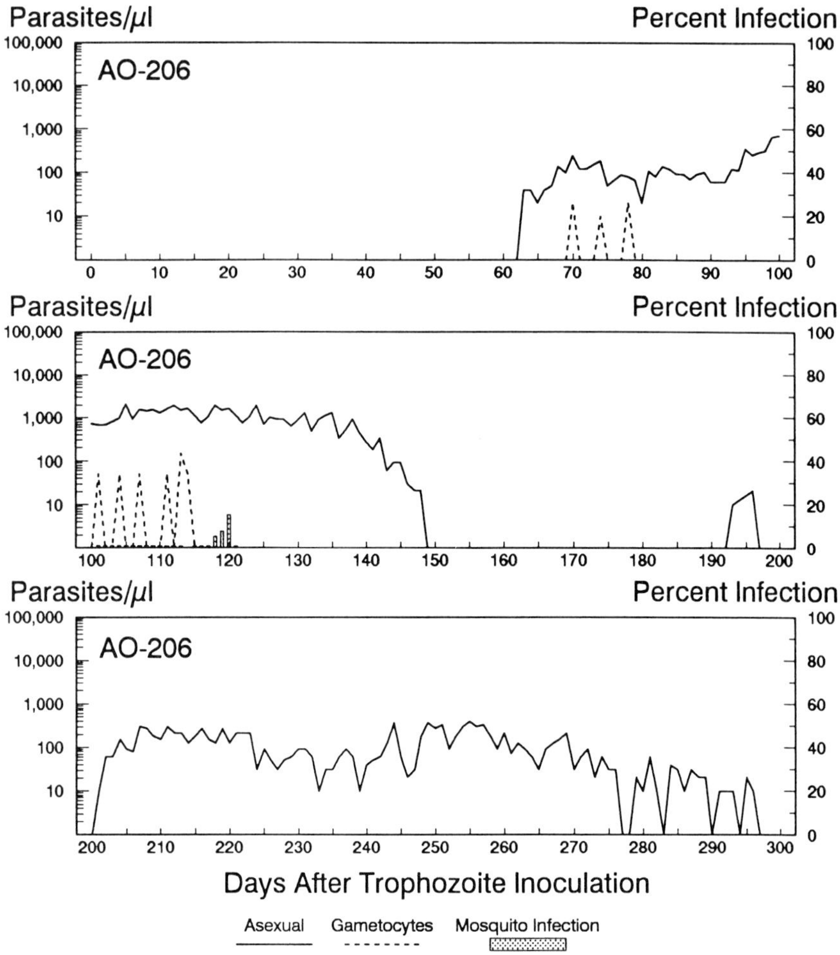

FIGURE 17 Prepatent period and course of parasitemia in a splenectomized *Aotus azarae boliviensis* monkey inoculated with the Uganda I/CDC strain of *Plasmodium malariae*.

following the inoculation of sporozoites either by the bites of 10 or more heavily infected mosquitoes or by the intravenous inoculation of sporozoites dissected from the salivary glands, large numbers of animals are needed to detect significant differences in protection. If the vaccine is only partially protective, 14 to 15 animals are required for each test group. Vaccines directed against the sporozoites of *P. vivax* can best be tested in *Saimiri* spp. since these animals are much more susceptible to sporozoite infection (Collins *et al.*, 1988c).

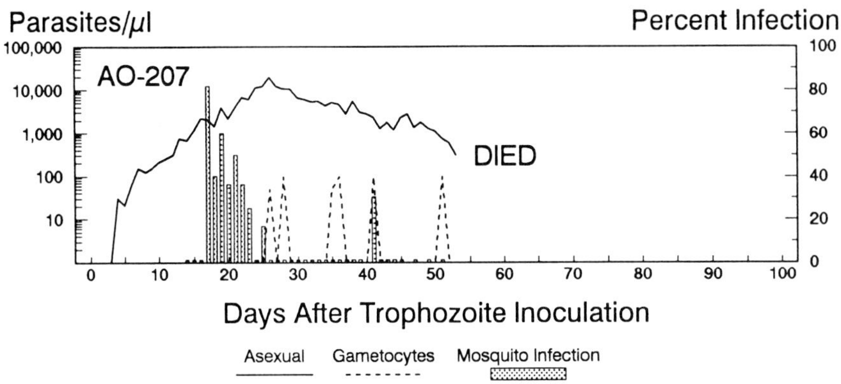

FIGURE 18 Prepatent period and course of parasitemia in a splenectomized *Aotus azarae boliviensis* monkey infected with the Uganda I/CDC strain of *Plasmodium malariae*.

B. Blood Stage Vaccines

Aotus lemurinus griseimembra, A azarae boliviensis, A. nancymae, and *A. vociferans* have been used in trials of vaccines directed against the blood stage of different strains of *P. falciparum* (Siddiqui, 1977; Siddiqui *et al.*, 1978, 1981, 1986; Reese *et al.*, 1978; Voller and Richards, 1968; Mitchell *et al.*, 1977; Collins *et al.*, 1988d; Ruebush *et al.*, 1990; Patarroyo *et al.*, 1987). Since the animals are highly susceptible to infection, both intact and splenectomized monkeys can be used. *Aotus* monkeys are therefore the nonhuman primates most suitable for testing the efficacy of *P. falciparum* blood stage vaccines. Vaccines directed against the blood stages of *P. vivax* could also be tested in any of the available species of *Aotus*.

C. Transmission-Blocking Vaccines

Because only *Aotus* monkeys that support the development of infective gametocytes are appropriate for testing the efficacy of transmission-blocking vaccines, splenectomized *A. lemurinus griseimembra* and *A. vociferans* are suitable for such studies with certain strains of *P. falciparum* (Kaslow *et al.*, 1993). These animals, as well as *A. azarae boliviensis* monkeys, are also suitable for the testing of *P. vivax* transmission-blocking vaccines.

V. SUMMARY

Aotus monkeys may be naturally infected with certain species of *Plasmodium*. Thus, *A. vociferans* may harbor *P. brasilianum*, a parasite of many species of monkeys from South America. Other monkeys may be experimentally infected with *P. simium, P. cynomolgi, P. fragile, P. knowlesi,* or *P. inui.* More importantly, splenectomized *Aotus* monkeys are susceptible to the human malaria parasites, *P. falciparum, P. malariae,* or *P. vivax.* Surprisingly, certain Old World monkey malaria parasites (*P. coatneyi, P. fieldi, P. simiovale,* and *P. gonderi*) cannot be adapted to *Aotus* monkeys. Different species of *Aotus* vary in their suitability as models for human malarias, but once an isolate has been successfully adapted to one species of *Aotus,* it can predictably be propagated and maintained in other species. A very important prerequisite in most cases is splenectomy. *Aotus lemurinus griseimembra, A. azarae boliviensis, A. nancymae,* and *A. vociferans* are highly susceptible to infection whether splenectomized or spleen-intact. By virtue of their small size and wide range of susceptibility to different species and developmental stages of human malaria parasites, these animals are attractive models for chemotherapeutic, biologic, and immunologic studies on sexual and asexual stages of human malaria parasites. Work at the CDC over the last three decades has unequivocally confirmed these data.

REFERENCES

Coatney, G. R., Collins, W. E., Warren, McW., and Contacos, P. G. (1971). "The Primate Malarias," U.S. Gov. Printing Office, Washington, DC.

Collins, W. E., Contacos, P. G., Stanfill, P. S., and Richardson, B. B. (1973a). Studies on human malaria in *Aotus* monkeys. I. Sporozoite transmission of *Plasmodium vivax* from El Salvador. *J. Parasitol.* **59,** 606–608.

Collins, W. E., Contacos, P. G., Jumper, J. R., Smith, C. S., and Skinner, J. C. (1973b). Studies on human malaria in *Aotus* monkeys. III. Exoerythrocytic stages of the Salvador II strain of *Plasmodium vivax. J. Parasitol.* **59,** 859–866.

Collins, W. E., Skinner, J. C., Richardson, B. B., Stanfill, P. S., and Contacos, P. G. (1974a). Studies of human malaria in *Aotus* monkeys. V. Blood-induced infections of *Plasmodium vivax. J. Parasitol* **60,** 393–398.

Collins, W. E., Stanfill, P. S., Richardson, B. B., and Smith, C. S. (1974b). Transmission of *Plasmodium fragile* to *Aotus trivirgatus* monkeys. *J. Parasitol.* **60,** 719–720.

Collins, W. E., Skinner, J. C., Richardson, B. B., and Stanfill, P. S. (1975a). Transmission of *Plasmodium cynomolgi* to *Aotus trivirgatus* monkeys. *J. Parasitol.* **61,** 146–148.

Collins, W. E., Contacos, P. G., and Richardson, B. B. (1975b). Studies on human malaria in *Aotus* monkeys. VI. Infectivity of *Plasmodium malariae* to different anophelines. *J. Parasitol.* **61,** 941–943.

Collins, W. E., Warren, McW., Skinner, J. C., Chin, W., and Richardson, B. B. (1977). Studies on the Santa Lucia (El Salvador) strain of *Plasmodium falciparum* in *Aotus trivirgatus* monkeys. *J. Parasitol.* **63,** 52–56.

Collins, W. E., Warren, McW., Skinner, J. C., Richardson, B. B., and Kearse, T. S. (1979). Effect of sequential infection with *Plasmodium vivax* and *P. falciparum* in the *Aotus trivirgatus* monkey. *J. Parasitol.* **65,** 605–608.

Collins, W. E., Warren, McW., Contacos, P. G., Skinner, J. C., Richardson, B. B., and Kearse, T. S. (1980). The Chesson strain of *Plasmodium vivax* in *Aotus* monkeys and anopheline mosquitoes. *J. Parasitol.* **66,** 488–497.

Collins, W. E., Skinner, J. C., Richardson, B. B., and Huong, A. Y. (1981). Development of *Plasmodium inui* in the *Aotus trivirgatus* monkey. *J. Parasitol.* **67,** 222–225.

Collins, W. E., Schwartz, I. K., Skinner, J. C., and Broderson, J. R. (1984). Studies on the Uganda I/CDC strain of *Plasmodium malariae* in Bolivian *Aotus* monkeys and different anophelines. *J. Parasitol.* **70,** 677–681.

Collins, W. E., Skinner, J. C., Broderson, J. R., Mehaffey, P., and Sutton, B. B. (1985a). Infection of *Aotus azarae boliviensis* monkeys with different strains of *Plasmodium vivax*. *J. Parasitol.* **71,** 239–243.

Collins, W. E., Skinner, J. C., Huong, A. Y., Broderson, J. R., Sutton, B. B., and Mehaffey, P. (1985b). Studies of a newly isolated strain of *Plasmodium brasilianum* in *Aotus* and *Saimiri* monkeys and different anophelines. *J. Parasitol.* **71,** 767–770.

Collins, W. E., Skinner, J. C., and Mehaffey, P. (1985c). Infection of *Aotus azarae boliviensis* monkeys with the RO strain of *Plasmodium cynomolgi*. *J. Parasitol.* **71,** 848–849.

Collins, W. E., Skinner, J. C., Pappaioanou, M., Ma, N. S.-F., Broderson, J. R., Sutton, B. B., and Stanfill, P. S. (1987a). Infection of *Aotus vociferans* (karyotype V) monkeys with different strains of *Plasmodium vivax*. *J. Parasitol.* **73,** 536–540.

Collins, W. E., Skinner, J. C., Pappaioanou, M., Broderson, J. R., Ma, N. S.-F., Stanfill, P. S., and Filipski, V. (1987b). Transmission of *Plasmodium simium* to *Aotus nancymae, A. vociferans, A. azarae boliviensis,* and *Saimiri sciureus boliviensis* monkeys. *J. Parasitol.* **73,** 653–655.

Collins, W. E., Skinner, J. C., Pappaioanou, M, Broderson, J. R., Ma, N. S.-F., Filipski, V., Stanfill, P. S., and Rogers, L. (1988a). Infection of Peruvian *Aotus nancymae* monkeys with different strains of *Plasmodium falciparum, P. vivax,* and *P. malariae*. *J. Parasitol.* **74,** 392–398.

Collins, W. E., Skinner, J. C.., Filipski, V., Wilson, C., Broderson, J. R., and Stanfill, P. S. (1988b). Transmission of the OS strain of *Plasmodium inui* to *Saimiri sciureus boliviensis* and *Aotus azarae boliviensis* monkeys by *Anopheles dirus* mosquitoes. *J. Parasitol.* **74,** 502–503.

Collins, W. E., Skinner, J. C., Pappaioanou, M., Broderson, J. R., Filipski, V. K., McClure, H. M., Strobert, E., Sutton, B. B., Stanfill, P. S., and Huong, A. Y. (1988c). Sporozoite-induced infections of the Salvador I strain of *Plasmodium vivax* in *Saimiri sciureus boliviensis* monkeys. *J. Parasitol.* **74,** 582–585.

Collins, W. E., Pappaioanou, M., Anders, R. F., Campbell, G. H., Brown, G. V., Kemp, D. J., Broderson, J. R., Coppel, R. L., Skinner, J. C., Procell, P. M., Favaloro, J. M., Corcoran, L. M., Ma, N. S.-F., Mitchell, G. F., and Campbell, C. C. (1988d). Immunization trials with the ring-infected erythrocyte surface antigen of *Plasmodium falciparum* in owl monkeys (*Aotus vociferans*). *Am. J. Trop. Med. Hyg.* **38,** 268–282.

Collins, W. E., Skinner, J. C., Broderson, J. R., Pappaioanou, M., Filipski, V., Sutton, B. B., Stanfill, P. S., Huong, A. Y., Roberts, J., and Wilson, C. (1989). The Uganda I/CDC strain of *Plasmodium malariae* in *Aotus lemurinus griseimembra* monkeys. *J. Parasitol.* **75,** 61–65.

DeGowin, R. L., and Power, R. D. (1965). Drug resistance of a strain of *Plasmodium falciparum* from Malaya. *Am. J. Trop. Med. Hyg.* **14,** 519–528.

Earle, W. E., and Perez, M. (1932). Enumeration of parasites in the blood of malarial patients. *J. Lab. Clin. Med.* **17,** 1124–1130.

Eyles, D. E. (1963). The species of simian malaria: Taxonomy, morphology, life cycle, and geographical distribution of the monkey species. *J. Parasitol.* **49,** 866–887.

Geiman, Q. M., and Meagher, M. J. (1967). Susceptibility of a New World monkey to *Plasmodium falciparum* from man. *Nature* (*London*) **215,** 437–439.

Geiman, Q. M., and Siddiqui, W. A. (1969). Susceptibility of a New World monkey to *Plasmodium malariae* from man. *Am. J. Trop. Med. Hyg.* **18,** 351–354.

Kaslow, D. C., Bathurst, I. C., Keister, D. B., Campbell, G. H., Adams, S., Morris, K. L., Sullivan, J. S., Barr, P. J., and Collins, W. E. (1993). Safety, immunogenicity, and *in vitro* efficacy of a muramyl tripeptide-based malaria transmission-blocking vaccine in an *Aotus nancymae* monkey model. *Vaccine Res.* **2,** 95–103.

Mitchell, G. H., Richards, W. H. G., Butcher, G. A., and Cohen, S. (1977). Merozoite vaccination of douroucouli monkeys against falciparum malaria. *Lancet* **1,** 1335–1338.

Patarroyo, M. E., Romero, P., Torres, M. L., Clavijo, P., Moreno, A., Martinez, A., Rodriguez, R., Guzman, F., and Cabezas, E. (1987). Induction of protective immunity against experimental infection with malaria using synthetic peptides. *Nature (London)* **328,** 629–632.

Porter, J. A., Jr. (1971). *Plasmodium vivax* in *Aotus trivirgatus. Am. J. Trop. Med. Hyg.* **20,** 535–538.

Porter, J. A. Jr., and Young, M. D. (1966). Susceptibility of Panamanian primates to *Plasmodium vivax. Mil. Med.* **131,** 952–958.

Reese, R. T., Trager, W., Jensen, J. B., Miller, D. A., and Tantravhi, R. (1978). Immunization against malaria with antigen from *Plasmodium falciparum* cultivated *in vitro. Proc. Natl. Acad. Sci. U.S.A.* **75,** 5655–5668.

Ruebush, T. K., II, Campbell, G. H., Moreno, A., Patarroyo, M. E., and Collins, W. E. (1990). Immunization of owl monkeys with a combination of *Plasmodium falciparum* asexual blood-stage synthetic peptide antigens. *Am. J. Trop. Med. Hyg.* **43,** 355–366.

Schmidt, L. H. (1973). Infections with *Plasmodium falciparum* and *Plasmodium vivax* in the owl monkey—Model systems for basic biological and chemotherapeutic studies. *Trans. R. Soc. Trop. Med. Hyg.* **67,** 446–474.

Schmidt, L. H. (1978a). *Plasmodium falciparum* and *Plasmodium vivax* infections the owl monkey (*Aotus trivirgatus*). I. The course of untreated infections. *Am. J. Trop. Med. Hyg.* **27,** 671–702.

Schmidt, L. H. (1978b). *Plasmodium falciparum* and *Plasmodium vivax* infections the owl monkey (*Aotus trivirgatus*). II Responses to chloroquine, quinine, and pyrimethamine. *Am. J. Trop. Med. Hyg.* **27,** 703–717.

Schmidt, L. H. (1978c). *Plasmodium falciparum* and *Plasmodium vivax* infections the owl monkey (*Aotus trivirgatus*). III. Methods employed in the search for new blood schizonticidal drugs. *Am. J. Trop. Med. Hyg.* **27,** 718–737.

Siddiqui, W. A. (1977). An effective immunization of experimental monkeys against a human malaria parasite, *Plasmodium falciparum. Science* **197,** 388–389.

Siddiqui, W. A., Schnell, J. V., and Geiman, Q. M. (1972). A model *in vitro* system to test the susceptibility of human malarial parasites to antimalarial drugs. *Am. J. Trop. Med. Hyg.* **21,** 392–399.

Siddiqui, W. A., Schnell, J. V., and Richmond-Crum, S. M. (1974). Susceptibility of a New World monkey (*Aotus trivirgatus*) to an Old World simian malarial parasite (*Plasmodium knowlesi*). *Trans. R. Soc. Trop. Med. Hyg.* **68,** 387–391.

Siddiqui, W. A., Taylor, D. W., Kan, S. C., Kramer, K., and Richmond-Crum, S. M. (1978). Vaccination of experimental monkeys against *Plasmodium falciparum*: A possible safe adjuvant. *Science* **201,** 1237–1239.

Siddiqui, W. A., Kan, S., Kramer, K., Case, S., and Palmer, K. (1981). Use of a synthetic adjuvant in an effective vaccination of monkeys against malaria. *Nature (London)* **289,** 64–66.

Siddiqui, W. A., Tam, L. Q., Kan, S., Kramer, K. J., Case, S. E., Palmer, K. L., and Hui, G. S. N. (1986). Induction of protective immunity to monoclonal-antibody-defined *Plasmodium falciparum* antigens requires strong adjuvant in *Aotus* monkeys. *Infect. Immun.* **52,** 314–318.

Voller, A., and Richards, W. H. G. (1968). An attempt to vaccinate owl monkeys (*Aotus trivirgatus*) against falciparum malaria. *Lancet* **2,** 1172–1174.

Ward, R. A., Rutledge, L. C., and Hickman, R. L. (1969). Cyclical transmission of Chesson vivax malaria in subhuman primates. *Nature (London)* **224,** 1126–1127.

Wellde, B. T., Diggs, C. L., and Anderson, S. (1979). Immunization of *Aotus trivirgatus* against *Plasmodium falciparum* with irradiated blood forms. *Bull. W. H. O.* **57,** 153–157.

Young, M. D., Porter, J. A., Jr., and Johnson, C. M. (1966). *Plasmodium vivax* transmitted from man to monkey to man. *Science* **153,** 1006–1007.

9

The Owl Monkey in Oncogenic Virus Research

Norval W. King
New England Regional Primate Research Center
One Pine Hill Drive
Southborough, Massachusetts 01772

I. INTRODUCTION

During the past two decades, the owl monkey (*Aotus* spp.) has been a useful animal model for studies of the mechanisms by which several DNA-containing viruses produce malignant neoplasms in nonhuman primates and other mammalian species. These studies of viral oncogenesis in owl monkeys have focused primarily on certain lymphotropic herpesviruses, that is, the gammaherpesviruses and selected polyomaviruses.

II. ONCOGENIC GAMMAHERPESVIRUSES

A. Herpesvirus saimiri

Interest in the owl monkey for studies of herpesvirus-induced neoplasia was initially spawned by the reports of Melendez *et al.* (1969a, 1969b) and Hunt *et al.* (1970), who demonstrated the occurrence of rapidly fatal leukemia and/or lymphoma in owl monkeys and cotton-top tamarins (*Saguinus oedipus*) experimentally inoculated with *Herpesvirus saimiri,* which is an indigenous virus of squirrel monkeys (*Saimiri sciureus*) that causes no apparent disease in this species either naturally or experimentally (Melendez *et al.*, 1969a; Falk *et al.*, 1973). These studies were the first to comfirm that a mammalian herpesvirus could induce lymphoma in a primate species, a fact that had been strongly suspected from the earlier work of Epstein *et al.* (1964), who isolated a lymphotropic herpesvirus of human origin, that is, Epstein–Barr virus (EBV), from African children with Burkitt's lymphoma.

Later, *H. saimiri* was also shown to cause lymphoma in a wide variety of other New World monkeys, including common marmosets (*Callithrix jacchus*) (Wright *et al.*, 1977), white-lipped (*Saguinus labiatus*), saddle-backed (*Saguinus fuscicollis*), and red and black tamarins (*Saguinus nigricollis*) (Wolfe *et al.*, 1971), cebus monkeys (*Cebus albifrons*) (Melendez *et al.*, 1970), spider monkeys (*Ateles geoffroyi*) (Hunt *et al.*, 1972a), and New Zealand white rabbits (*Oryctolagus* sp.) (Daniel *et al.*, 1974; Hunt *et al.*, 1975). Initially, the cotton-top tamarin was the species of choice for studies with *H. saimiri* because of its smaller size and the extremely short interval between inoculation and the development of leukemia and lymphoma in this species. When the cotton-top tamarin was declared an endangered species in 1976, the owl monkey emerged as the best alternative primate model for studies with *H. saimiri* and related herpesviruses. In addition to causing lymphoma and leukemia in owl monkeys under experimental conditions, *H. saimiri* was subsequently also isolated from owl monkeys that died of spontaneously occurring lymphoma (Hunt *et al.*, 1973; Rabin *et al.*, 1975). In these cases it was assumed that the owl monkeys had acquired their *H. saimiri* infection from squirrel monkeys, although this was not confirmed. Horizontal transmission of *H. saimiri* from infected squirrel monkeys to uninfected owl monkeys was subsequently demonstrated experimentally by simply housing the two species in the same cage separated only by a fine-mesh screen (Barahona *et al.*, 1975). Transmission of the virus under these conditions was thought to have occurred via oropharyngeal secretions, which were shown to contain large quantities of virus. Other studies with *H. saimiri* have focused on defining the molecular basis of malignant transformation of lymphoid cells by this virus. For these studies, peripheral blood lymphocytes from common marmosets (*Callithrix jacchus*), cotton-top tamarins (*Saguinus oedipus*), and human beings, all of which can be immortalized *in vitro* by *H. saimiri,* have largely replaced the owl monkey for assessment of the transforming ability of *H. saimiri*-encoded oncoproteins (Desrosiers *et al.*, 1985, 1986; Biesinger *et al.*, 1992).

I. Clinical Disease The clinical course of disease in owl monkeys inoculated with *H. saimiri* can vary considerably but generally takes one of three patterns. Some animals die within 40 days of a widely disseminated lymphoma of a reticulum cell type that often is associated with extensive necrosis and replacement of vital organs (Hunt *et al.*, 1976). Animals that survive 50–150 days generally have a less aggressive form of lymphocytic lymphoma with multiple organ involvement and an associated lymphocytic leukemia. Those animals that survive for longer periods usually have a localized form of well-differentiated lymphocytic lymphoma often without leukemia. Clinically, affected animals present with generalized lymphadenopathy, splenomegaly, and a variable degree of hepatomegaly that may be preceded or followed by lymphocytic leukemia (Cicmanec *et al.*, 1974). Peripheral leukocyte counts may exceed 100,000 cells/mm^3, of which 90% are lymphocytes. A significant number of the circulating lymphoid cells are atypical lymphocytes or prolymphocytes (Fig. 1). Smears of bone marrow aspirates also contain increased numbers of lymphocytes and roughly a normal complement of erythropoietic and granulopoietic elements. Terminally, affected animals become lethargic and anorectic, lose weight, and may be anemic. Neurologic signs may be seen in animals with brain involvement.

2. Pathologic Features

a. Gross Findings Animals dying of *H. saimiri*-induced lymphoma have moderate to marked enlargement of lymph nodes, spleen, and occasionally the thymus due to focal, diffuse, or nodular infiltration of gray-white, homogeneous tissue that displaces the normal parenchyma. There may be focal hemorrhages scattered throughout the neoplastic infiltrates as well. Similar infiltrates may be present in the liver, kidneys, adrenals, lungs, and Peyer's patches of the intestine and impart a mottled appearance to the natural and cut surfaces of these organs. In some animals, the neoplastic infiltrates occur as large solid masses that invade and distort the normal tissues or organs (Fig. 2).

b. Microscopic Findings The cells comprising the neoplastic infiltrates vary considerably in their degree of differentiation in individual owl monkeys and in other species. In the most aggressive form of *H. saimiri*-induced lymphoma, the neoplastic cells are large and pleomorphic and resemble reticulum cells or histiocytes. Many are polygonal to stellate and have small to moderate amounts of granular eosinophilic cytoplasm and a single, large, round to ovoid nucleus with leptochromatic chromatin and one to two prominent nucleoli (Fig. 3). Other cells within these infiltrates are smaller, round to ovoid, with small amounts of cytoplasm and pleomorphic nuclei that vary from round to indented to angular in shape. Mitotic activity is high. This aggressive reticulum cell or histiocytic form of lymphoma may have extensive necrosis of tumor cells. In animals with a more protracted clinical course, the neoplastic infiltrates are often composed of more differentiated cells resembling lymphocytes. These form solid sheets or masses

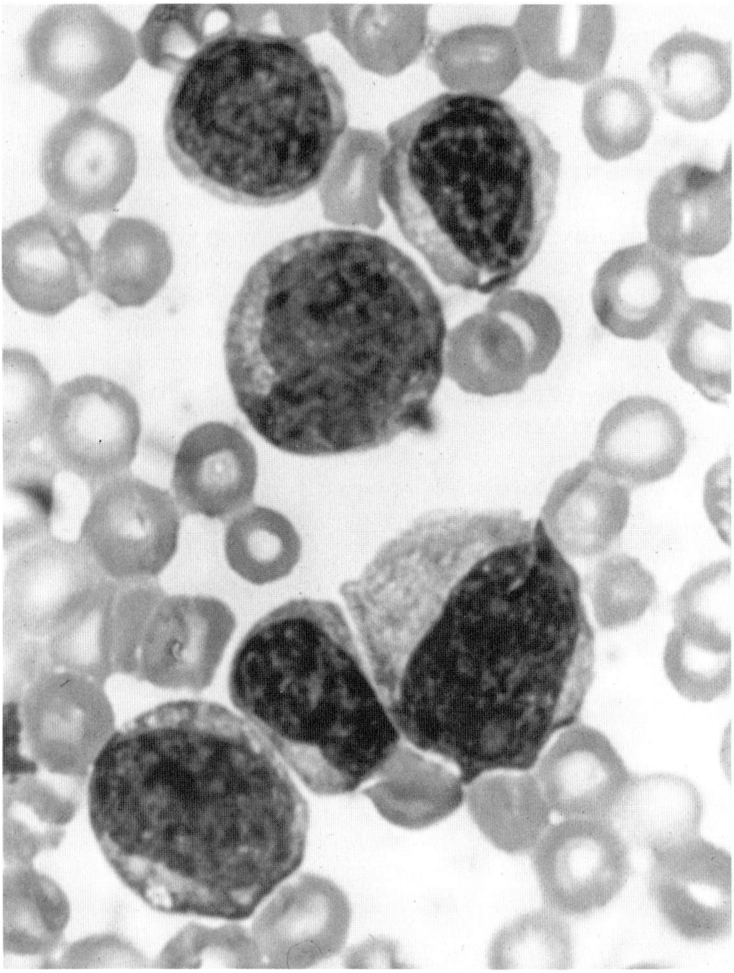

FIGURE 1 Photomicrograph of a portion of a peripheral blood smear of an owl monkey with *Herpesvirus saimiri*-induced leukemia. Note the large atypical lymphocytes and prolymphocytes. ×2000, Wright–Giemsa stain.

that invade and replace the parenchyma of multiple organs or occasionally only affect a single organ. This lymphocytic form of *H. saimiri*-induced lymphoma may be associated with leukemia that is reflected by the increased number of mononuclear cells in the vasculature of affected organs. The neoplastic cells comprising the tumor infiltrates and leukemia are polyclonal and of T lymphocyte origin (Wright *et al.*, 1976; Wallen *et al.*, 1973, 1974; Fleckenstein and Desrosiers, 1982). Ultrastructurally, the neoplastic cells have large bizarre-shaped nuclei and contain no demonstrable herpes-

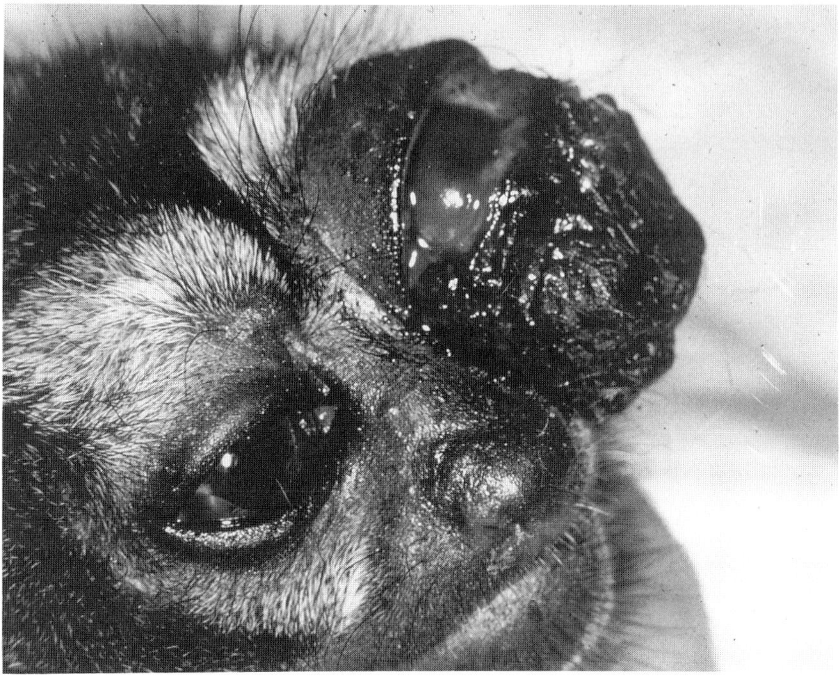

FIGURE 2 An owl monkey with a large *H. saimiri*-induced lymphoma in the left retro-orbital space causing compression and displacement of the eye.

virus virions (Fig. 4). After being maintained in cell culture for 7–10 days, a small percentage of the neoplastic cells will exhibit replicating viral particles in their nuclei.

The distribution of the neoplastic infiltrates within different organs is somewhat consistent. In lymph nodes, spleen, and thymus the neoplastic infiltrates are generally diffuse and replace most or all of the normal parenchyma and may infiltrate the surrounding adipose tissue. In the liver, the infiltrates are consistently found in the portal tracts but may extend into the hepatic lobules (Figs. 5 and 6). In the kidney they are usually confined to the cortex, whereas in the adrenals the corticomedullary junction is most often involved (Fig. 7). Pulmonary infiltrates usually surround pulmonary arteries and in the intestine the infiltrates arise within lymphoid follicles.

3. Etiology and Pathogenesis

Herpesvirus saimiri is a lymphotropic virus classified in the subfamily Gammaherpesvirinae of the family Herpesviridae. It occurs as an extremely common infection of squirrel monkeys (*Saimiri sciureus*), where it persists

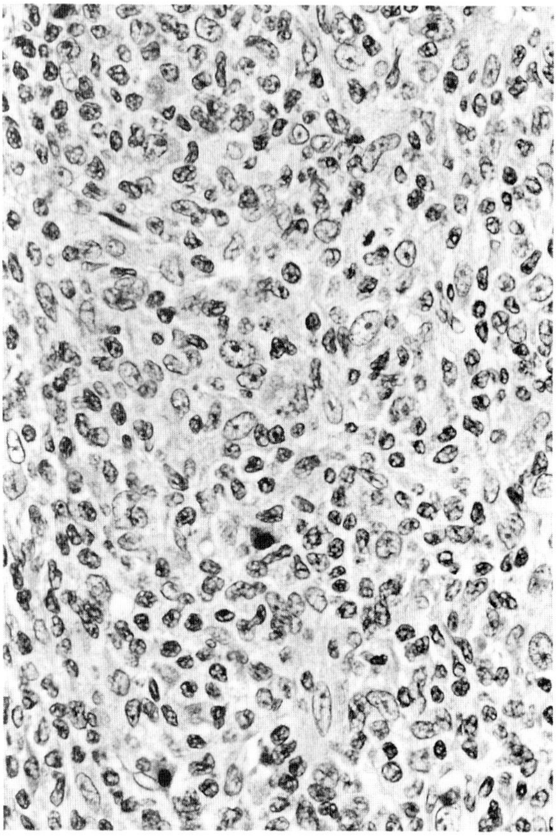

FIGURE 3 A portion of a lymphomatous infiltrate in which there is a mixed population of neoplastic reticulum cells and lymphoblasts. ×400, H & E.

in T lymphocytes without causing apparent disease (Falk *et al.*, 1973). The squirrel monkey is regarded as the natural or reservoir host for this virus. Strains of *H. saimiri* have been divided into three different subgroups (A, B, and C) based on the extent of sequence divergence at the left end of L-DNA (Medveczky *et al.*, 1984). Strains belonging to subgroups A and C are highly oncogenic *in vivo* and will immortalize peripheral blood lymphocytes of common marmosets (*Callithrix jacchus*) *in vitro*, whereas subgroup B will not (Desrosiers *et al.*, 1986). These three subgroups also differ in their oncogenic potential in rabbits (Medveczky *et al.*, 1989). The left terminus of L-DNA of strains in subgroup A and C contains sequences that encode proteins responsible for *in vitro* transformation, that is, transforming pro-

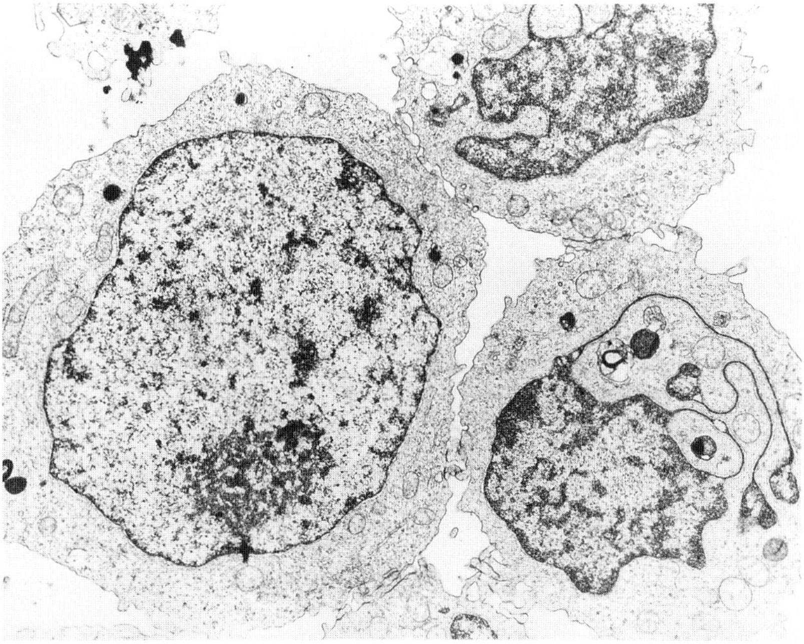

FIGURE 4 Electron micrograph of portions of three neoplastic lymphoid cells. Note the bizarre nuclear protrusions in some of the cells and the absence of herpesvirus virions in the tumor cells. ×4000.

teins (Desrosiers *et al.*, 1986). In the case of *H. saimiri*, strain 11, this protein has been termed "saimiri transformation-associated protein" or STP-A11 (Jung *et al.*, 1991). It is a polypeptide with a highly hydrophobic stretch of amino acids at the carboxyl terminus, an acidic hydrophilic amino terminus, and a sequence appropriate for formation of a metal binding domain. *Herpesvirus saimiri*, strain 488, a subgroup C virus, has two open reading frames (ORF-1 and ORF-2) in a position and orientation in the viral genome similar to that of the STP-A11 gene. Except for the hydrophobic, putative transmembrane domain at the carboxyl terminus, limited sequence identity exists between STP-A11, ORF-1, and ORF-2. ORF-2 contains a stretch of 54 amino acids with collagenlike repeat motif, a feature seen only in scattered areas of STP-A11 protein (Jung *et al.*, 1991). ORF-2 of strain 488 has been shown to encode for a phosphoprotein with transforming and tumor-inducing activity and thus it has been renamed STP-C488 (Jung and Desrosiers, 1991, 1992). STP-A11 and STP-C488 are unique oncogenes that do not have sequence homology with other known viral oncogenes or cellular proto-oncogenes.

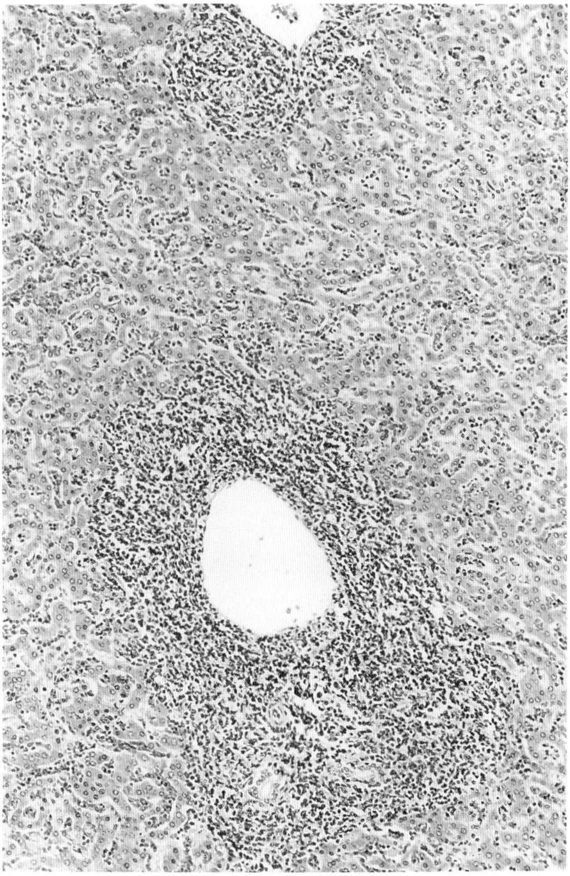

FIGURE 5 Photomicrograph of a portion of liver in an *H. saimiri*-infected owl monkey in which there are extensive infiltrates of neoplastic lymphocytes in the portal tracts and filling the sinusoids. ×125, H & E.

B. Herpesvirus ateles

The epidemiology and biological properties of *Herpesvirus ateles* have many elements in common with *H. saimiri*. *Herpesvirus ateles* was originally isolated from primary kidney cell cultures of a clinically healthy black-handed spider monkey (*Ateles geoffroyi*) (Melendez *et al.*, 1972a), and subsequently from the peripheral blood lymphocytes of healthy black spider monkeys (*Ateles paniscus*) (Falk *et al.*, 1974). Spider monkeys are commonly infected with *H. ateles* but do not develop clinical disease. For this reason, they are regarded as the natural host for this agent. Like *H. saimiri, H. ateles* also induces lymphoma that may be associated with leukemia when

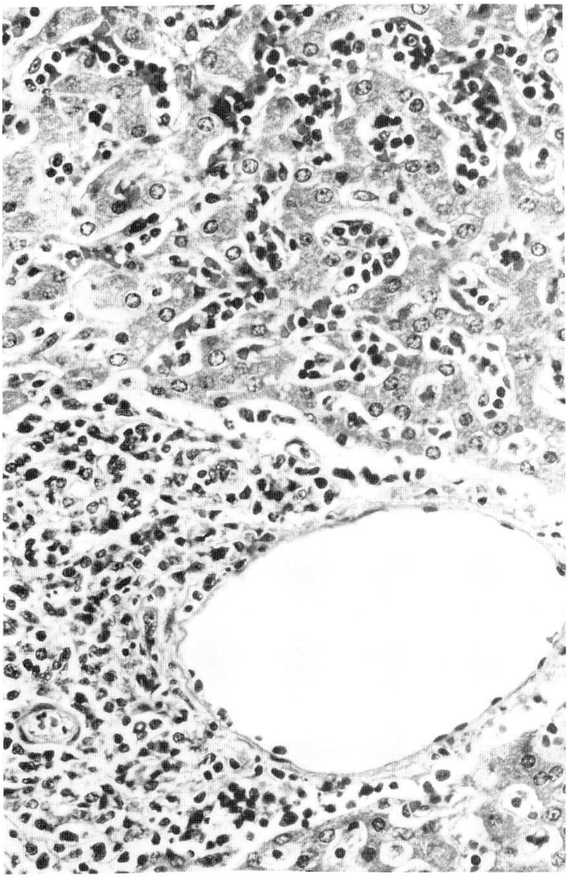

FIGURE 6 Higher magnification of liver showing neoplastic lymphocytes in the portal area and leukemic cells in the sinusoids. ×400, H & E.

inoculated into cotton-top tamarins (*Saguinus oedipus*), white-lipped tamarins (*S. labiatus*), saddle-backed tamarins (*S. fuscicollis*), black and red tamarins (*Saguinus nigricollis*), owl monkeys (*Aotus* spp.) (Melendez *et al.*, 1972a,b; Hunt *et al.*, 1972c, 1978), and the ACCRB strain of rabbits (Daniel *et al.*, 1978). New Zealand white rabbits are relatively resistant to *H. ateles*-induced lymphomas. *Herpesvirus ateles* was isolated several years after *H. saimiri* and because its biology appears to be similar to that of *H. saimiri*, it has not been as thoroughly studied.

I. Clinical Disease Owl monkeys experimentally infected with *H. ateles* develop lymphomas within 3–4 weeks. No noteworthy clinical signs were observed in affected animals. Interestingly, owl monkeys with *H. ateles*-

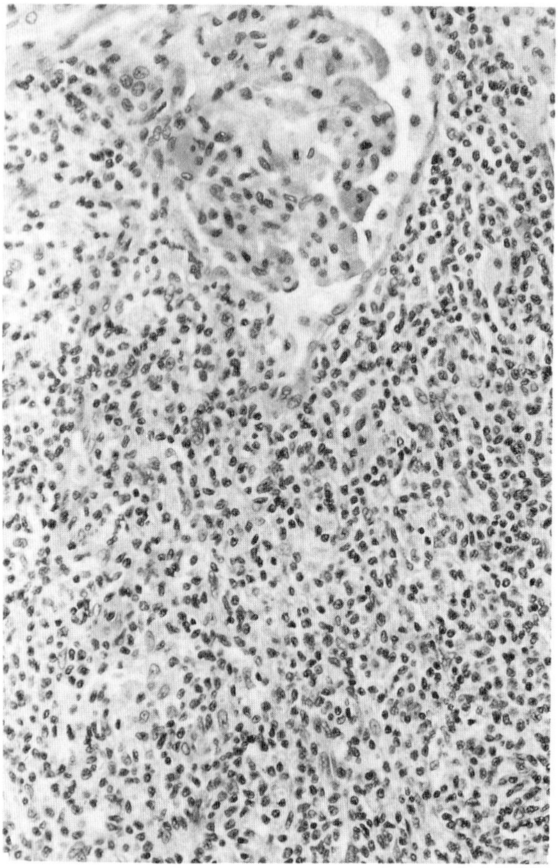

FIGURE 7 Photomicrograph of a portion of renal cortex of an owl monkey with *H. saimiri*-induced lymphoma. Sheets of neoplastic lymphocytes surround the glomerulus and replace the adjacent renal tubules. ×250, H & E.

induced lymphoma do not become leukemic and exhibit no consistent hematologic abnormalities.

2. Pathologic Features

a. Gross Findings Animals dying with *H. ateles*-induced lymphoma have diffuse enlargement of lymph nodes and spleen and nodular infiltrates in the kidneys. Other organs exhibit no obvious gross alterations.

b. Microscopic Findings Microscopically, the lymphomatous infiltrates are confined to lymph nodes, spleen, bone marrow, and kidneys. Other organs, including the liver, are not affected. The neoplastic cells are pleomorphic but this is not as pronounced as one sees in *H. saimiri*-induced lymphoma in this species. In general, the cells resemble lymphoblasts rather than reticulum cells or lymphocytes. They are round to ovoid and have a small amount of eosinophilic cytoplasm and round, leptochromatic nuclei. Immunologic studies have shown that these neoplastic cells are of T lymphocyte origin. Mitotic figures are frequent throughout the infiltrates. The lymph nodes and spleen are diffusely infiltrated by solid sheets of these neoplastic cells, which replace the normal follicular and sinus architecture. The bone marrow also contains diffuse or nodular infiltrates of neoplastic lymphoblasts. In the kidneys, the tumor cells form discrete nodular masses in the cortex and medulla in contrast to the diffuse infiltrates that one sees in *H. saimiri*-induced lymphoma in this species. These nodules appear to grow by expansion and compression of adjacent parenchyma rather than being diffusely infiltrative. There is no consistent pattern of distribution of the nodules within the kidneys of affected animals, and in some individuals much of the renal cortex may be spared. Other organs, including the liver, are not affected.

3. Etiology and Pathogenesis *Herpesvirus ateles*, like *H. saimiri*, is a member of the gammaherpesvirus subfamily of herpesviruses. Serological surveys indicate that it occurs commonly as a nonpathogenic infection of spider monkeys (*Ateles* spp.) in their natural habitats, the rain forests of South America. Like *H. saimiri*, *H. ateles* has a specific tropism for T lymphocytes of both its natural and experimentally susceptible hosts. Although the two viruses share a number of biological and molecular properties, they are genetically different agents (Fleckenstein *et al.*, 1978). Unlike *H. saimiri*, there are no published reports describing the genetic basis of T lymphocyte transformation by *H. ateles*, hence this remains to be determined.

C. Epstein–Barr Virus

Epstein–Barr virus (EBV) is a pathogenic, B-lymphotropic gammaherpesvirus of human origin. It is the cause of infectious mononucleosis in humans and has been strongly linked to two human neoplastic diseases that are endemic in certain parts of the world, that is, Burkitt's lymphoma in Africa and nasopharyngeal carcinoma in Southeast Asia, North and East Africa, the Mediterranean region, and in the Eskimo population in Alaska and Greenland (de-Thé, 1982). Epstein *et al.* (1964) first isolated EBV from cell cultures derived from Burkitt's lymphoma, a malignant B cell lymphoma that affects primarily children in Africa. It is clear in those areas where these two neoplastic diseases are endemic that other environmental factors besides EBV are involved in their pathogenesis because EBV occurs commonly in other parts of the world where these neoplasms are rare. In efforts to confirm the oncogenicity of EBV, a number of different species of nonhuman primates have been innoculated with it. Epstein *et al.* (1973)

described the occurrence of a fatal reticuloproliferative disease in one of three owl monkeys inoculated with an EBV-producing cell line of Burkitt's lymphoblasts. The authors were careful not to conclude that the lesion was a lymphoma. To this author's knowledge, this finding has not been reproduced. Miller *et al.* (1977) were the first to provide conclusive evidence that EBV is oncogenic for cotton-top tamarins (*Saguinus oedipus*), in which it produces B cell lymphomas. The following is a summary of the findings in the one *Aotus* with the experimentally induced EBV-associated reticuloproliferative disease.

1. Clinical Disease The one *Aotus* that died of reticuloproliferative disease survived 14 weeks after being inoculated intraperitoneally with EBV-containing human Burkitt's lymphoma cells. Cervical lymphadenopathy was noted during the terminal stages of disease.

2. Pathologic Features

a. Gross Findings The cervical lymph nodes were enlarged, measured 1.2 cm in diameter, and on cut-surface appeared whitish-gray with red flecks. The axillary, inguinal, pancreatic, and right renal nodes were also enlarged and the latter contained a cyst. The thymus and tracheobronchial and mesenteric lymph nodes were normal sized. The spleen was enlarged but a prominent follicular pattern was preserved.

b. Microscopic Findings The enlarged lymph nodes, thymus, kidney, liver, uterine cervix, adrenals, and bone marrow all contained focal or diffuse infiltrates of a mixed population of cells that included large and small lymphocytes, lymphoblasts, reticulum cells, plasma cells, and scattered eosinophils. Mitotic figures were numerous, particularly in the enlarged lymph nodes. The lymphoid follicles of the spleen were hyperplastic with prominent germinal centers. The question that has not been resolved in this one case is whether the changes noted were simply an immune response to the transplantation of human cells or whether it represented an unusual neoplastic disease.

3. Etiology and Pathogenesis As mentioned earlier, the basis for the unusual reticuloproliferative disease in the one owl monkey receiving EBV-containing transformed human B cells is not known. EBV has been shown to cause *in vitro* transformation of B lymphocytes of humans as well as a number of nonhuman primate species (Robinson and Miller, 1982).

III. POLYOMAVIRUSES

Polyomaviruses are a family of small DNA-containing viruses that contain members that infect human and many other mammalian species. A number of them have been associated with a wide variety of neoplasms (hence their name) under natural and experimental conditions. Humans are the natural host for two polyomaviruses referred to by the initials of the patients from whom they were originally isolated, that is, JC virus (JCV) and BK virus (BKV). Infection with JCV and BKV occurs commonly during childhood, is not associated with clinical illness, and persists in a latent state for life. Clinical illness associated with these two human polyomaviruses occurs only in patients whose immune systems become severely compromised, that is, patients receiving immunosuppressive therapy, AIDS patients, and patients with other debilitating diseases such as cancer. In these patients, reactivation of the their latent JCV or BKV infection may result in clinical illness and death. Reactivation of BKV in immunocompromised patients has been associated with a viral tubulointerstitial nephritis with excretion of virus in the urine. JCV, on the other hand, is the cause of a severe demyelinating disease referred to as progressive multifocal leukoencephalopathy (PML) that affects the white matter of the brain. JCV has also been shown to induce a variety of tumors of the nervous system in hamsters, including medulloblastoma, undifferentiated neuroectodermal tumor, glioblastoma, ependymoma, pineocytoma, neuroblastoma, and meningioma (London *et al.*, 1978). Interest in polyomaviruses as a potential cause of brain tumors in humans prompted experimental studies of the oncogenicity of these viruses in nonhuman primates.

JC Virus

In 1978, London *et al.* inoculated 14 Colombian owl monkeys intracerebrally, subcutaneously, and intravenously with high doses of JCV. Six of the 14 animals received immunosuppressive doses of cyclophosphamide and prednisone continuously from 10 weeks prior to inoculation to 10 weeks after inoculation. Other *Aotus* in this study were inoculated with BKV and SV40 (a macaque polyomavirus) but did not develop neoplasms.

1. Clinical Disease Sixteen months after inoculation, one of the JCV-infected owl monkeys that had not been immunosuppressed developed opisthotonic posturing that was thought to be the result of an inner ear infection. After no apparent response to treatment it was euthanatized. Twenty-five months after inoculation a second JCV-infected owl monkey who also had not been immunosuppressed developed a mild right hemiparesis and hemisensory deficit. It was also euthanatized. In a subsequent publication from this same laboratory, reference is made to an additional eight JVC-infected owl monkeys that developed brain tumors over a 36-month period but the pathologic descriptions of these tumors are not given (Miller *et al.*, 1983). This indicates that greater than 30% of the animals infected with JCV developed such tumors.

2. Pathologic Features

a. Gross Findings In the first monkey, coronal sections of the brain revealed two discrete masses. One was a 10.0 mm, round, hemorrhagic mass located in the thalamus of the left cerebral hemisphere. The other was a 3.0-mm, round gray mass located in the left rostral brainstem and involving the midbrain and pons. In the second monkey, the left cerebral hemisphere was swollen and on cut section contained a 15.0 × 9.0 mm, nonhemorrhagic mass deep in the cerebral hemisphere involving the thalamus and extending to the midline.

b. Microscopic Findings Microscopically the two masses in the brain of the first monkey to die were consistent with grade 3 and grade 4 (glioblastoma multiforme) astrocytoma. They were composed of pleomorphic cells with pleomorphic nuclei and many mitotic figures. There were extensive necrosis and hemorrhage present in the larger tumor that were not present in the smaller one. The microscopic features in the neoplasm in the second animal varied. In some areas, the appearance and arrangement of the tumor cells were consistent with a neuroblastoma, whereas in other areas they had features of an astrocytoma. Mitotic figures were common in it as well. The neoplasms in both animals were regarded as malignant.

3. Etiology and Pathogenesis A total of six JCV-induced brain tumors in *Aotus* have been analyzed for the presence of JCV genome in the tumor virus DNA (Miller *et al.*, 1983). These studies revealed that all tumor cells contained integrated copies of the complete JCV genome and that in all but one of the tumors integration was at one or only a limited number of sites, indicating a clonal origin of the tumors. In all but one of the tumors there was tandem, head-to-tail integration of two or more copies of the genome.

REFERENCES

Barahona, H., Melendez, L. V., Hunt, R. D., Forbes, M., Fraser, C. E. O., and Daniel, M. D. (1975). Experimental horizontal transmission of *Herpesvirus saimiri* from squirrel monkeys to an owl monkey. *J. Infect. Dis.* **132,** 694–697.

Biesinger, B., Muller-Fleckenstein, I., Simmer, B., Lang, G., Wittmann, S., Platzer, E., Desrosiers, R. C., and Fleckenstein, B. (1992). Stable growth transformation of human T lymphocytes by *Herpesvirus saimiri. Proc. Natl. Acad. Sci. U.S.A.* **89,** 3116–3119.

Cicmanec, J. L., Loeb, W. F., and Valerio, M. G. (1974). Lymphoma in owl monkeys (*Aotus trivirgatus*) inoculated with *Herpesvirus saimiri*: Clinical, hematologic and pathologic findings. *J. Med. Primatol.* **3,** 8–17.

Daniel, M. D., Melendez, L. V., Hunt, R. D., King, N. W., Anver, M., Fraser, C. E. O., Barahona, H., and Baggs, R. B. (1974). *Herpesvirus saimiri.* VII. Induction of malignant lymphoma in New Zealand white rabbits. *J. Natl. Cancer Inst.* (*U.S.*) **53,** 1803–1807.

Daniel, M. D., Hunt, R. D., King, N. W., and Ingalls, J. K. (1978). The oncogenicity of *Herpesvirus ateles* in rabbits. *IARC Publ.* **24,** 213 (abstr.).

Desrosiers, R. C., Bakker, A., Kamine, J., Falk, L. A., Hunt, R. D., and King, N. W. (1985). A region of the *Herpesvirus saimiri* genome required for oncogenicity. *Science* **228,** 184–187.

Desrosiers, R. C., Silva, D. P., Waldron, L. M., and Letvin, N. L. (1986). Nononcogenic deletion mutants of *Herpesvirus saimiri* are defective for *in vitro* immortalization. *J. Virol.* **57,** 701–705.

de-Thé, G. (1982). Epidemiology of Epstein–Barr virus and associated diseases in man. *In* "The Herpesviruses" (B. Roizman, ed.), Vol. 1, pp. 25–103. Plenum, New York and London.

Epstein, M. A., Achong, B. G., and Barr, Y. M. (1964). Virus particles in cultured lymphoblasts from Burkitt's lymphoma. *Lancet* **1,** 252–253.

Epstein, M. A., Hunt, R. D., and Rabin, H. (1973). Pilot experiments with EB virus in owl monkeys (*Aotus trivirgatus*). I. Reticuloproliferative disease in an inoculated animal. *Int. J. Cancer* **12,** 309–318.

Falk, L. A., Nigida, S., Deinhardt, F., Cooper, R. W., and Hernandez-Camacho, J. I. (1973). Oral excretion of *Herpesvirus saimiri* in captive squirrel monkeys and incidence of infection in feral squirrel monkeys. *J. Natl. Cancer Inst.* (*U.S.*) **51,** 1987–1989.

Falk, L. A., Nigida, S., Deinhardt, F., Wolfe, L. G., Cooper, R. W., and Hernandez-Camacho, J. I. (1974). *Herpesvirus ateles*: Properties of an oncogenic herpesvirus isolated from circulating lymphocytes of spider monkeys (*Ateles* sp.). *Int. J. Cancer* **14,** 473–482.

Fleckenstein, B., and Desrosiers, R. C. (1982). *Herpesvirus saimiri*. *In* "The Herpesviruses" (B. Roizman, ed.), Vol. 1, pp. 251–332. Plenum, New York and London.

Fleckenstein, B., Bornkamm, G. W., Mulder, C., Werner, F.-J., Daniel, M. D., Falk, L. A., and Delius, H. (1978). *Herpesvirus ateles* DNA and its homology with *Herpesvirus saimiri* nucleic acid. *J. Virol.* **25,** 361–373.

Hunt, R. D., Melendez, L. V., King, N. W., Gilmore, C. E., Daniel, M. D., Williamson, M. E., and Jones, T. C. (1970). Morphology of a disease with features of malignant lymphoma in marmosets and owl monkeys inoculated with *Herpesvirus saimiri*. *J. Natl. Cancer Inst.* (*U.S.*) **44,** 447–465.

Hunt, R. D., Melendez, L. V., King, N. W., and Garcia, F. G. (1972a). *Herpesvirus saimiri* malignant lymphoma in spider monkeys: A new susceptible host. *J. Med. Primatol.* **1,** 114–128.

Hunt, R. D., Melendez, L. V., Garcia, F. G., and Trum, B. F. (1972c). Pathologic features of *Herpesvirus ateles* lymphoma in cotton-topped marmosets (*Saguinus oedipus*). *J. Natl. Cancer Inst.* (*U.S.*) **49,** 1631–1639.

Hunt, R. D., Garcia, F. G., Barahona, H. H., King, N. W., Fraser, C. E. O., and Melendez, L. V. (1973). Spontaneous *Herpesvirus saimiri* lymphoma in an owl monkey. *J. Infect. Dis.* **127,** 723–725.

Hunt, R. D., Daniel, M. D., Baggs, R. B., Blake, B. J., Silva, D., DuBose, D., and Melendez, L. V. (1975). Clinicopathologic characterization of *Herpesvirus saimiri* malignant lymphoma in New Zealand white rabbits. *J. Natl. Cancer Inst.* (*U.S.*) **54,** 1401–1412.

Hunt, R. D., Blake, B. J., and Daniel, M. D. (1976). *Herpesvirus saimiri* lymphoma in owl monkeys (*Aotus trivirgatus*): Susceptibility, latent period, hematologic picture and course. *Theriogenology* **6,** 139–151.

Hunt, R. D., Barahona, H., and Daniel, M. D. (1978). *Herpesvirus ateles* malignant lymphoma in owl monkeys: A new susceptible primate. *In* "Advances in Comparative Leukemia Research" (D. S. Yohn and P. Bentvelzlen, eds.), pp. 198–200. Elsevier/North-Holland, Amsterdam.

Jung, J. U., and Desrosiers, R. C. (1991). Identification and characterization of the *Herpesvirus saimiri* oncoprotein STP-C488. *J. Virol.* **65,** 6953–6960.

Jung, J. U., and Desrosiers, R. C. (1992). *Herpesvirus saimiri* oncogene STP-C488 encodes a phosphoprotein. *J. Virol.* **66,** 1777–1780.

Jung, J. U., Trimble, H. J., King, N. W., Biesinger, B., Fleckenstein, B. W., and Desrosiers,

R. C. (1991). Identification of transforming genes of subgroup A and C strains of *Herpesvirus saimiri*. *Proc. Natl. Acad. Sci. U.S.A.* **88,** 7051–7055.

London, W. T., Houff, S. A., Madden, D. L., Fuccillo, D. A., Gravell, M., Wallen, W. C., Palmer, A. E., Sever, J. L., Padgett, B. L., Walker, D. L., Zu Rhein, G. M., and Ohashi, T. (1978). Brain tumors in owl monkeys inoculated with a human polyomavirus (JC virus). *Science* **201,** 1246–1249.

Medveczky, P., Szomolanyi, E., Desrosiers, R. C., and Mulder, C. (1984). Classification of *Herpesvirus saimiri* into three groups based on extreme variation in a DNA region required for oncogenicity. *J. Virol.* **52,** 938–944.

Medveczky, M. M., Szomolanyi, E., Hesselton, R., DeGrand, D., Geck, P., and Medveczky, P. G. (1989). *Herpesvirus saimiri* strains from three DNA subgroups have different oncogenic potential in New Zealand white rabbits. *J. Virol.* **63,** 3601–3611.

Melendez, L. V., Daniel, M. D., Garcia, F. G., Fraser, C. E. O., Hunt, R. D., and King, N. W. (1969a). *Herpesvirus saimiri.* I. Further characterization studies of a new virus from the squirrel monkey. *Lab. Anim. Care* **19,** 372–377.

Melendez, L. V., Hunt, R. D., Daniel, M. D., Garcia, F. G., and Fraser, C. E. O. (1969b). *Herpesvirus saimiri.* II. Experimentally induced malignant lymphoma in primates. *Lab. Anim. Care* **19,** 378–386.

Melendez, L. V., Hunt, R. D., Daniel, M. D., Fraser, C. E. O., Garcia, F. G., and Williamson, M. E. (1970). Lethal reticuloproliferative disease induced in *Cebus albifrons* monkeys by *Herpesvirus saimiri*. *Int. J. Cancer* **6,** 431–435.

Melendez, L. V., Hunt, R. D., Daniel, M. D., Blake, B. J., and Garcia, F. G. (1971). Acute lymphocytic leukemia in owl monkeys inoculated with *Herpesvirus saimiri*. *Science* **171,** 1161–1163.

Melendez, L. V., Castellano, H., Barahona, H. H., Daniel, M. D., Fraser, C. E. O., and Garcia, F. G. (1972a). Two new herpesviruses from spider monkeys (*Ateles geoffroyi*). *J. Natl. Cancer Inst. (U.S.)* **49,** 233–238.

Melendez, L. V., Hunt, R. D., King, N. W., Barahona, H. H., Daniel, M. D., Fraser, C. E. O., and Garcia, F. G. (1972b). *Herpesvirus ateles,* a new lymphoma virus of monkeys. *Nature (London), New Biol.* **235,** 182–184.

Miller, G., Shope, T., Coope, D., Waters, L., Pagano, J., Bornkamm, G. W., and Henle, W. (1977). Lymphoma in cotton-top marmosets after inoculation with Epstein–Barr virus: Tumor incidence, histologic spectrum, antibody responses, demonstration of viral DNA, and characterization of viruses. *J. Exp. Med.* **145,** 948–967.

Miller, N. R., London, W., Padgett, B. L., Walker, D. L., and Wallen, W. C. (1983). The detection of JC viral genome in owl monkey tumors. *In* "Polyomaviruses and Human Neurological Diseases, pp. 271–288. Liss, New York.

Rabin, H., Neubauer, R. H., Pearson, G. R., Cicmanec, J. L., Wallen, J. L., Loeb, W. C., and Valerio, M. G. (1975). Spontaneous lymphoma associated with *Herpesvirus saimiri* in owl monkeys. *J. Natl. Cancer Inst. (U.S.)* **54,** 499–502.

Robinson, J. E., and Miller, G. (1982). Biology of lymphoid cells transformed by Epstein–Barr virus. *In* "The Herpesviruses" (B. Roizman, ed.), Vol 1, pp. 151–207. Plenum, New York and London.

Wallen, W. C., Neubauer, R. H., Rabin, H., and Cicmanec, J. L. (1973). Nonimmune rosette formation by lymphoma and leukemia cells from *Herpesvirus saimiri*-infected owl monkeys. *J. Natl. Cancer Inst. (U.S.)* **51,** 967–974.

Wallen, W. C., Neubauer, R. H., and Rabin, H. (1974). *In vitro* immunological characteristics of lymphoid cells derived from owl monkeys infected with *Herpesvirus saimiri*. *J. Med. Primatol.* **3,** 41–53.

Wolfe, L. G., Falk, L. A., and Deinhardt, F. (1971). Oncogenicity of *Herpesvirus saimiri* in marmoset monkeys. *J. Natl. Cancer Inst. (U.S.)* **47,** 1145–1162.

Wright, J., Falk, L. A., Collins, D., and Deinhardt, F. (1976). Brief communication: Mononuclear fraction carrying *Herpesvirus saimiri* in persistently infected squirrel monkeys. *J. Natl. Cancer Inst.* (*U.S.*) **57,** 959–962.

Wright, J., Falk, L. A., Wolfe, L. G., Ogden, J., and Deinhardt, F. (1977). Susceptibility of common marmosets (*Callithrix jacchus*) to oncogenic and attenuated strains of *Herpesvirus saimiri*. *J. Natl. Cancer Inst.* (*U.S.*) **59,** 1475–1478.

10

Ophthalmologic Research in the Owl Monkey

Thomas E. Ogden
Doheny Eye Institute
University of Southern California
School of Medicine
Los Angeles, California 90033

I. THE EYE OF *AOTUS:* STRUCTURE

A. The Globe

The owl monkey has been a popular subject for ophthalmologic research because its eye is large relative to its body size and the pupil, when fully dilated, seems nearly as large as the eye at the equator. The cornea occupies one third of the globe (Detwiler, 1941; Walls, 1942), the ciliary body is displaced posteriorly, and the anterior chamber is very deep; this geometry results in a spectacular view of the fundus through the dilated pupil.

The dimensions of the owl monkey eye have been presented by Jones (1965) from studies of 12 adult animals (*Aotus trivirgatus*). Mean equatorial diameter is 16.4 mm; mean optical axis is 16.3 mm; length of retina along the horizontal meridian is 27 mm; and retinal area is 325 mm^2. Jones calculated the equivalent of 1° of visual angle to be 128 μm on the retina. This figure is probably low because the extent of the visual field was overestimated. Ogden (1975) estimated that 1° of visual angle covered 150 μm on the retina, based on anatomic measurements. Webb and Kaas (1976) arrived at the same figure from measurements of the visual fields of cortical neurons.

Differences between the owl monkey and rhesus eye are apparent in the ratio of the anterior chamber to vitreous body (1 : 3.8 in the owl monkey and 1 : 7.1 in rhesus) and in the percentage of the eye volume occupied by the ciliary body (1% in the owl monkey and 0.25% in rhesus) (Detwiler, 1941).

Photographs of the fundus of the owl monkey eye have been presented in numerous publications (Brown *et al.*, 1984; Constable *et al.*, 1975; Klöti, 1967; Luxenberg, 1979; Parrish *et al.*, 1982; Pomerantzeff *et al.*, 1984; Radius, 1980a; Radius and Anderson, 1978, 1979b, 1980a,b, 1981a,b; Tsai and Smith, 1975; Wieder *et al.*, 1981). Some of these publications also show fluorescein angiograms of the owl monkey eye (Radius and Anderson, 1980a,b, 1981a,b; Tsai and Smith, 1975). The fundus of the owl monkey is similar to that of other primates, with vascular arcades above and below the area centralis (Wolin and Massopust, 1967). The center of the macula (or fovea), however, is vascularized, at least in some specimens (Ferraz de Oliveira and Ripps, 1968). The retinal vascular pattern is variable but, compared with that of rhesus, usually shows more prominent papillomacular vessels. The nerve fiber layer is easily discerned and shows the usual radial nasal and arcuate temporal pattern.

Most observers have described the owl monkey as afoveate (Jones, 1965; Ferraz de Oliveira and Ripps, 1968), but in my experience about one in ten of these animals has a funduscopic foveal reflex, and an occasional animal shows evidence of a rudimentary (shallow) foveal pit on histologic examination (Jones, 1965). When the retina is removed for a retinal whole-mount preparation, the macula is readily apparent and often shows a small central depression (Ogden, 1983b).

B. The Retina

1. Receptors The debate over whether or not the owl monkey has cones was ended by the electron microscopic studies of Murray and associates (1973), who demonstrated typical rods and cones, although the cones were few in number. Ogden (1974) confirmed this observation with Gogli studies. As in all primates, the cone perikaryon abuts the outer limiting membrane. The outer segments are short and tapered and measure approximately $4 \times 10\ \mu m$. The slender cone axons expand into a typical cone pedicle with several prominent telodendria (Fig. 1, structure C). The pedicles are about $6\ \mu m$ in diameter, contain numerous synaptic ribbons and vesicles, and have horizontal cell dendrites and bipolar cell processes in invaginating triadic complexes (Murray *et al.*, 1973; Ogden *et al.*, 1982). Horizontal cell axon terminals do not contact the cones.

Owl monkey rods also are typical of those of most primates, but are more numerous in the area centralis than are those in diurnal primates. Receptor density maps are used to show variations in receptor density across the retina. Such studies in the owl monkey reveal rod density to vary from 450,000 rods/mm^2 centrally to 100,000/mm^2 peripherally across the retina, with the highest density in the area centralis (Ogden, 1975). Wikler and Rakic (1990), in a similar study, found densities of 100,000 to 325,000 rods/mm^2. Cone density across the retina shows little variation (2000 to 7000/mm^2; Ogden, 1975; Wikler and Rakic, 1990). The specimen studied by Wikler and Rakic (1990) was of interest because the area of highest cone density did not correspond to the anatomic center of the area centralis as determined by the vascular pattern. The high density of rods in the owl monkey is possible because the rods are very slender, resembling macaque foveal cones. The small rod spherules are arranged in stacks in the outer plexiform layer, and are invaginated by horizontal cell axon terminals and the dendrites of rod bipolar cells (Fig. 1, structure RB). The receptor outer segments of the owl monkey are immunoreactive for alpha transducin (van Veen *et al.*, 1986), a G-protein thought to be involved in phototransduction.

2. Horizontal Cells All horizontal cells observed by Ogden (1974) with the Golgi procedure possessed a well-defined axon. Several of the peripherally located cells were stained completely. These were very large cells with an axon that was thin near the perikaryon but that increased to 5 to $6\ \mu m$ in diameter near the terminal, which was as much as 1 mm from the cell body in some instances. The basal dendrites of these peripherally located horizontal cells numbered from two to six and were thick, irregular trunk-

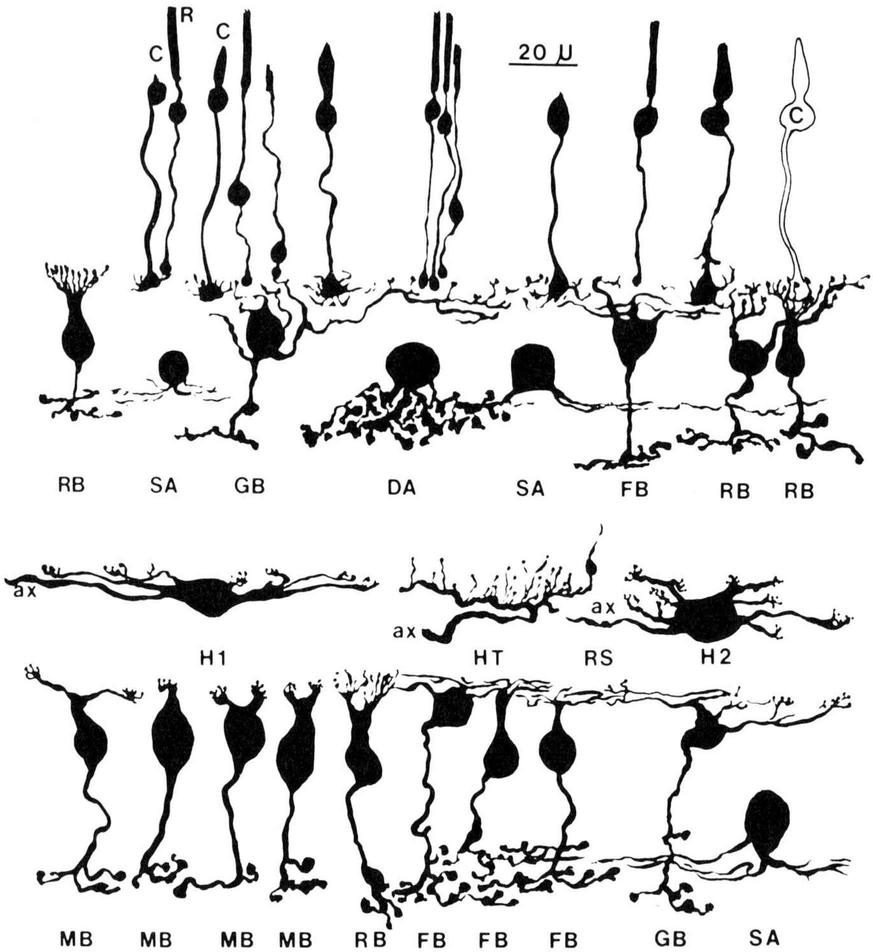

FIGURE I Camera lucida drawings of retinal neurons of the owl monkeys as revealed by the Golgi–Colonnier stain. All the cells were observed near the area centralis in retinal cross sections. ax, axon of a horizontal cell; C, cone; DA, diffuse amacrine; FB, flat bipolar; GB, giant bipolar; H1, peripheral horizontal cell; H2, central horizontal cell; HT, axon terminal of a horizontal cell; MB, midget bipolar; R, rod; RB, rod bipolar. Calibration, 20 μm. Reprinted with permission from Ogden (1974, Fig. 1).

like processes. Secondary processes were much finer, and the dendritic synaptic terminals, like clusters of grapes, rose from minute stems, as in most other forms (Fig. 1, structure H1). The horizontal cells varied gradually in number of contacts and size from center to periphery: centrally located cells were very compact, with the largest extent of dendritic spread measuring about 40 μm (Fig. 1, structure H2); the peripheral cells were as large as 150 μm. Central cells were observed to contact 16 to 31 cones, whereas peripheral cells contacted from 21 to 43 cones. This relative consistency of cone contacts across the retina despite a 10:1 variation in horizontal cell dendritic spread may relate to the relative constancy of cone density. The axon terminals were richly arborized and also varied in size across the retina (Fig. 1, structure HT).

3. Bipolar Cells The bipolar cells of the owl monkey resemble those of other primates, with the exception that the owl monkey has a very large, diffuse bipolar cell called the giant bipolar (Fig. 1, structure GB), with its processes in all sublaminae of the inner plexiform layer. Rod bipolars are very numerous, as would be expected, and their moplike dendrites contact as many as 200 rods. Their axon is slender, about 1 μm in diameter, and ends in a bulbous terminal enlargement that spans the inner plexiform layer (IPL). The rod bipolars of the owl monkey closely resemble those of rhesus, with the exception that the dendritic spread is about half the size of that of rhesus; this is in keeping with the high density of rods in the owl monkey. (Fig. 1, structure RB). Midget, flat and invaginating cone bipolars are virtually identical in the owl monkey and rhesus macaque (Fig. 1, structures MB and FB) (Ogden, 1974).

4. Amacrine Cells Amacrine cells of the owl monkey are easily classified morphologically as diffuse or stratified. The diffuse cells do not spread far laterally and have thickly branched, bushlike terminals in all sublayers of the inner plexiform layer (Fig. 1, structure DA). The stratified cells have a few slender processes that extend various distances within one or two sublayers of the IPL (Fig. 1, structure SA). Presumed amacrine cells of the owl monkey have been shown to stain positively for NADPH diaphorase (Sandell, 1985).

5. Retinal Ganglion Cells My Golgi preparations of the owl monkey retina contained many partially stained ganglion cells but the material was not complete enough to permit a detailed study of these cells, and I am not aware of any other such study in the literature. I can say with certainty that the general classes of cells found in other primates are also present in the owl monkey. Cells are of various sizes with dendrites that ramify in one or more sublayers of the IPL. Dendritic spread varies from tightly restricted to extensive. Cells in the area centralis are smaller than those in the periphery. Webb and Kaas (1976) studied retinal whole mounts of the owl monkey to determine cell soma size; they noted small- to medium-sized cells with a density that decreased from 10,000/mm^2 centrally to 300/mm^2 in the extreme periphery. A distinct population of larger cells was found primarily in the periphery, extending out from about 2 mm of the macula (Fig. 2). Webb and Kaas created an isodensity map (Fig. 3) that shows moderate elongation of isodensity lines in the horizontal meridian,

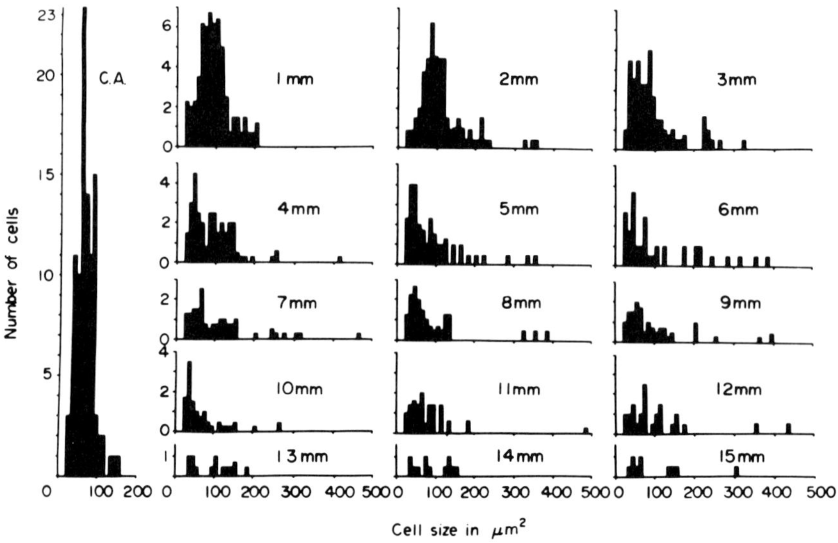

FIGURE 2 Histograms showing the number of ganglion cells of different sizes at 1-mm intervals from the central area (C.A.). The population of small cells declines rapidly in the first few millimeters. Populations of larger cells begin appearing at 2 mm eccentricity. Counts were made from retinal whole mounts of the owl monkey retina. Reproduced with permission from Webb and Kaas (1976, Fig. 7).

which is similar to the distribution of receptors noted by Ogden (1975). The projection of ganglion cells to the lateral geniculate nucleus of the thalamus was described by Sherman *et al.* (1976).

6. Nerve Fiber Layer The nerve fiber layer of the owl monkey retina has been the subject of intense study as a primate model in an exquisitely photogenic eye. The nerve fiber layer is an important clinical indicator of disease of the optic nerve, particularly glaucoma, and the owl monkey has served as a model for several studies relevant to clinical glaucoma, described in the following. Glaucoma is characterized clinically by elevation of intraocular pressure; this may be associated with atrophy of the arcuate nerve fibers and progressive loss of the paracentral visual field, resulting in blindness when the macula becomes involved. Accordingly, patients with ocular hypertension are monitored closely for evidence of progressive optic atrophy; if this is detected, corrective surgery may be necessary. Loss of arcuate fibers and thinning of the supero- and inferotemporal optic disc rim, or loss of visual field in the Bjerrum area corresponding to the arcuate fibers, suggest uncontrolled glaucoma. In fact, some clinicians feel that loss of arcuate fibers is the most sensitive indicator of progression of optic atrophy secondary to glaucoma.

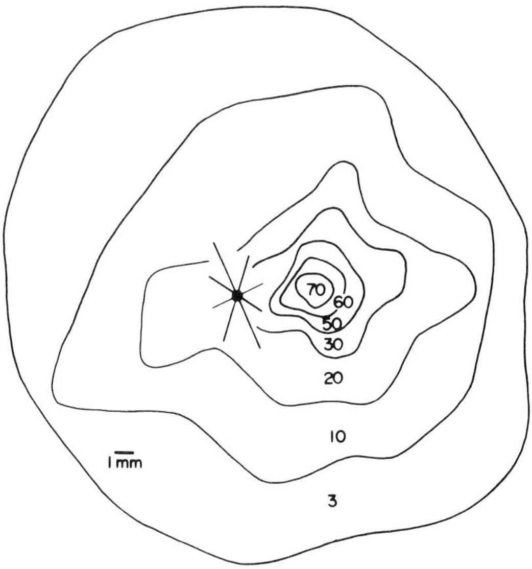

FIGURE 3 Diagram of ganglion cell density in the owl monkey retina. Lines represent isodensity contours of ganglion cell concentration. Numbers indicate the lowest number of cells found in 0.01-mm^2 sample areas in each contour. The solid circle represents the optic disc, and the lines radiating from it show where the thickness of the nerve fiber layer obscured the underlying ganglion cells in the retinal whole mount. The isodensity lines are fairly concentric, with a slight horizontal elongation outside the central retina. Reproduced with permission from Webb and Kass (1976, Fig. 3).

The nerve fiber layer of the owl monkey retina appears to be striated with reflective wedge-shaped spokes radiating from the optic disc and separated from each other by dark lines. Radius and Anderson (1979b,c) confirmed the general belief that the reflective areas are indeed bundles of nerve fibers; they showed also that the dark lines correspond to the end feet of Müller cells, which separate the bundles. Loss of fibers appears as a wedge-shaped dark area. Those investigators destroyed small areas of the peripheral retina of the owl monkey, in the distribution of the arcuate fibers, and documented the development of fiber drop-out using fundus photography, histology, and electron microscopy.

The nerve fiber layer of the owl monkey also has been the subject of studies concerning primate retinotopic organization. Placement of ganglion cell axons within the individual nerve fiber bundles that cross the retina en route to the optic disc has been the subject of controversy. These bundles contain many thousands of axons and may be over 200 μm thick and 40 μm wide at the disc margin (Ogden, 1978; Radius, 1980a). In the growing eye, the central retina develops first, and retinal enlargement occurs by means of growth around the periphery. The axons of newly developed

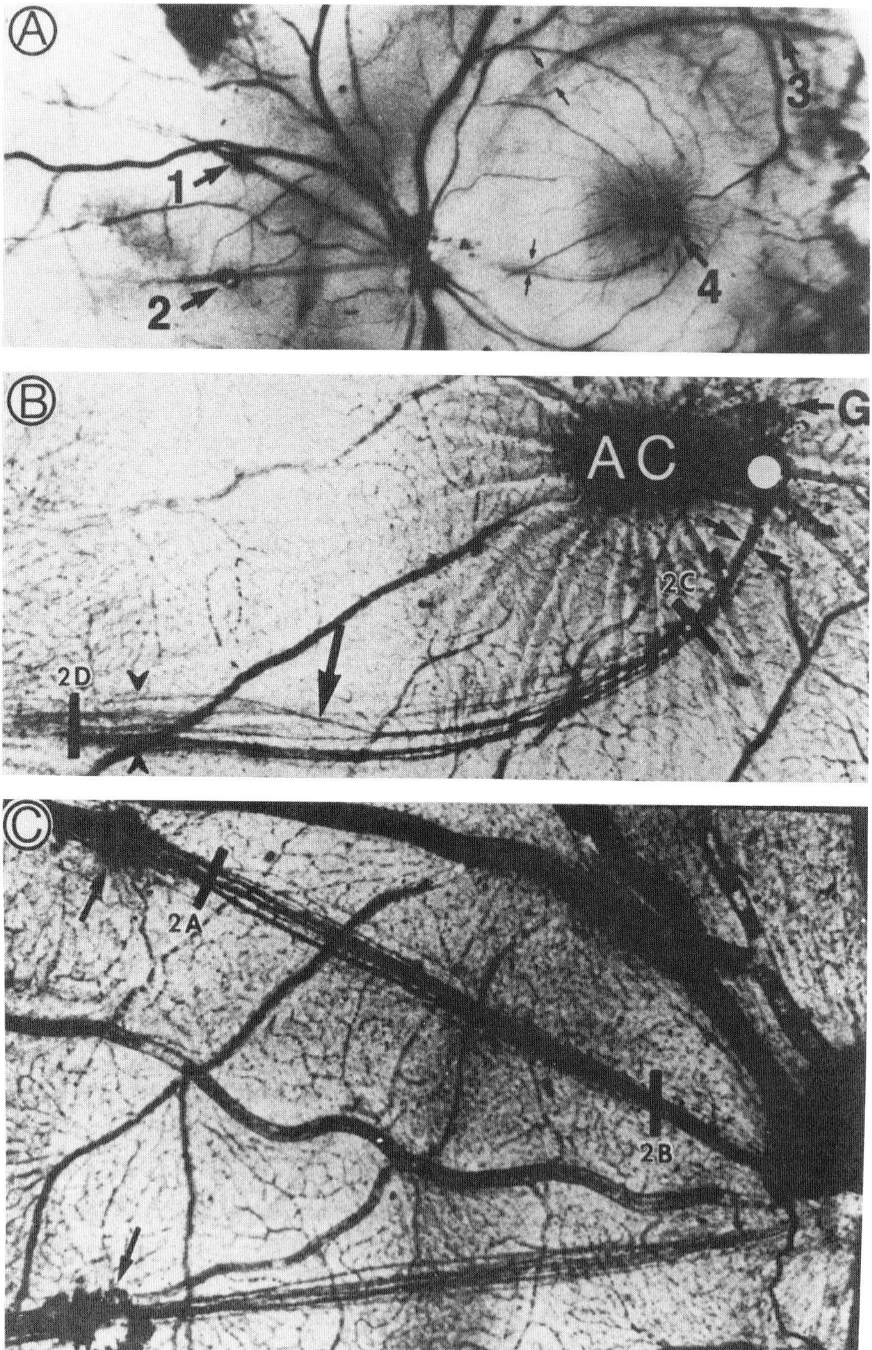
A
1
2
3
4
B
AC
G
2C
2D
C
2A
2B

ganglion cells join those of older cells located in a wedge-shaped bundle and converging on the disc. Most of the parent ganglion cells of the axons of a bundle are located beneath the bundle in a well-defined horizontal retinotopy, but there is the possibility of vertical retinotopy within a bundle. As axons join a bundle, they could be retinotopically segregated on the scleral or the vitreal margin of the bundle or dispersed randomly within it if no vertical retinotopy were present. Since the earliest visual loss in glaucoma involves peripheral rather than central ganglion cells, the position of the peripheral fibers within the bundles might be important from the standpoint of fiber susceptibility to injury in glaucoma or, for instance, following laser treatment for diabetic retinopathy.

Radius and Anderson (1979b,c) attempted to determine vertical retinotopy within nerve fiber bundles by making small peripheral laser burns and tracing the degenerating fibers. Their results were confounded by a combination of data from both rhesus macques and owl monkeys, and perhaps also by uncertainties associated with experiments involving degenerating nerve fibers. They concluded that nerve fibers show little if any lateral dispersion; that is, fibers join a bundle and stay within it as they pass to the disc. They concluded also that axons of more peripheral ganglion cells are segregated on the scleral surface of the bundles, and that shorter axons of more centrally located cells are segregated on the vitreal part of the bundles.

The study was repeated in both cynomolgus macaques and owl monkeys using intraretinal microinjections of horseradish peroxidase (Ogden, 1983a,b). The peroxidase labeled a few axons of a bundle with an electron-dense, optically opaque reaction product, thereby permitting unambiguous identification of the path of fibers from the injection site to the disc. The studies in macaques (Ogden, 1983b) proved that peripheral fibers are segregated in the vitreal, not the scleral portions of the bundles, which contradicted the earlier results of Radius and Anderson (1979b,c). The explanation for these conflicting results became clear, however, when the peroxidase experiments were repeated in the owl monkey (Ogden, 1983a). These experiments revealed important differences between the highly specialized nocturnal owl monkey and the diurnal macaques; the retina of the latter is assumed to be similar to that of humans. Unlike the macaques, significant lateral dispersion of fibers was found among bundles in the temporal retina (Figs. 4A and 4B; opposed arrows) but not the nasal retina (Fig. 4C) of the

FIGURE 4 Photomicrographs of horseradish peroxidase-injected retinal whole mounts of the owl monkey. Four microinjections (A: 1,2,3,4) were delivered into the nerve fiber layer. The labeled temporal fibers, injections 3 and 4, become widely dispersed (opposed arrowheads) as they approach the optic disc (see also B). The nasal fibers, injections 1 and 2, are seen at higher magnification in C and show no lateral dispersion. AC, area centralis. Calibration: A, 4 mm; B, C, 1 mm. Reproduced with permission from Ogden (1983b, Fig. 1).

owl monkey; peripheral temporal fibers were segregated, as described by Radius and Anderson but incorrectly attributed to the macaque, in the scleral portions of the bundles. An unexpected finding was that the nasal fibers of the owl monkey had no vertical retinotopy; the peripheral nasal fibers were dispersed randomly within the bundles at the disc margin. The retinotopy of the owl monkey thus differed significantly from that of the mascaque. These studies are important in demonstrating the danger of extrapolating data obtained from the owl monkey to macaques and to humans. The owl monkey is a popular subject for study of problems with clinical significance because its fundus is easily visible, but the results of such studies can be misleading.

The thickness and glial content of the nerve fiber layer have also been the subjects of study. Like the diurnal primates, including humans, the nerve fiber layer of the owl monkey is thickest at the supero- and infero-temporal margins of the disc, is virtually absent in the central portion of the area centralis corresponding to the fovea, is relatively thin between the area centralis and the disc, and shows thickened "arcuate ridges" above and below the area centralis at about the position of the temporal vascular arcades (Radius, 1980a; Ogden, 1978). The major difference between the owl monkey and the rhesus macaque is the thickness of the nerve fiber layer, which is from three to five times thicker in the rhesus macaque than in the owl monkey (Radius, 1980a).

Astrocytes are common in the primate nerve fiber layer, where their numbers are proportional to the thickness of the layer. The owl monkey, like other primates, has two major categories of retinal astrocyte; an elongated astrocyte that has processes that parallel the nerve fiber bundles (Fig. 5) and a stellate astrocyte with processes that conspicuously cross the bundles. The owl monkey differs from the diurnal primates in having a "giant" astrocyte in the far periphery of its retina (Ogden, 1978).

II. THE EYE OF *AOTUS:* FUNCTION

A. Spectral Sensitivity

Early studies of the spectral sensitivity of the owl monkey failed to reveal evidence of cone activity. Brown and Watanabe (1962) recorded the electroretinogram (ERG) with vitreal leads and the local ERG (LERG) with intraretinal microelectrodes. The shape of the volume-conducted receptor potential, isolated from the other ERG components by compression of the retinal arteries, showed no evidence of cone activity, and was called a "pure rod receptor potential." Jones and Jacobs (1963) also failed to find evidence of cone activity in the corneal ERG of the owl monkey, but did find a Purkinje shift in the flicker ERG, suggesting the presence of function-

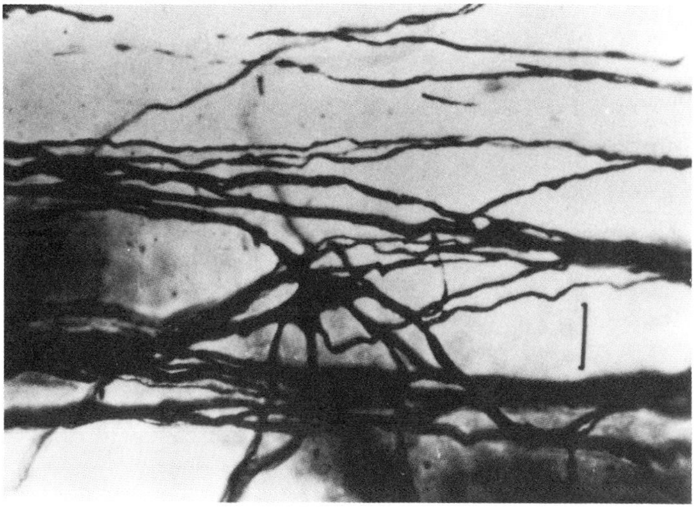

FIGURE 5 Photomicrograph of a retinal whole mount of the owl monkey, stained with the Golgi–Colonnier method. The soma of a fibrous astrocyte is seen suspended between two bundles of nerve fibers that the astrocyte processes join. Calibration: 10 μm. Reproduced with permission from Ogden (1978, Fig. 6D).

ing cones. Jones (1965) studied spectral responses of lateral geniculate nucleus neurons and found three classes of cells with response maxima to 500, 530, and 560 nm stimulation, respectively. He concluded that "more than one pigment" was responsible for the groups but found no evidence for color opponency. Hamasaki (1967) and Hamasaki and colleagues (1969) studied dark adaptation in the owl monkey using the ERG. They noted a well-defined "rod–cone break" in the sensitivity curve, which supported their light microscopic finding of scattered cones throughout the retina. Ripps and Vaughan (1969), however, failed to find evidence of photopic (cone) input in either the ERG or visual evoked response (VER) of the owl monkey. The spectral response curves that they obtained matched that of rhodopsin.

An early behavioral study of the spectral sensitivity of the owl monkey reported a well-defined Purkinje shift, a scotopic spectral sensitivity similar to that of other mammals (Fig. 6), and a photopic sensitivity with a peak at about 530 nm (Jacobs, 1977a). An explicit test of color vision that involved training the monkeys to make color discriminations was carried out, and it was tentatively concluded that the color vision of the owl monkey might be similar to that of a human diagnosed as a severely protonomalous trichromat.

The photopigments and spectral sensitivity of the owl monkey have also been investigated by Jacobs *et al.* (1993) for two reasons. First, in the past

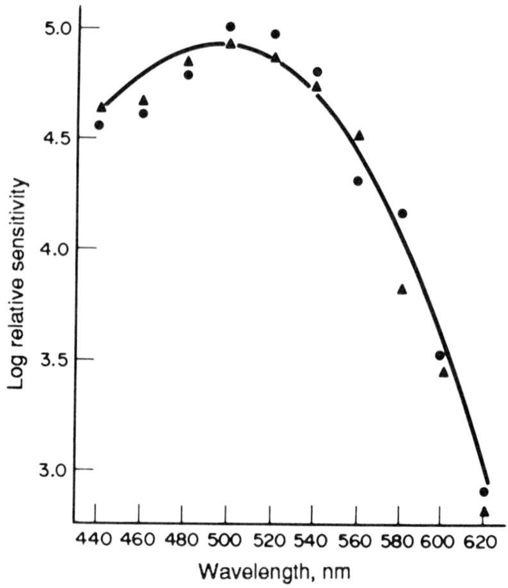

FIGURE 6 Graph of the scotopic spectral sensitivity of the owl monkey, based on an increment-threshold discrimination task. The symbols represent two subjects. The solid line shows the sensitivity of a nomogram photopigment having a 500-nm peak. Reproduced with permission from Jacobs (1977a, Fig. 2).

decade it has become clear that platyrrhine monkeys have a striking photopigment (and color vision) polymorphism, but it was not known whether the owl monkey shares this trait with the diurnal New World monkeys. Second, it has also recently become clear that there are many instances where rods may contribute usable signals in direct tests of color vision. Putative contributions from rods would have been difficult to avoid in the earlier behavioral studies. In the new investigation, Jacobs and colleagues measured the photopigment complements of owl monkeys using electroretinographic flicker photometry, a procedure that is particularly good at isolation of cone signals. In addition to the usual indication of the presence of a rod pigment, evidence was found of a single cone pigment with a maximum sensitivity at about 543 nm (Fig. 7). No hint of the presence of other cone pigments was found in a series of chromatic adaptation tests, nor was there a cone pigment polymorphism in owl monkeys similar to that now found in other platyrrhines. Parallel behavioral tests conducted under more stringent adaptation conditions than those reported previously (Jacobs, 1977a) verified that the owl monkey seems to have only two photopigments: a rod pigment and a single cone pigment (Jacobs *et al.*, 1993).

The owl monkey is unique among primates in apparently possessing only a single cone photopigment. All primate retinas for which there is

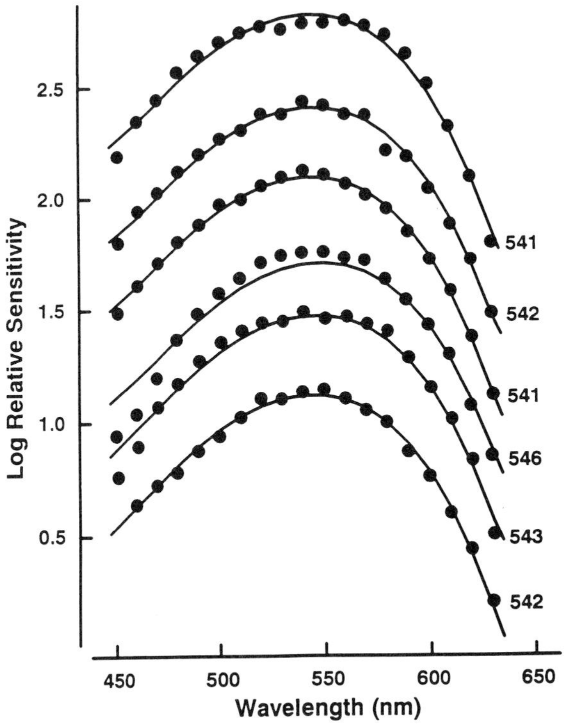

FIGURE 7 Spectral sensitivity functions for the cone pigment of the owl monkey. Each function summarizes the results obtained from electroretinogram flicker photometric measurements for one monkey. The solid circles are the average photometric equations. The continuous line is the curve for the best-fitting visual pigment absorption function; the computed peak of this function (in nm) is given to the right of each curve. The individual functions have been arbitrarily positioned on the ordinate. Reproduced with permission of G. H. Jacobs (Jacobs *et al.*, 1993).

current evidence contain a short-wavelength-sensitive (SWS) photopigment and one or more middle- or long-wavelength-sensitive photopigments. To determine the basis of this difference between the owl monkey and other anthropoids, Jacobs *et al.* (1993) carried out a hybridization analysis to search for the presence of a gene that might code for a cone opsin appropriate for producing a SWS photopigment. They discovered, surprisingly, that the owl monkey has such a gene. Why the apparent SWS-pigment gene is not translated into a functional product is uncertain, but it may be instructive to note that there is a parallel case in human color-defective vision where point mutations that produce amino acid substitutions in the SWS cone opsin render the pigment malfunctional (Weitz *et al.*, 1992). (The author gratefully acknowledges that this section on spectral sensitivity and Fig. 7 were contributed by Dr. G. H. Jacobs.)

B. Acuity

Visual acuity of the owl monkey was incorrectly claimed to be "not much below that of the human" by Kluver (cited by Polyak, 1957). Ordy and Samorajski (1968) tested minimum separable acuity using a bar pattern and behavioral testing. Eight minutes of visual angle was resolved by all animals, and 4 minutes was resolved by some. Jacobs (1977b) used more sophisticated behavioral methods, namely, sinusoidal gratings of variable contrast and spatial frequency in a forced choice discrimination task, and found acuity to be limited to 1.5 to 2.0 cycles/degree in dim light, and 10 cycles/degree under photopic conditions (Fig. 8). This acuity is inferior to that of the human in photopic conditions, but superior in scotopic conditions. Jacobs also tested minimum separable acuity in the owl monkey and found responses at 3 minutes of arc, which is somewhat better than that reported by Ordy and Samorajski (1968). The owl monkey eye is remarkable

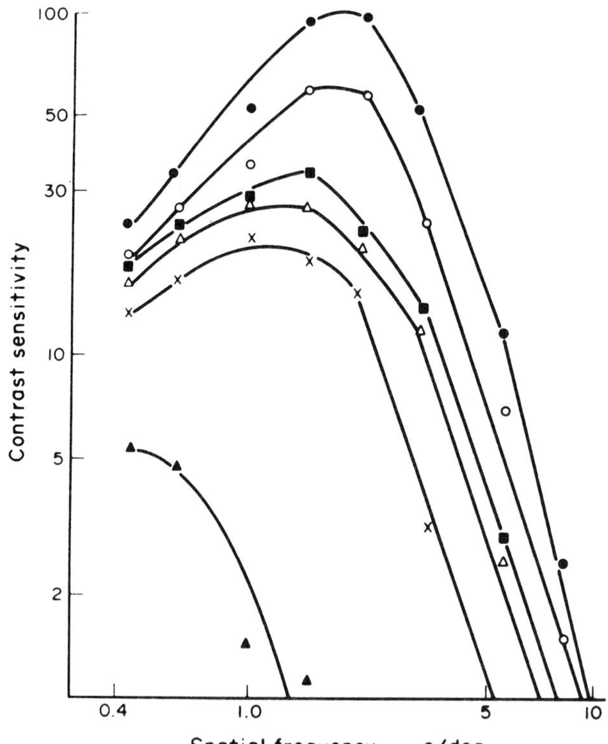

FIGURE 8 Spatial contrast sensitivity functions for the owl monkey as a function of stimulus luminance level. The top function was measured for a mean grating luminance of 11.4 cd/m^2; the bottom five functions are for successive log-attenuations below that level. Reproduced with permission from Jacobs (1977b, Fig. 3).

in showing relatively little variation in spatial resolution over a wide range of light levels, having a 5 : 1 variation, as compared with the 30 : 1 variation seen in humans. The superior acuity of the owl monkeys can be attributed to the high density and small size of its rods (Ogden, 1975). The photopic spatial resolution of the owl monkey, 3 minutes of arc, corresponds to about 7.5 μm on the retina (Jones, 1965), which is less than the average cone separation of 10 μm observed by Ogden (1975).

C. Electrophysiology

The first careful description of the electroretinogram of the owl monkey is that of Brown and Watanabe (1962), who used both intravitreal and intraretinal electrodes to record the response. Their interest focused on the possibility that the owl monkey has so few cones that the response could be considered to be purely rod driven. The owl monkey ERG, recorded from the vitreous, resembled that of the diurnal primate recorded under scotopic conditions, and they found no evidence of cone function. With intraretinal recording, however, they found a well-defined c-wave that was not evident in their vitreous recording, probably owing to masking by the large negative rod receptor potential. They isolated the receptor potential from the b-wave by blocking the retinal circulation.

Jones and Jacobs (1963) used corneal electrodes to record the ERG of the owl monkey, so their data resemble those obtained with vitreal recording by Brown and Watanabe (1962). They obtained an approximation of the ICI scotopic spectral sensitivity curve (see the foregoing) using dim lights of different wavelengths flickering at 4 Hz. A photopic function was obtained with stimuli delivered at 12 to 20 Hz, providing the first functional evidence that this retina contains cones.

Hamasaki (1967) complemented previous studies of the owl monkey retinal anatomy with careful corneal recording of the ERG under controlled scotopic and photopic conditions. He found distinct differences in waveforms and was able to distinguish a photopic b-wave. His paper provides excellent descriptions of the variation in waveform during dark adaptation and as a function of stimulus intensity (Fig. 9). These recordings clearly show the presence of a DC component, a c-wave, and a d-wave.

Newman (1987a,b) performed important studies of Müller cell function in the owl monkey. Müller cells were isolated from the retina and their response was studied in terms of membrane polarization following local potassium exposure. The results support the suggestion that the Müller cell provides a path for potassium to pass form the extracellular space to the inner retinal capillaries.

Ripps and Vaughn (1969) used the ERG to determine the spectral sensitivity of the owl monkey (see earlier) and compared this function at retinal

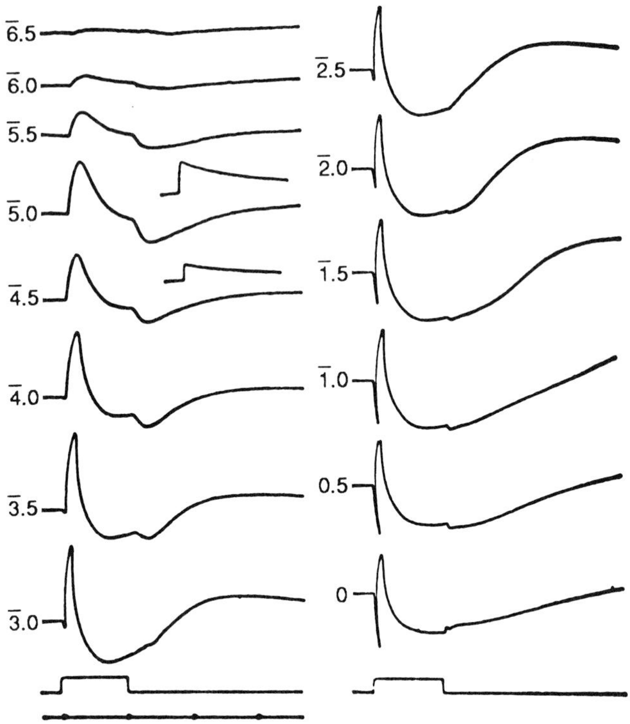

FIGURE 9 Tracings of the corneal electroretinogram of the owl monkey obtained with a wide range of stimulus intensities in the dark-adapted animal. Calibration: 250 μV. Note change in sensitivity between I = 5.0 and 4.5. Reproduced with permission from Hamasaki (1967, Fig. 3).

and cortical levels, the latter determined by recording the visual evoked response.

Hamasaki *et al.* (1969) used the ERG of the owl monkey to assess the loss and restoration of function of the retina following total detachment and reattachment. This study involved comparison of dark adaptation curves and intensity/response functions. Their results indicated total loss of function within 7 days of total detachment; evidence of recovery could be seen as soon as 5 hr following reattachment surgery, and the process was complete in about 12 weeks.

D. Axoplasmic Transport

The owl monkey has been used for numerous studies of axoplasmic transport relevant to attempts to understand the pathogenesis of glaucoma and papilledema. Glaucoma is a blinding disease characterized by optic nerve atrophy, and most, but not all, cases are associated with elevated intraocular pressure. In normal neurons, including retinal ganglion cells, there is continuous movement of proteins and organelles back and forth between the cell body and the axon terminal. This orthograde and retrograde flow has a slow and fast component. When an axon section becomes ischemic or is compressed, flow is blocked through the segment and intracellular material accumulates on either side of the blockage, resulting in axonal swelling. A number of studies have suggested that a block of axoplasmic flow at the level of the lamina cribrosa of the optic disc occurs in glaucoma and in papilledema. This phenomenon has been studied in the owl monkey in which a mild degree of elevated intraocular pressure has been induced experimentally (Quigley and Anderson, 1976, 1977; Quigley *et al.*, 1979; Radius, 1983).

Retinal necrosis was observed when intraocular pressure was elevated to a level higher than that of systolic pressure (Anderson and Davis, 1975; Parrish *et al.*, 1982). Optic nerve damage caused by such extreme elevations of intraocular pressure proved to be reversible if the pressure was relieved within 8 hr (Radius and Anderson, 1981a). Axonal swelling due to impairment of the slow component of transport was induced in the owl monkey eye by lowering intraocular pressure (hypotony); this caused papilledema or swelling of the optic nerve head (Radius and Anderson, 1980b).

Impairment of fast axonal transport in the optic nerve of the owl monkey was induced by occlusion of the short posterior ciliary artery. The impairment occurred in a region of focal ischemia at the lamina cribrosa and could be distinguished from impairment of both fast and slow components caused by ocular hypertension (Radius, 1980b). The ultrastructural correlates of interruption of fast orthograde and retrograde transport were described subsequently by Radius and Anderson (1981b) in owl monkeys that had been subjected to occlusion of the central retinal artery. In many of these studies, both owl monkeys and macaques were used and no significant species differences were found.

E. Retinal Circulation

Retinal circulation time was studied in owl monkeys to establish the reproducibility of a new two-point fluorophotometer based on the fluorescein dye-dilution technique. Retinal circulation time (delay between arterial and venous appearance of the dye) in anesthetized animals was about 2.2 sec. (Eberli *et al.*, 1979).

Vortex vein angiograms were obtained from the owl monkey by Tsai and Smith (1975) for the first direct demonstration that the retinal and choroidal circulations of the primate are completely separate, although the optic disc is perfused from both circulations.

Optic disc vascular morphology and function were studied by Anderson and Braverman (1976). The microvasculature was continuous with and similar to that of the retina in that the capillaries of the disc were not fenestrated, fluorescein was retained, and pericytes were rare, in contradistinction to choroidal capillaries.

III. *AOTUS* AS AN EXPERIMENTAL MODEL

The primary use of the owl monkey in ophthalmologic research has been as an experimental model for the study of various pathologic conditions: retinal ischemia, optic nerve atrophy, ocular trauma, drug effects, endophthalmitis, ocular cell transplantation, photic or ionizing radiation injury to the retina, ocular trauma, and retinal detachment/proliferative vitreoretinopathy. The rationale for the selection of the owl monkey for these studies seems to be based on the ease of their use and the magnificent view afforded of their retina, rather than on this species being suited to the particular study based on scientific reasons.

A. Ischemia

Ischemia has been the subject of several studies: experimental embolization of choroidal and retinal arteries was investigated by Klöti (1967), Algvere and Balazs (1974), and Algvere (1976,1977). The procedure resulted in areas of necrosis of the retinal pigment epithelium, the photoreceptors, or the full thickness of the retina. In other experiments, the lateral short posterior ciliary circulation was occluded with the result that axonal flow was blocked in the optic nerve at the level of the lamina cribrosa (Radius, 1980b); carotid ligation did not affect axonal blockade that had been induced by increased intraocular pressure (Radius *et al.*, 1980); and retinal degeneration was caused by acute retinal ischemia induced by elevation of the intraocular pressure (Anderson and Davis, 1975).

B. Optic Nerve Atrophy

The pathophysiology of optic nerve atrophy was studied using fluorescein angiography and histology of the nerve fiber layer following photocoagulation (Radius and Anderson, 1978,1979a,b).

C. Ocular Trauma

Ocular trauma effects were studied in owl monkeys following injection of whole blood into the posterior chamber; observations indicate that this procedure does not damage the retina (Bennett *et al.*, 1975). Experimental commotio retinae, induced by blunt trauma and studied by light and electron microscopy, did cause degeneration of the outer segments of photoreceptors (Sipperley *et al.*, 1978).

D. Pharmacological Studies

Drug studies were done to evaluate the toxic effects of micanazole on ocular structures. Intravitreal doses of up to 80 μg caused no obvious retinopathy (Tolentino *et al.*, 1982).

E. Endophthalmitis

Endophthalmitis due to *Toxocara canis* infestation was studied in the owl monkey model by Luxenberg (1979).

F. Cell Transplantation

Transplantation of cultured human retinal pigment epithelial (RPE) cells into the subretinal space (Gouras *et al.*, 1984,1985) and into the vitreous (Grierson *et al.*, 1986) has been studied from the standpoint of restoration of function of the transplanted cells. Autotransplantation of RPE cells into the vitreous resulted in proliferation and metaplasia of the transplanted cells (Mandelcorn *et al.*, 1975).

G. Retinal Detachment

Retinal detachment and proliferative vitreoretinopathy (PVR), a serious complication of detachment, have been the subject of many studies in the owl monkey, which would seem to be the ideal primate model (van Horn *et al.*, 1977; Tolentino *et al.*, 1978). Retinal detachment methods (Machemer and Norton, 1968), histology (Machemer, 1968a), electron microscopy (Kroll and Machemer, 1968), reattachment techniques (Machemer, 1986b), electron microscopy of the reattached retina (Kroll and Machemer, 1969), electrophysiology (Hamasaki *et al.*, 1969), photoreceptor protein renewal in the detached retina (Machemer and Kroll, 1971) and in the reattached retina (Kroll and Machemer, 1971), protein metabolism (Machemer and Buettner, 1972), the early receptor potential (Charles and Machemer, 1974a), and photoreceptor degeneration (Charles and Machemer, 1974b) in the detached retina have been studied. Retinal reattachment using intraocular gas tamponade also has been evaluated in the owl monkey (Anderson, 1975; Peters *et al.*, 1985).

H. Irradiation Injury

Irradiation injury to the retina as the result of exposure to a variety of light sources has been studied exhaustively in the owl monkey. Light from an intraocular fiber optic operative source (Fuller *et al.*, 1978,1980; Meyers and Bonner, 1982), proton irradiation (Constable *et al.*, 1975; Gragoudas *et al.*, 1979), krypton laser irradiation (Ducrey *et al.*, 1981; Hultsch and Ducrey, 1979; Wieder *et al.*, 1981; Pomerantzeff *et al.*, 1983,1984), YAG laser irradiation (Brown *et al.*, 1984), and argon laser irradiation (Pomerantzeff *et al.*, 1976) have been evaluated in the owl monkey model.

Other studies in the owl monkey have described naturally occurring retinal detachment (Aaberg and Machemer, 1970), massive periretinal proliferation, an analog of PVR (Laqua and Machemer, 1975b), glial cell proliferation in PVR (Laqua and Machemer, 1975a; Laqua, 1975), hemorrhagic uveal edema (Aaberg, 1975) and choroidal edema (Aaberg and Maggliano, 1979) associated with retinal detachment, and silicone retinopathy caused by the intravitreal injection of silicone oil (Mukai *et al.*, 1975). That the owl monkey model of PVR has clinical relevance is suggested by the similarity of its pathology to that of the human condition (van Horn *et al.*, 1977; Machemer *et al.*, 1978).

The popularity of the owl monkey as a model of primate ocular disease has declined dramatically since 1985 owing to a number of factors, including an increased realization that the survival of this species may be in question. There are concerns about the propriety of any large-scale experiment involving primates, and particularly those that employ the owl monkey. Also, all primate experimentation has been reduced because of rapidly escalating animal costs. Although members of the genus *Aotus* are not on the endangered species list, they are certainly in short supply and their use in vision research should be restricted to those studies directed to the evaluation of their unique adaptations for nocturnal vision, or for which there is no alternative animal model available.

REFERENCES

Aaberg, T. M. (1975). Experimental serous and hemorrhagic uveal edema associated with retinal detachment surgery. *Invest. Ophthalmol.* **14,** 243–246.

Aaberg, T. M., and Machemer, R. (1970). Correlation of naturally occurring detachments with long-term retinal detachment in the owl monkey. *Am. J. Ophthalmol.* **69,** 640–650.

Aaberg, T. M., and Maggliano, J. M. (1979). Choroidal edema associated with retinal detachment repair: Experimental and clinical correlation. *Mod. Probl. Ophthalmol.* **20,** 6–15.

Algvere, P. (1976). Retinal detachment and pathology following experimental embolization of choroidal and retinal circulation. *Albrecht von Graefes Arch. Klin. Exp. Ophthalmol.* **201,** 123–134.

Algvere, P. (1977). Dissociation of the vitreo-retinal junction following experimental embolization of retinal circulation. *Albrecht von Graefes Arch. Klin. Exp. Ophthalmol.* **201,** 229–235.

Algvere, P., and Balazs, E. A. (1974). Experimental retinal detachment in owl monkeys. Effect of intravitreal hyaluronidase injection and embolization of the choroidal and retinal circulation. *Mod. Probl. Ophthalmol.* **12,** 152–166.

Anderson, D. R. (1975). Sulfur hexafluoride in owl monkey vitreous cavity. *Am. J. Ophthalmol.* **79,** 67–76.

Anderson, D. R., and Braverman, S. (1976). Reevaluation of the optic disk vasculature. *Am. J. Ophthalmol.* **82,** 165–174.

Anderson, D. R., and Davis, E. B. (1975). Sensitivities of ocular tissues to acute pressure-induced ischemia. *Arch. Ophthalmol. (Chicago)* **93,** 267–274.

Bennett, T. O., Peyman, G. A., and Vlchek, J. K. (1975). Intravitreous injection of autologous blood in primates. *Can. J. Ophthalmol.* **10,** 248–254.

Brown, G. C., Green, W. R., Shaw, H. G., Woodlief, N., and Gruber, M. (1984). Effects of the Nd:YAG laser on the primate retina and choroid. *Ophthalmology (Rochester, Minn.)* **91,** 1397–1405.

Brown, K. T., and Watanabe, K. (1962). Rod receptor potential from the retina of the night monkey. *Nature (London)* **196,** 547–550.

Charles, S., and Machemer, R. (1974a). Experimental retinal detachment in the owl monkey. 10. The early receptor potential. *Am. J. Ophthalmol.* **77,** 509–512.

Charles, S., and Machemer, R. (1974b). Experimental retinal detachment in the owl monkey. 11. The role of light in photoreceptor atrophy. *Am. J. Ophthalmol.* **78,** 233–235.

Constable, I. J., Koehler, A. M., and Schmidt, R. A. (1975). Proton irradiation of simulated ocular tumors. *Invest. Ophthalmol.* **14,** 547–555.

Detwiler, S. R. (1941). The eye of the owl monkey (*Nyctipithecus*). *Anat. Rec.* **80,** 233–241.

Ducrey, N., Pomerantzeff, O., and Schneider, J. (1981). Utilisation du laser a krypton. Données experiméntales. *Bull. Mem. Soc. Fr. Ophtalmol.* **93,** 95–99.

Eberli, B., Riva, C. E., and Feke, G. T. (1979). Mean circulation time of fluorescein in retinal vascular segments. *Arch. Ophthalmol. (Chicago)* **97,** 145–148.

Ferraz de Oliveira, L., and Ripps, H. (1968). The "area centralis" of the owl monkey (*Aotus trivirgatus*). *Vision Res.* **8,** 223–228.

Fuller, D., Machemer, R., and Knighton, R. W. (1978). Retinal damage produced by intraocular fiber optic light. *Am. J. Ophthalmol.* **85,** 519–537.

Fuller, D., Machemer, R., and Knighton, R. W. (1980). Retinal damage produced by intraocular fiber optic light. *Vision Res.* **20,** 1055–1072.

Gouras, P., Flood, M. T., and Kjeldbye, H. (1984). Transplantation of cultured human retinal cells to monkey retina. *An. Acad. Bras. Cienc.* **56,** 431–443.

Gouras, P., Flood, M. T., Kjeldbye, H., Bilek, M. K., and Eggers, H. (1985). Transplantation of cultured human retinal epithelium to Bruch's membrane of the owl monkey's eye. *Curr. Eye Res.* **4,** 253–265.

Gragoudas, E. S., Zakoz, N. Z., Albert, D. M., and Constable, I. J. (1979). Long-term observation of proton-irradiated monkey eyes. *Arch. Ophthalmol. (Chicago)* **97,** 2184–2191.

Grierson, I., Boulton, M., Hiscott, P., Hitchins, C., Gilbert, D., and McLeod, D. (1986). Human retinal pigment epithelial cells in the vitreous of the owl monkey. *Exp. Eye Res.* **43,** 491–502.

Hamasaki, D. I. (1967). An anatomical and electrophysiological study of the retina of the owl monkey, *Aotus trivirgatus*. *J. Comp. Neurol.* **130,** 163–173.

Hamasaki, D. I., Machemer, R., and Norton, E. W. D. (1969). Experimental retinal detachment in the owl monkey. VI. The ERG of the detached and reattached retina. *Albrecht von Graefes Arch. Klin. Exp. Ophthalmol.* **177,** 212–221.

Hultsch, E., and Ducrey, N. (1979). Effect of argon and krypton laser irradiation on the normal owl monkey vitreous. *Mod. Probl. Ophthalmol.* **20,** 184–187.

Jacobs, G. H. (1977a). Visual capacities of the owl monkey (*Aotus trivirgatus*). I. Spectral sensitivity and color vision. *Vision Res.* **17,** 811–820.

Jacobs, G. H. (1977b). Visual capacities of the owl monkey (*Aotus trivirgatus*). II. Spatial contrast sensitivity. *Vision Res.* **17,** 821–825.

Jacobs, G. H., Deegan, J. F., II, Neitz, J., Crognale, M. A., and Neitz, M. (1993). Photopigments and color vision in the nocturnal monkey, *Aotus. Vision Res.* **33,** 1773–1783.

Jones, A. E. (1965). The retinal structure of *Aotus trivirgatus* the owl monkey. *J. Comp. Neurol.* **125,** 19–28.

Jones, A. E., and Jacobs, G. H. (1963). Electroretinographic luminosity functions of the *Aotus* monkey. *Am. J. Physiol.* **204,** 47–50.

Klöti, R. (1967). Experimental occlusion of retinal and ciliary vessels in owl monkeys. I. Technique and clinical observations of selective embolism of the central retinal artery system. *Exp. Eye Res.* **6,** 393–399.

Kroll, A. J., and Machemer, R. (1968). Experimental retinal detachment in the owl monkey. III. Electron microscopy of retina and pigment epithelium. *Am. J. Ophthalmol.* **66,** 410–427.

Kroll, A. J., and Machemer, R. (1969). Experimental retinal detachment in the owl monkey. V. Electron microscopy of the reattached retina. *Am. J. Ophthalmol.* **67,** 117–130.

Kroll, A. J., and Machemer, R. (1971). Experimental retinal detachment in the owl monkey. VIII. Photoreceptor protein renewal in early retinal redetachment. *Am. J. Ophthalmol.* **72,** 356–366.

Laqua, H. (1975). Massive periretinal proliferation (MPP). IV. Pre- and subretinal proliferation of glial tissue in experimental retinal detachment. *Mod. Probl. Ophthalmol.* **15,** 235–245.

Laqua, H., and Machemer, R. (1975a). Glial cell proliferation in retinal detachment (massive periretinal proliferation). *Am. J. Ophthalmol.* **80,** 602–618.

Laqua, H., and Machemer, R. (1975b). Clinical-pathological correlation in massive periretinal proliferation. *Am. J. Ophthalmol.* **80,** 913–929.

Luxenberg, M. N. (1979). An experimental approach to the study of intraocular *Toxocara canis. Trans. Am. Ophthalmol. Soc.* **77,** 542–602.

Machemer, R. (1968a). Experimental retinal detachment in the owl monkey. II. Histology of retina and pigment epithelium. *Am. J. Ophthalmol.* **66,** 396–410.

Machemer, R. (1968b). Experimental retinal detachment in the owl monkey. IV. The reattached retina. *Am. J. Ophthalmol.* **66,** 1075–1091.

Machemer, R., and Buettner, H. (1972). Experimental retinal detachment in the owl monkey. IX. Radioautographic study of protein metabolism. *Am. J. Ophthalmol.* **73,** 377–389.

Machemer, R., and Kroll, A. J. (1971). Experimental retinal detachment in the owl monkey. VII. Photoreceptor protein renewal in normal and detached retina. *Am. J. Ophthalmol.* **71,** 690–695.

Machemer, R., and Norton, E. W. D. (1968). Experimental retinal detachment in the owl monkey. I. Methods of production and clinical picture. *Am. J. Ophthalmol.* **66,** 388–396.

Machemer, R., van Horn, D., and Aaberg, T. M. (1978). Pigment epithelial proliferation in human retinal detachment with massive periretinal proliferation. *Am. J. Ophthalmol.* **85,** 181–191.

Mandelcorn, M. S., Machemer, R., Fineberg, E., and Hersch, S. B. (1975). Proliferation and metaplasia of intravitreal retinal pigment epithelium cell autotransplants. *Am. J. Ophthalmol.* **80,** 227–237.

Meyers, S. M., and Bonner, R. F. (1982). Retinal irradiance from vitrectomy endoilluminators. *Am. J. Ophthalmol.* **94,** 26–29.

Mukai, N., Lee, P. F., Oguri, M., and Schepens, C. L. (1975). A long-term evaluation of silicone retinopathy in monkeys. *Can. J. Ophthalmol.* **10,** 391–402.

Murray, R. G., Jones, A. E., and Murray, A. (1973). Fine structure of photoreceptors in the owl monkey. *Anat. Rec.* **175,** 673–695.

Newman, E. A. (1987a). Regulation of potassium levels by Müller cells in the vertebrate retina. *Can. J. Physiol. Pharmacol.* **65,** 1028–1032.

Newman, E. A. (1987b). Distribution of potassium conductance in mammalian Müller (glial) cells: A comparative study. *J. Neurosci.* **7,** 2423–2432.

Ogden, T. E. (1974). The morphology of retinal neurons of the owl monkey *Aotus*. *J. Comp. Neurol.* **153,** 399–428.

Ogden, T. E. (1975). The receptor mosaic of *Aotus trivirgatus:* Distribution of rods and cones. *J. Comp. Neurol.* **163,** 193–202.

Ogden, T. E. (1978). Nerve fiber layer astrocytes of the primate retina: Morphology, distribution and density. *Invest. Ophthalmol. Visual Sci.* **17,** 499–510.

Ogden, T. E. (1983a). Nerve fiber layer of the macaque retina: Retinotopic organization. *Invest. Ophthalmol. Visual Sci.* **24,** 85–98.

Ogden, T. E. (1983b). Nerve fiber layer of the owl monkey retina: Retinotopic organization. *Invest. Ophthalmol. Visual Sci.* **24,** 265–269.

Ogden, T. E., Pietantoni, R., and Citron, M. C. (1982). Studies of horizontal cell processes in frogs and primates. *In* "The Structure of the Eye" (J. G. Hollyfield, ed.), pp. 141–149. Elsevier, North-Holland, Amsterdam.

Ordy, J. M., and Samorajski, T. (1968). Visual acuity and ERG-CFF in relation to the morphologic organization of the retina among diurnal and nocturnal primates. *Vision Res.* **8,** 1205–1225.

Parrish, R., Gass, J. D. M., and Anderson, D. R. (1982). Outer retina ischemic infarction—A newly recognized complication of cataract extraction and closed vitrectomy. Part 2. An animal model. *Ophthalmology (Rochester, Minn.)* **89,** 1472–1477.

Peters, M. A., Abrams, G. W., Hamilton, L. H., Burke, J. M., and Schreiber, T. M. (1985). The non-expansile, equilibrated concentration of perfluoropropane gas in the eye. *Am. J. Ophthalmol.* **100,** 831–839.

Polyak, S. L. (1957). "The Vertebrate Visual System." Univ. of Chicago Press, Chicago.

Pomerantzeff, O., Kaneko, H., Donovan, R. H., Schepens, C. L., and McMeel, J. W. (1976). Effect of the ocular media on the main wavelengths of argon laser emission. *Invest. Ophthalmol.* **15,** 70–77.

Pomerantzeff, O., Eng, D., Wang, G. J., Pankratov, M., and Schneider, J. (1983). Time and location analysis of lesion formation in photocoagulation. *Arch. Ophthalmol. (Chicago)* **101,** 954–957.

Pomerantzeff, O., Timberlake, G., Wang, G. J., Pankratov, M. M., and Schneider-Goren, J. (1984). Automation in krypton laser photocoagulation. *Invest. Ophthalmol. Visual Sci.* **25,** 711–719.

Quigley, H. A., and Anderson, D. R. (1976). The dynamics and location of axonal transport blockade by acute intraocular pressure elevation in primate optic nerve. *Invest. Ophthalmol.* **15,** 606–616.

Quigley, H. A., and Anderson, D. R. (1977). Distribution of axonal transport blockade by acute intraocular pressure elevation in the primate optic nerve head. *Invest. Ophthalmol. Visual Sci.* **16,** 640–644.

Quigley, H. A., Guy, J., and Anderson, D. R. (1979). Blockade of rapid axonal transport. Effect of intraocular pressure elevation in primate optic nerve. *Arch. Ophthalmol. (Chicago)* **97,** 525–531.

Radius, R. L. (1980a). Thickness of the retinal nerve fiber layer in primate eyes. *Arch. Ophthalmol. (Chicago)* **98,** 1625–1629.

Radius, R. L. (1980b). Optic nerve fast axonal transport abnormalities in primates. Occurrence after short posterior ciliary artery occlusion. *Arch. Ophthalmol. (Chicago)* **98,** 2018–2022.

Radius, R. L. (1983). Pressure induced fast axonal transport abnormalities and the anatomy at the lamina cribrosa in primate eyes. *Invest. Ophthalmol. Visual Sci.* **24,** 343–346.

Radius, R. L., and Anderson, D. R. (1978). Retinal ganglion cell degeneration in experimental optic atrophy. *Am. J. Ophthalmol.* **86,** 673–679.

Radius, R. L., and Anderson, D. R. (1979a). The mechanism of disc pallor in experimental optic atrophy. A fluorescein angiographic study. *Arch. Ophthalmol. (Chicago)* **97,** 532–535.

Radius, R. L., and Anderson, D. R. (1979b). The histology of retinal nerve fiber layer bundles and bundle defects. *Arch. Ophthalmol. (Chicago)* **97,** 948–950.

Radius, R. L., and Anderson, D. R. (1979c). The course of axons through the retina and optic nerve head. *Arch. Ophthalmol. (Chicago)* **97,** 1154–1158.

Radius, R. L., and Anderson, D. R. (1980a). Fast axonal transport in early experimental disc edema. *Invest. Ophthalmol. Visual Sci.* **19,** 158–168.

Radius, R. L., and Anderson, D. R. (1980b). Breakdown of the normal optic nerve head blood–brain barrier following acute elevation of intraocular pressure in experimental animals. *Invest. Ophthalmol. Visual Sci.* **19,** 245–255.

Radius, R. L., and Anderson, D. R. (1981a). Reversibility of optic nerve damage in primate eyes subjected to intraocular pressure above systolic blood pressure. *Br. J. Ophthalmol.* **65,** 661–672.

Radius, R. L., and Anderson, D. R. (1981b). Morphology of axonal transport abnormalities in primate eyes. *Br. J. Ophthalmol.* **65,** 767–777.

Radius, R. L., Schwartz, E. L., and Anderson, D. R. (1980). Failure of unilateral carotid artery ligation to affect pressure-induced interruption of rapid axonal transport in primate optic nerves. *Invest. Ophthalmol. Visual Sci.* **19,** 153–157.

Ripps, H., and Vaughan, H. G., Jr. (1969). The spectral sensitivity of evoked potentials from the retina and cortex of nocturnal and diurnal monkeys. *Vision Res.* **9,** 895–907.

Sandell, J. H. (1985). NADPH diaphorase in the mammalian inner retina. *J. Comp. Neurol.* **238,** 466–472.

Sherman, M. S., Wilson, J. R., Kaas, J. H., and Webb, S. V. (1976). X- and Y-cells in the dorsal lateral geniculate nucleus of the owl monkey (*Aotus trivirgatus*). *Science* **192,** 475–477.

Sipperley, J. O., Quigley, H. A., and Gass, V. D. M. (1978). Traumatic retinopathy in primates. The explanation of commotio retinae. *Arch. Ophthalmol. (Chicago)* **96,** 2267–2273.

Tolentino, F. I., Refojo, M. F., Liu, H. S., Schepens, C. L., and Freeman, H. M. (1978). Intravitreous silicone balloon: an experimental study. *Ophthalmic Surg.* **9,** 73–80.

Tolentino, F. I., Foster, C. S., Lahav, M., Liu, L. H. S., and Rabin, A. R. (1982). Toxicity of intravenous miconazole. *Arch. Ophthalmol. (Chicago)* **100,** 1504–1509.

Tsai, E. N. C., and Smith, J. L. (1975). Experimental vortex-choroidal angiograms. *Arch. Ophthalmol. (Chicago)* **93,** 198–206.

van Horn, D. L., Aaberg, T. M., Machemer, R., and Fenzl, R. (1977). Glial cell proliferation in human retinal detachment with massive periretinal proliferation. *Am. J. Ophthalmol.* **84,** 383–393.

van Veen, T., Ostholm, T., Gierschik, P., Spiegel, A., Somers, R., Korf, H. W., and Klein, D. C. (1986). Alpha-transducin immunoreactivity in retinae and sensory pineal organs of adult vertebrates. *Proc. Natl. Acad. Sci. U.S.A.* **83,** 912–916.

Walls, G. L. (1942). "The Vertebrate Eye and its Adaptive Radiation." Cranbrook Press, Bloomfield Hills, MI.

Webb, S. V., and Kaas, J. H. (1976). The sizes and distribution of ganglion cells in the retina of the owl monkey, *Aotus trivirgatus. Vision Res.* **16,** 1247–1254.

Weitz, C. J., Went, L. N., and Nathans, J. (1992). Human tritanopia associated with a third amino acid substitution in the blue sensitive visual pigment. *Am. J. Hum. Genet.* **52,** 444–446.

Wieder, M., Pomerantzeff, O., and Schneider, J. (1981). Retinal vessel photocoagulation: A quantitative comparison of argon and krypton laser effects. *Invest. Ophthalmol. Visual Sci.* **20,** 418–424.

Wikler, K. C., and Rakic, P. (1990). Distribution of photoreceptor subtypes in the retina of diurnal and nocturnal primates. *J. Neurosci.* **10,** 3390–3401.

Wolin, L. R., and Massopust, L. C., Jr. (1967). Characteristics of the ocular fundus in primates. *J. Anat.* **101,** 693–699.

11

The Functional Organization of Visual Cortex in Owl Monkeys

John Allman and Richard Jeo
Division of Biology
California Institute of Technology
Pasadena, California 91125

Martin Sereno
Department of Cognitive Science
University of California at San Diego
La Jolla, California 92093

I. MULTIPLE MAPS OF THE VISUAL FIELD IN VISUAL CORTEX

The cerebral cortex of the owl monkey is less fissured than that of most higher primates. Because of this feature, Allman and Kaas selected the owl monkey for their studies of the functional organization of primate visual cortex that began in 1968 (Allman and Kaas, 1971a,b, 1974b, 1975, 1976). These studies led to the discovery of the large expanse of visually responsive cortex that could be subdivided into many visual areas on the basis of receptive field mapping.

An example of the mapping of a cortical visual area is illustrated in Fig. 1. In the experiment, the anesthetized monkey was placed at the center of a translucent plastic hemisphere. The electrical activity of neurons was recorded with microelectrodes in the visual cortex while moving simple visual stimuli on the surface of the plastic hemisphere. The receptive field obtained from each cortical recording site was the area on the hemisphere that when stimulated resulted in an increase in the electrical spiking activity detected by listening to an audio monitor. The receptive fields obtained in this experiment are depicted as broad ovals in the perimeter chart representing the contralateral half of the visual field that is illustrated in the right side of Fig. 1. In this particular experiment the microelectrode was advanced parallel to the medial wall of the occipital lobe, which is illustrated in the lower left part of Fig. 1. Thus, in penetration 1, electrode sites A through D were recorded as the microelecrode advanced and the corresponding visual receptive fields 1A through 1D were mapped. There was a steady progression of receptive fields as the microelectrode was advanced. When the receptive fields for a series of penetrations were plotted, they formed a "map" of the visual field, which is illustrated in the upper left part of Fig. 1. The map is a representation of the contralateral half of the visual field. Figure 1 is the map of the Medial Visual Area, abbreviated as M, which is unique among cortical visual areas because of its relatively large representation of the more peripheral parts of the visual field, which makes it convenient for illustrative purposes. For the other cortical visual areas the mapping principles are the same, but there is a much greater emphasis on the central part of the visual field, which sustains higher acuity.

This investigation has continued for the past 25 years. Sereno, McDonald, and Allman have recently produced the map illustrated in Fig. 2 based on a large number of receptive fields recorded from throughout the visual cortex. The delineation of the inferotemporal areas was based on the anatomical studies of Weller and Kaas (1985, 1987). In Fig. 2, in the upper right-hand corner there is a dorsolateral view of the owl monkey brain with the anatomical location of the areas depicted. The main body of Fig. 2 is an unfolded and flattened version of the owl monkey visual cortex. The star indicates the representation of the center of the visual field, the line of circles indicates the representation of the vertical midline of the visual field, and the line of black squares indicates the representation of the horizontal meridian. The vertical meridian forms the border between the

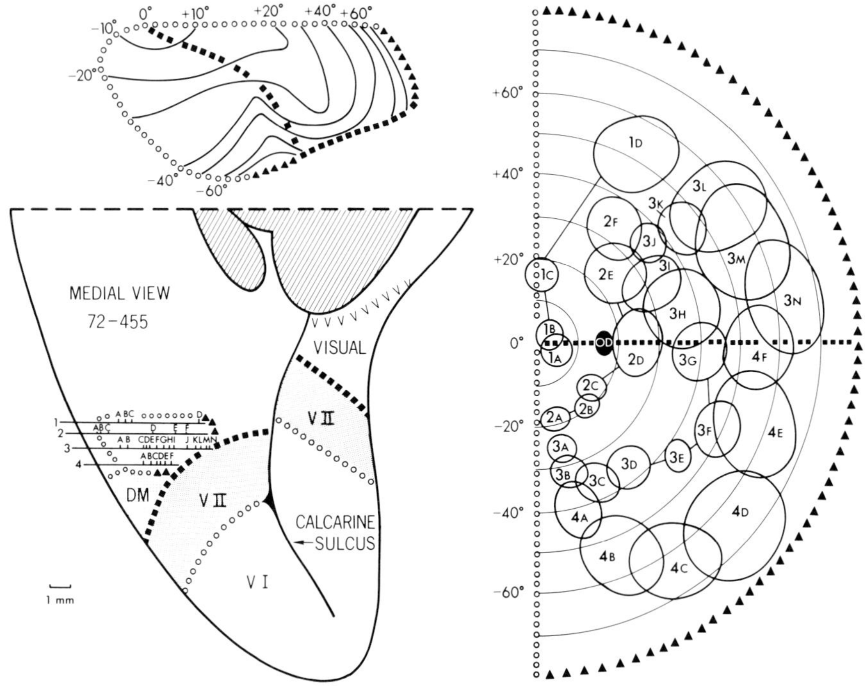

FIGURE 1 Microelectrode recording and receptive field data for the Medial Visual Area in owl monkey 72–455. The diagram on the lower left is a view of the posterior half of the medial wall of cerebral cortex of the left hemisphere with the brain stem and cerebellum removed. Anterior is up and dorsal is to the left in this diagram. Microelectrode penetrations are numbered, and recording sites are indicated by short bars denoted by letters. The corresponding receptive fields are shown in the perimeter chart on the right. In the upper left is an expanded map of the visuotopic organization of the Medial Visual Area. The circles indicate the representation of the vertical meridian (midline) of the visual field; the squares indicate the horizontal meridian of the contralateral half of the visual field; the triangles indicate the temporal periphery of the contralateral hemifield. V I is the First Visual Area, denoted at V1 in the text; V II is the Second Visual Area, denoted as V2 in the text; DM is the Dorsomedial Visual Area; OD indicates the projection of the optic disk or blind spot. Reproduced from Allman and Kaas (1976), with permission from the American Association for the Advancement of Science.

First Visual Area (V1) and the Second Visual Area (V2), and the border between the Ventral Posterior (VP) and Ventral Anterior (VA) Visual areas, as well as many other visual areas. The horizontal meridian forms the border between V2 and a series of adjacent "third tier" visual areas: VP; the Dorsal Lateral Posterior, DLp; the Dorsal Intermediate Area, DI; the Dorsal Medial Area, DM; and finally, area M. The horizontal meridian also forms the border of many other areas.

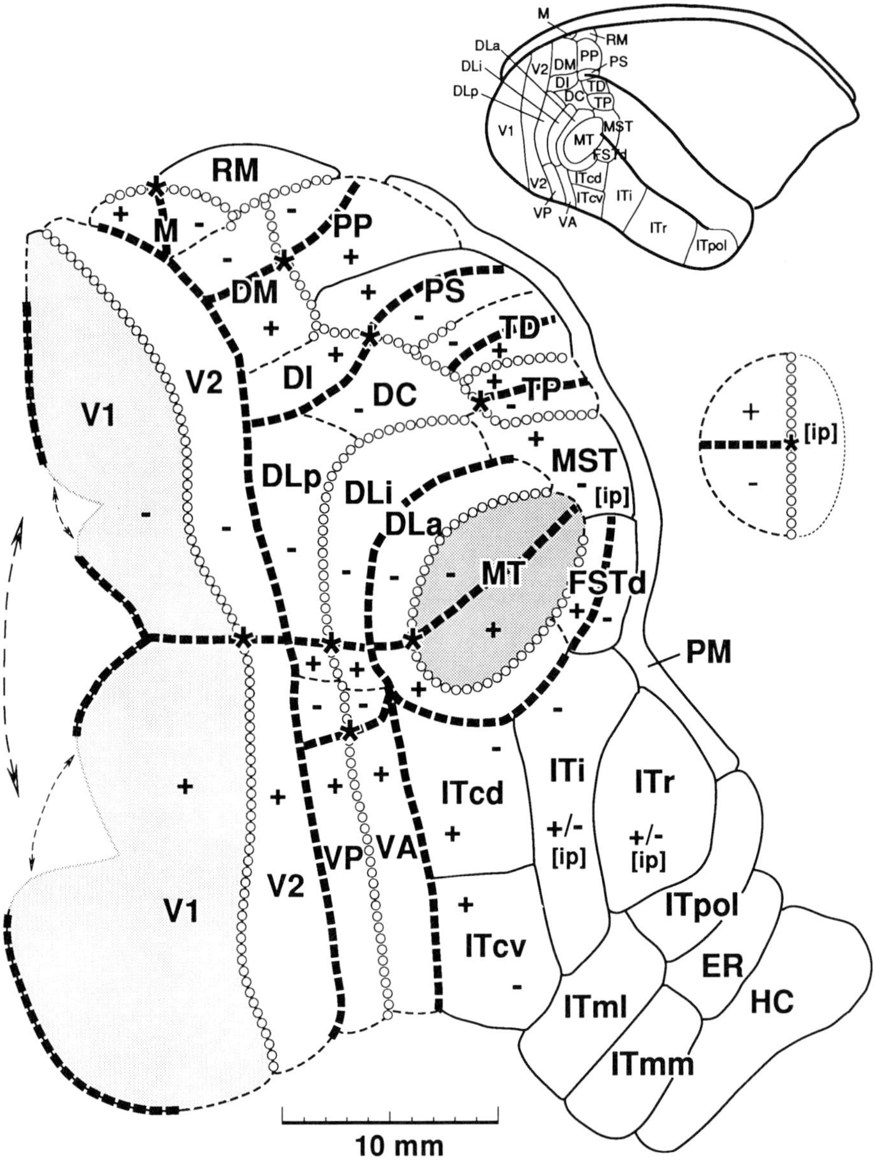

FIGURE 2 Visual cortical areas in the owl monkey (M. Sereno, C. McDonald, and J. Allman, unpublished data). Area boundaries are based on Allman and Kaas (1971a,b, 1974a,b, 1975, 1976), Weller and Kaas (1985, 1987), and Sereno *et al.* (1987; Sereno and Allman, 1991). They were drawn using a cytochrome oxidase-stained flat-mounted hemisphere as a template. A cut was made in V1 in the depths of the calcarine sulcus to allow the cortex to lie flat. Abbreviations: DC, dorsocentral; DI, dorsointermediate; DLa, dorsolateral anterior; DLi, dorsolateral intermediate; DLp, dorsolateral posterior; DM, dorsomedial; ER, entorhinal; FSTd, fundus superior temporal—dorsal division; ITcd, caudal inferotemporal—dorsal division; ITcv, caudal inferotemporal—ventral division; ITi, intermediate inferotemporal; ITml, medial inferotemporal—lateral division; ITmm, medial inferotemporal—medial division; ITpol, polar inferotemporal; ITr, rostral inferotemporal; MT, middle temporal; PM, polymodal; PP, poste-

II. HISTOLOGICAL FEATURES OF CORTICAL VISUAL AREAS

Figures 3 and 4 are histological sections stained for myelinated fibers through the visual cortex. Figure 3 is a horizontal section from V1 to MT. V1 is characterized by the densely stained stria of Gennari (SG), which is the anatomical basis for the term striate cortex. V2 is characterized by a slightly denser myelin staining in the lower cortical layers. MT is characteristically densely stained for myelin. Figure 4 is a coronal section through MT, which stains out in bold contrast to the adjacent visual areas. The underlying part of V1 in the calcarine sulcus is located below MT.

The corpus callosum connects the two hemispheres and carries the fiber connections that unify the representation of the two halves of the visual field. Figure 5 illustrates the distribution of corpus callosum projections indicated by the dots. Areas V1 and MT are shaded. There is a definite clustering of projections along the border between V1 and V2, and to a lesser extent along the border of MT. The corpus callosum unifies the two halves of the visual field and its termination in these areas approximately corresponds to the representation of the vertical midline of the visual field. This correspondence with the vertical midline also is true for the second discrete band on the ventral surface about 5 mm anterior to striate cortex. Figure 6 illustrates the receptive fields recorded near the vertical midline of the visual field and the callosal band running across the ventral occipital cortex. This band led to the discovery of the twin ventral areas VP and VA.

Figure 7 illustrates a section parallel to the surface of flattened visual cortex stained for the energy-producing enzyme, cytochrome oxidase, from the work of Tootell *et al.* (1985). Area V1 is to the right in this figure and is filled with small dark spots. These are the cytochrome oxidase-rich "puffs," "patches," or "blobs" (Wong-Riley and Carroll, 1984; Horton, 1984; Livingstone and Hubel, 1984). In diurnal monkeys, these cytochrome oxidase-rich spots are clearly related to color processing (Livingstone and Hubel, 1984; Tootell *et al.*, 1988; Ts'o and Gilbert, 1988), but their presence in nocturnal owl monkeys suggests that they have some other, more basic function. Allman and Zucker (1990) have suggested that they are involved in the continuous analysis of luminosity and thus have greater energy requirements than the surrounding lightly stained zones, which are involved in the analysis of statistically rarer events, such as a specific orientation or disparity. Area V2 is characterized by a series of stripes abutting on the V1 border. In the left side of Fig. 7, area MT is an oval of dense

---

rior parietal; PS, posterior sylvian; RM, rostromedial; ST, superior temporal; TD, temporodorsal; TP, temporoparietal; VA, ventroanterior; VP, ventroposterior; V1, First Visual Area; V2, Second Visual Area.

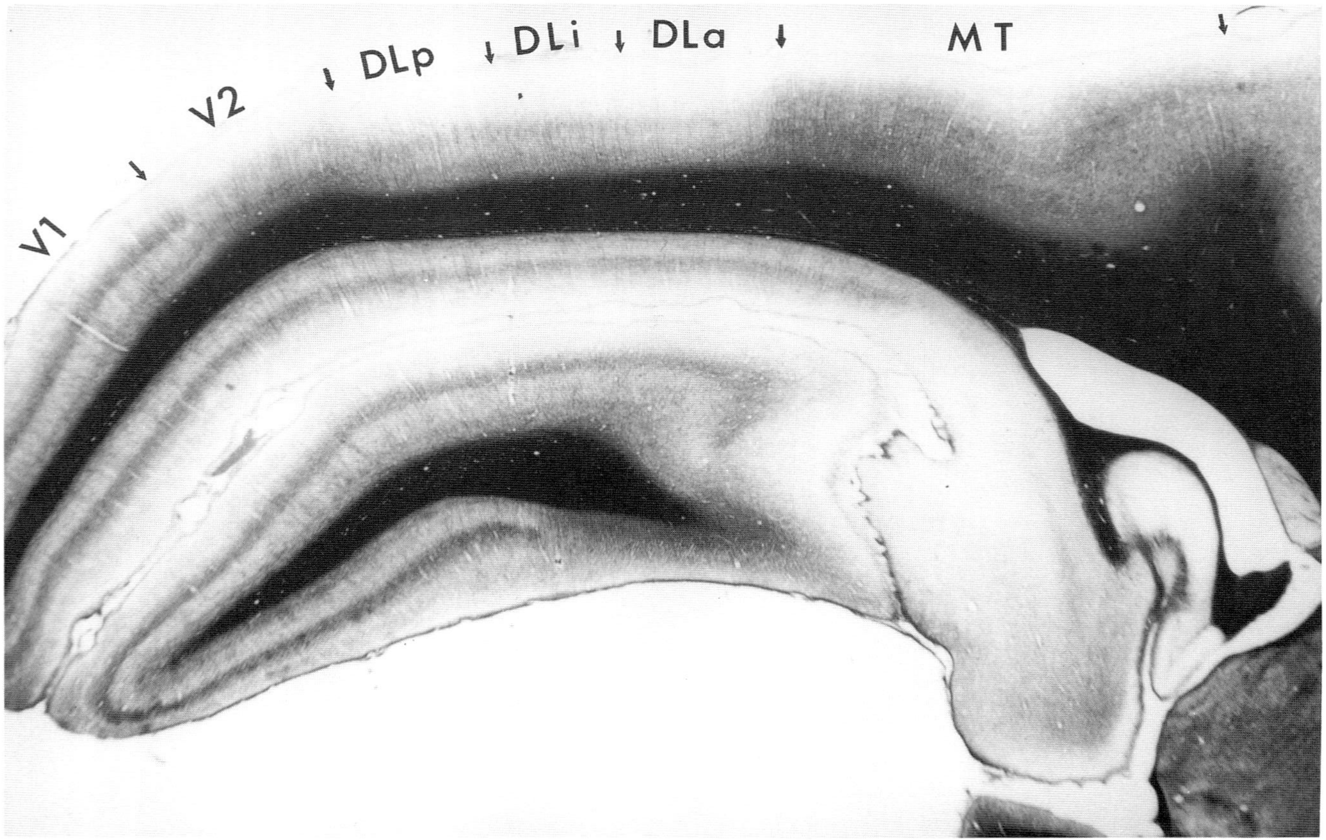

FIGURE 3 An oblique section between horizontal and sagittal through the visual cortex of the owl monkey stained with the Gallyas technique for fibers. Abbreviations are as in Fig. 2.

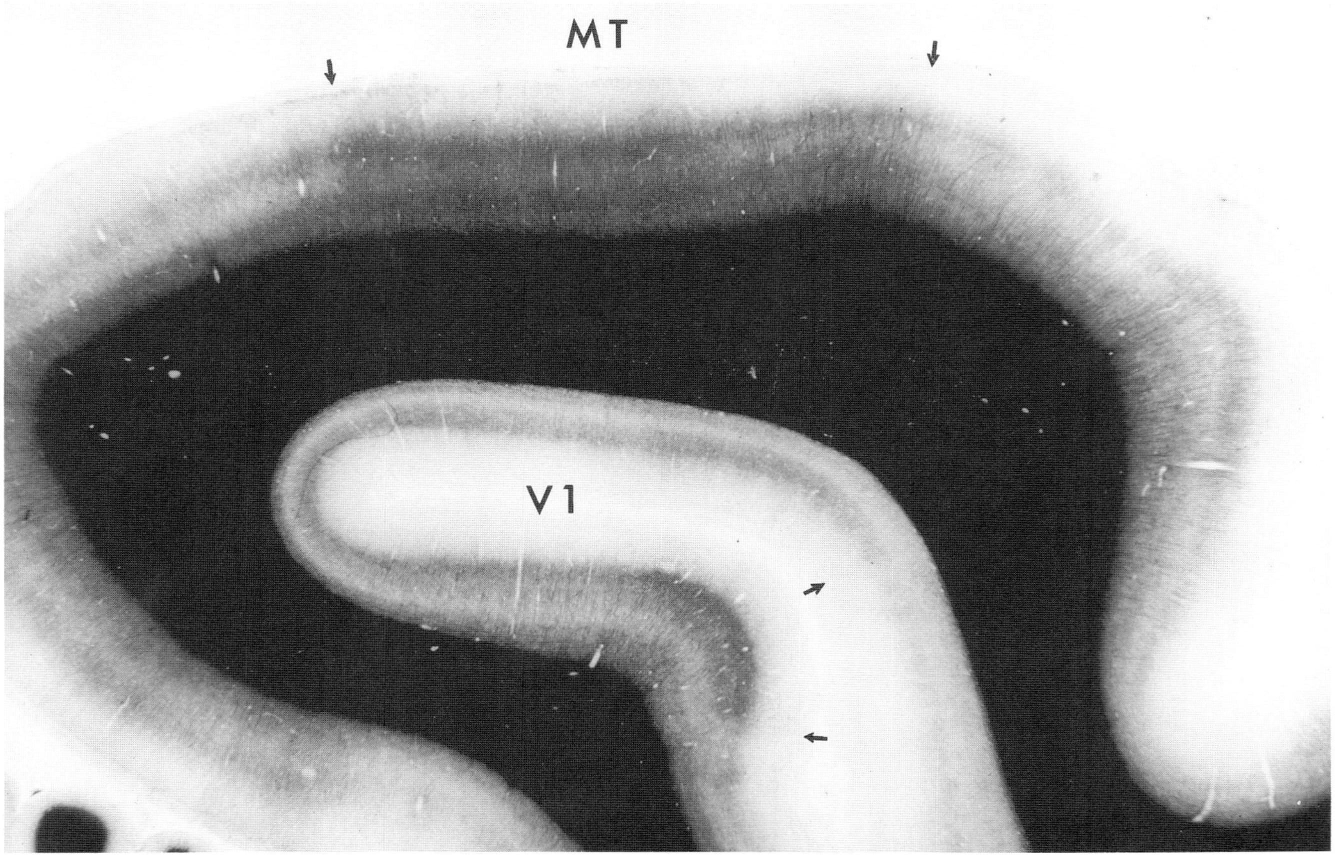

FIGURE 4 A coronal section through area MT and adjacent structure in the owl monkey stained with the Heidenhain technique for fibers.

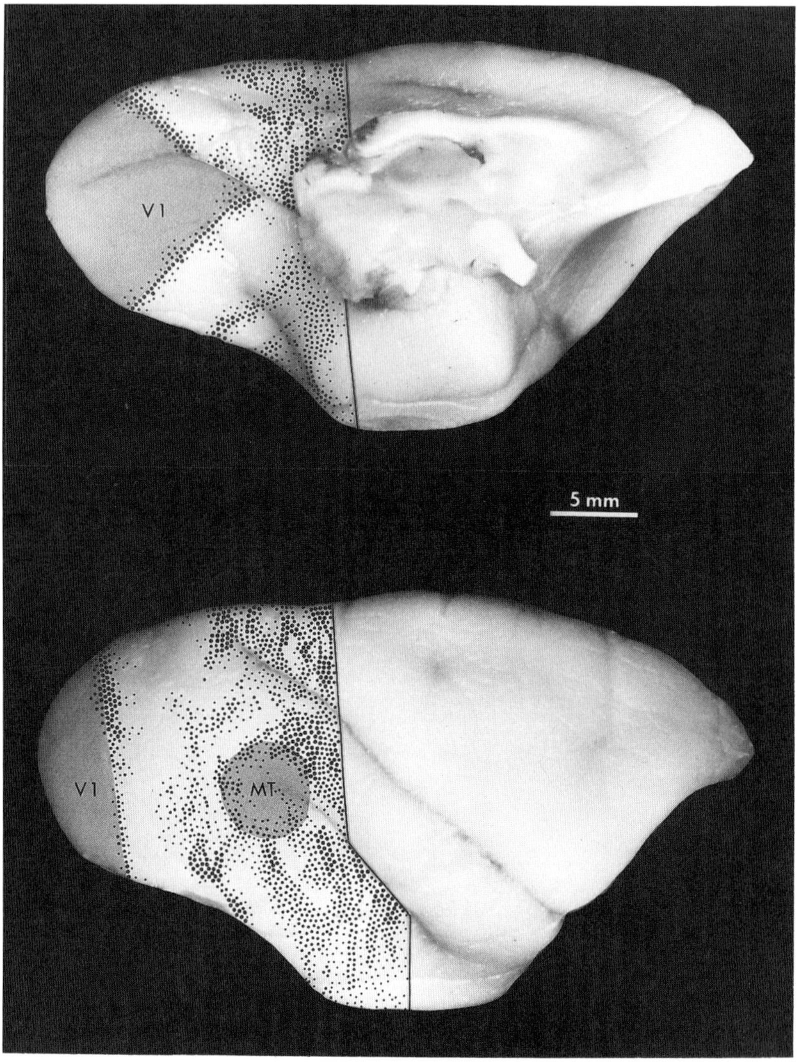

FIGURE 5 Distribution of corpus callosum terminals in the visual cortex of the owl monkey. Posterior is to the left. Dorsolateral (top) and ventromedial (bottom) views of the cortex of an owl monkey. The posterior shaded region is V1 and the oval shaded region is MT. The location of corpus callosum terminals is indicated by the dots; the terminals were plotted in only the posterior half of the cortex. Reproduced from Newsome and Allman (1980) with permission.

cytochrome oxidase staining, which is probably related to the high energetic requirements inherent in processing temporarlly varying stimuli (see the following). Within the densely stained oval are islands of lower activity. The work of Born and Tootell (1992) suggests that the antagonistic surround mechanisms described in the next section are stronger in the lightly stained zones within MT.

III. NEURAL RESPONSES IN THE CORTICAL VISUAL AREAS

A. Overview

Figure 8 summarizes some of the other important response properties of neurons in these areas. In each case the size of the square indicates the relative strength of the response property as expressed in the area. Neurons in area DL are significantly less orientation selective than in MT, DM, or M, but DL neurons are much more dimensionally selective (i.e., sensitive to stimulus shape). Area MT neurons are more sensitive to direction of motion than DL neurons, which in turn are more sensitive to directionality than neurons in DM or M. Neurons in MT are much more sensitive to the motion of fields of random dots than are neurons in other areas. Area DL contains the greatest emphasis on the representation of the central visual field, whereas area M has the least emphasis on central vision.

B. Responses to Flashed and Moving Stimuli

Figure 9 illustrates the summed histogram responses of populations of neurons recorded from areas MT, DL, M, and DM, when an optimally oriented stationary bar of light was presented. (This work was done before Sereno, McDonald, and Allman discovered the subdivisions of Dl.) The histograms for area MT have an initial response latency of 33 msec with a strong transient peak response at 59 msec, followed by a trough at about 175 msec and then a modest rebound in activity. Area M has a similar histogram with a slightly slower initial response at 42 msec. Area DL has a much slower initial response at 63 msec, a peak at 84 msec, and a much larger tonic level of activity. Area DM is intermediate between DL and M both anatomically and in its histogram. These data indicate that areas MT and M have mainly a fast, transient response to visual stimuli, whereas area DL has a much slower and more sustained response. These response histograms reflect a fundamental difference between areas MT and M, which are involved in the analysis of motion, and area DL, which is more involved in the analysis of stimulus form.

Figure 10 illustrates histograms of the directionality of neurons in these four areas. The higher the index, the greater the differential sensitivity of

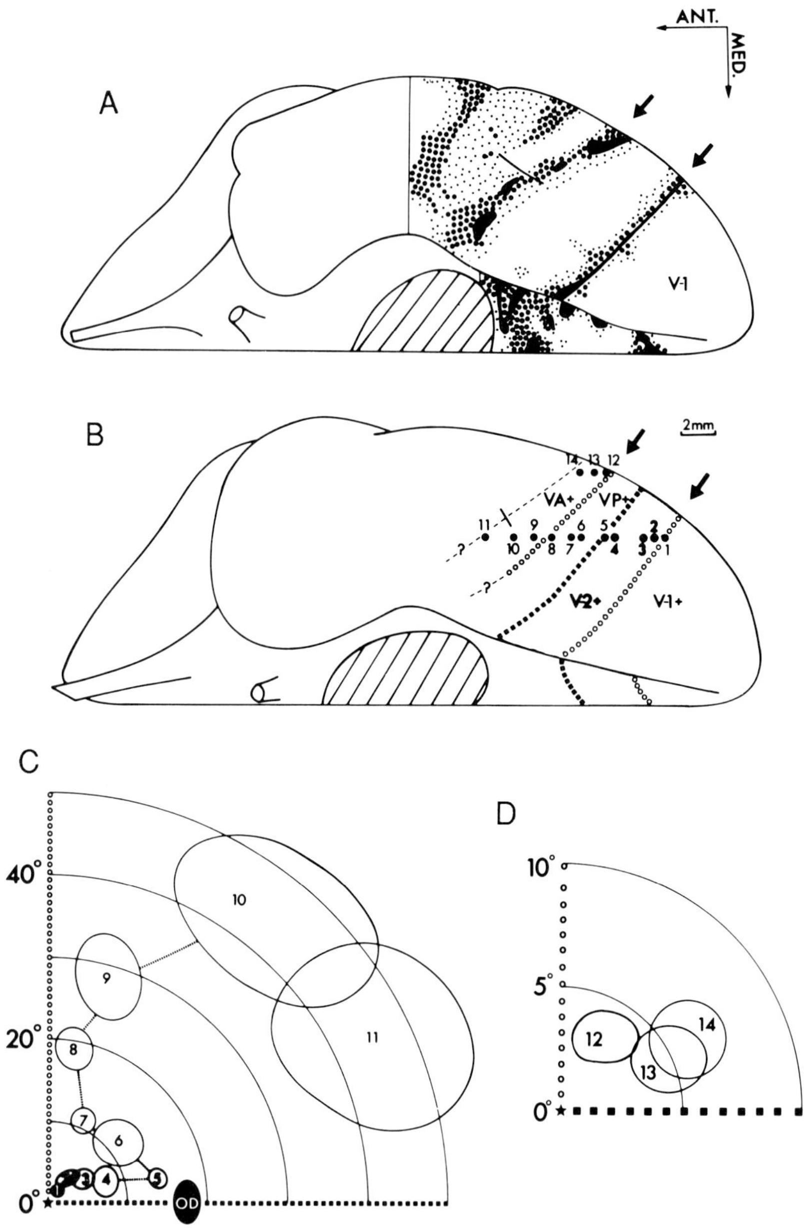

FIGURE 6 Comparison of anatomical results with physiological data for the ventral surface of owl monkey visual cortex. (A) The pattern of degeneration on the ventral surface of owl monkey 78-3. Heavy degeneration is solid black, moderate degeneration is represented by large dots, and light degeneration by the small dots. (B) Electrophysiological recording sites (1 through 14) on the ventral surface of owl monkey 72-343. The location of the recording sites is based on a histological reconstruction of the electrode tracks superimposed on a photograph of the ventral surface taken postmortem. Open circles denote the vertical meridian representations and the solid squares signify the horizontal meridian representation. Dashed lines represent uncertain boundaries. (C) Receptive fields for recording sites 1 through 11.

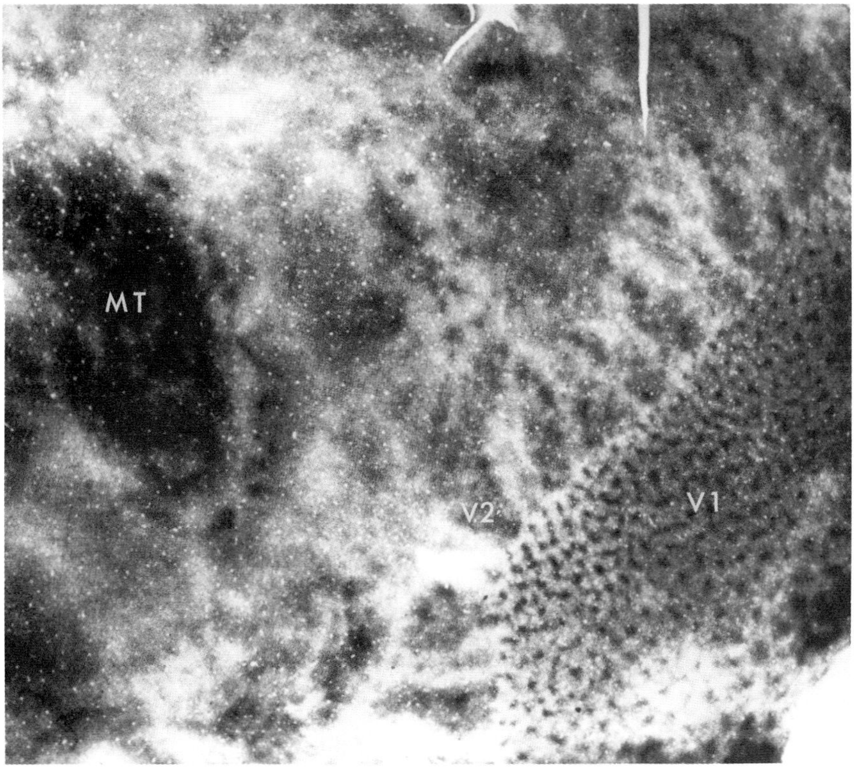

FIGURE 7 Portion of a flattened piece of visual cortex from an owl monkey, sectioned parallel to the surface and stained for cytochrome oxidase activity. Posterior is to the right; dorsal is up. The posterior region containing the fairly regular array of dark spots is V1; the adjacent narrow region containing transverse stripes is V2. The dark-stained oval to the left is MT. Reproduced from Tootell *et al.* (1985) with permission.

The black receptive field is in V1, the gray-shaded receptive fields are in V2, and the unshaded receptive fields are in VP and VA. Note the progression of receptive fields from the hortizontal meridian (recording site 5) at the V2–VP border to the vertical meridian (recording site 8) at the VP–VA border. (D) Receptive fields for recording sites 12–14. Note the change in scale between C and D. Recording site 12, near the VP–VA boundary, yielded a receptive field very near the vertical meridian. The physiological representations of the vertical meridian illustrated in B appear remarkably similar to the distinct bands of callosal degeneration in A (solid arrow). The data of C and D also illustrate the progression from the representation of the center-of-gaze laterally to the representation of the periphery medially. Abbreviations as in Fig. 2. Reproduced from Newsome and Allman (1980) with permission.

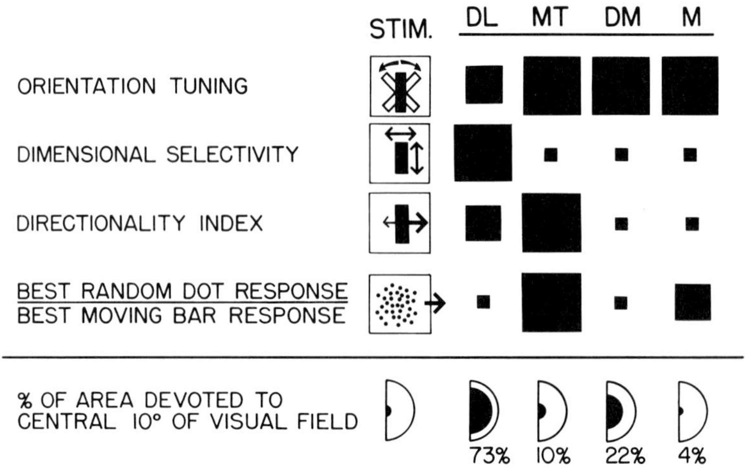

FIGURE 8 Functional specializations in the extrastriate cortical visual areas of the owl monkey based on Baker *et al.* (1981) and Petersen *et al.* (1980). The functional attributes are listed on the left; a pictoral representation of the functional attribute is illustrated in the leftmost column. The four columns labeled DL, MT, DM, and M indicate the relative strength of the functional attribute in that area by the magnitude of the square. The relative representation of the central 10 degrees of the visual field is indicated in the horizontal column in the bottom of the figure.

the neurons to direction of motion. Note than in MT the neurons cluster around a directionality index of 0.9 to 1.0, indicating that they are very sensitive to direction of motion.

Figure 11 illustrates the responses recorded sequentially from a series of 12 neurons in a single vertical penetration perpendicular to the surface of MT. On the left are responses to different directions of bar motion; on the right are responses to flashed bars of different orientations. Note that the first 11 neurons responded optimally to the horizontally oriented flashed bar and that the 12th neuron was inhibited by horizontally oriented flashed bars. The microelectrode was passing down a cortical column defined by preference for horizontal bar stimuli. It is interesting to note that the first 6 neurons in this penetration preferred a horizontal bar moving downward, and that then there was an abrupt shift to horizontal bars moving upward until the last neuron, in which the preferred stimulus was a vertical bar moving to the left. These data suggest that the basic columnar system in MT is for orientation and that grafted upon this is a second system for direction of motion. From the data presented in Fig. 11 it is also possible to interpret the directional preference to be related to cortical lamination, however, there is no other evidence to support the notion of laminar specialization for direction.

C. Effects of Motion Adaptation in Area MT

The presence of adjacent neuron populations specialized for detecting opposite directions of motion suggests a possible mechanism for the "waterfall illusion" or "motion aftereffect." When an observer gazes steadily at an array of stimuli moving in a particular direction for a few minutes, and then the stimulus movement stops, the observer will see the stationary array appear to move in the opposite direction. Thus motion perception may be the product of the precise balance between oppositely tuned populations of neurons. Figure 12 illustrates the influence of adapting motion with random dot backgrounds on the responses of MT neurons. The top two histograms in Fig. 12B were the responses to bar motion in the opposite and preferred direction of motion when the random dot background was stationary. When the background random dots moved in the opposite to the preferred direction of the cell for 20 sec before being tested with a bar moving in the preferred direction, there was a substantial enhancement of the response. When the background random dots moved in the preferred direction of the cell prior to testing with bar moving in the preferred direction, the response was substantially reduced. When the two effects are combined, directionally selective neurons show a large effect from prior motion adaptation.

D. Responses from beyond the Classical Receptive Field

The true receptive fields for MT neurons are very much larger than the classical receptive fields mapped against a stationary background. The stimuli presented in the nonclassical field strongly and selectively influence the responses to stimuli presented within the classical receptive field. For example, in the left half of Fig. 13, the classical receptive field was the small central square containing moving dots. The directional tuning curve below this shows the responses elicited by different directions of movement of random dot array. The optimal direction of motion was horizontal to the right (zero degrees). When this central square was stimulated with random dots moving in the optimal direction and simultaneously the surrounding field was stimulated with moving dots, there was substantial suppression of the response by background movement in the direction that was optimal for the center, and there was considerable enhancement of the response by background movements in the direction opposite to the optimal direction. It should be noted that no observable response could be obtained from this neuron by stimulating the surround by itself. Thus the response from the classical receptive field was contingent on what was happening elsewhere in the visual field in a directionally antagonistic way. This is probably an important mechanism for the integration of local and global motion in visual

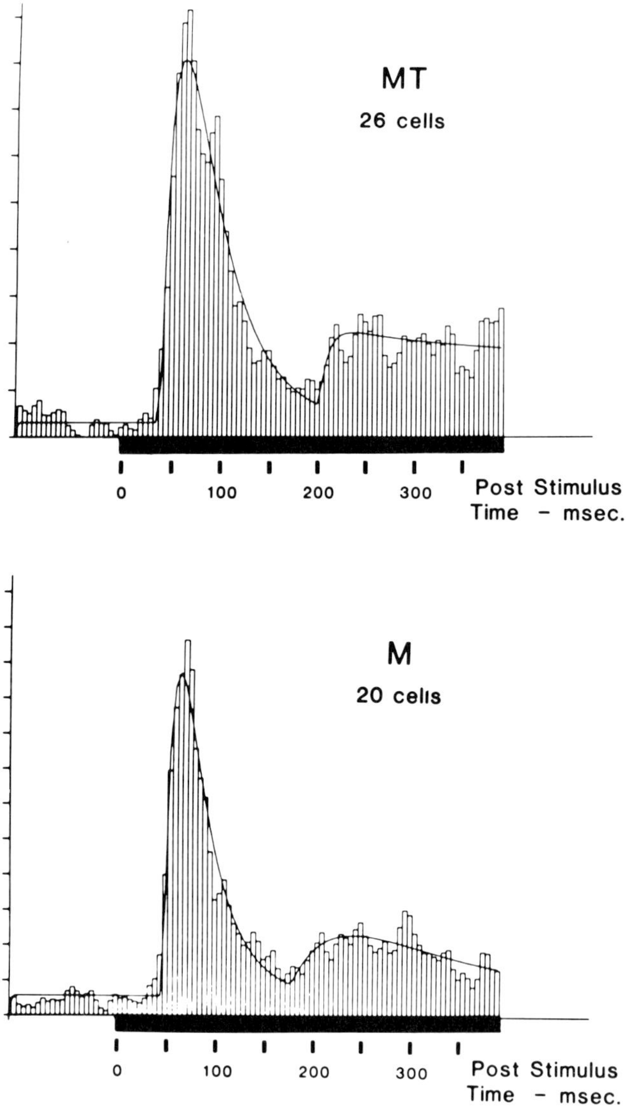

FIGURE 9 Panels display the averaged peristimulus histogram for the response to an optimally oriented flashed bar in the four visual areas sampled. The histogram bins are 5 msec wide. Time since stimulus onset is represented on the horizontal axis with responses before 0 msec being spontaneous activity. The vertical axis was normalized to the peak response for each panel. Superimposed on each histogram is the best-fitting curve obtained by using a random search algorithm. Reproduced from Petersen *et al.* (1988) with permission.

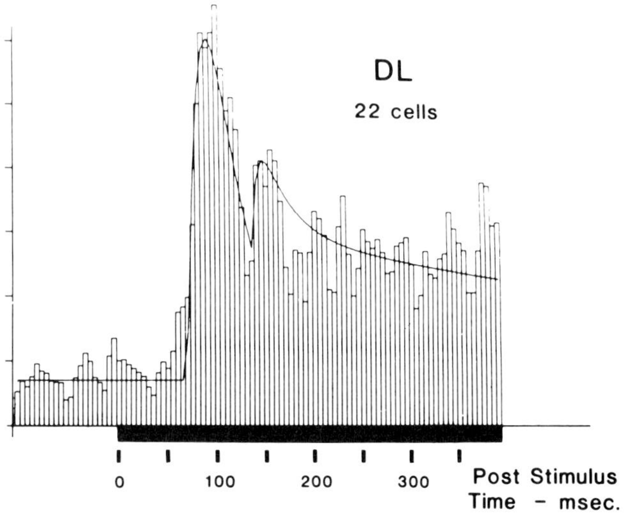
DL
22 cells
0
100
200
300
Post Stimulus
Time - msec.

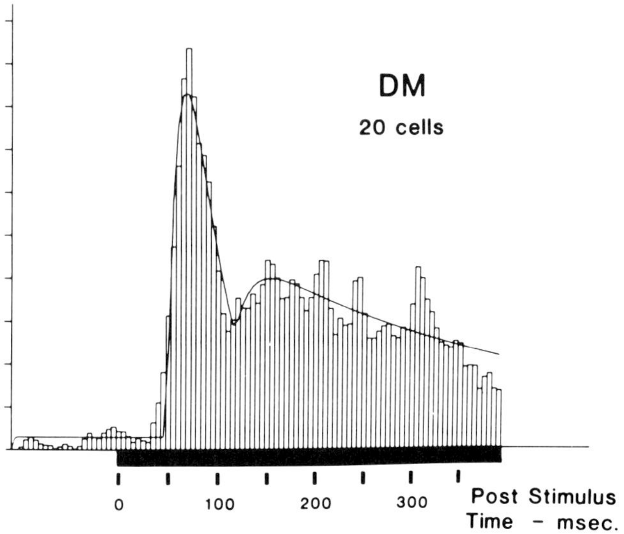
DM
20 cells
0
100
200
300
Post Stimulus
Time - msec.

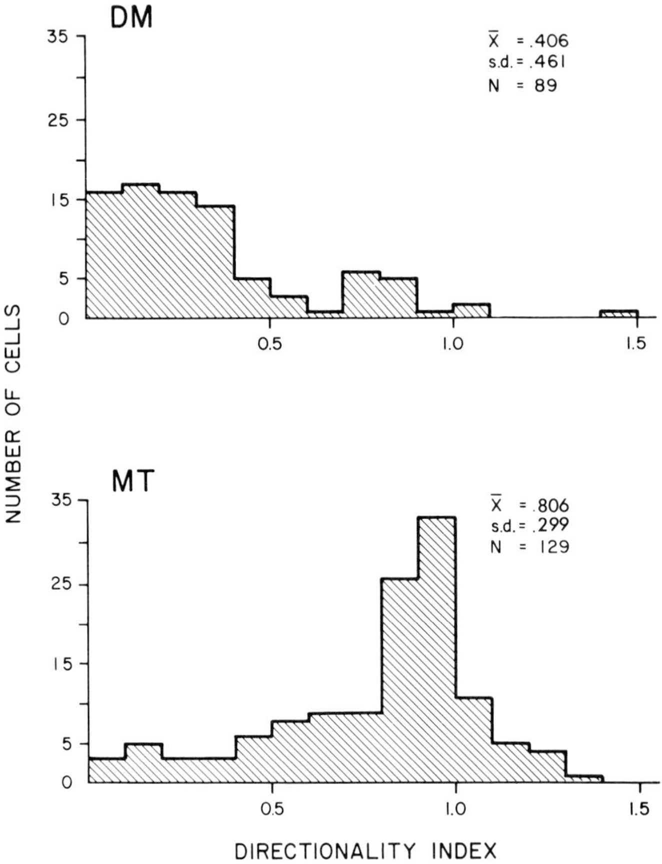

FIGURE 10 Distribution of directionality indices for neurons recorded from DM, DL, MT, and M. The directionality index was (1 minus opposite direction response divided by the best direction response). Reproduced from Baker *et al.* (1981) with permission.

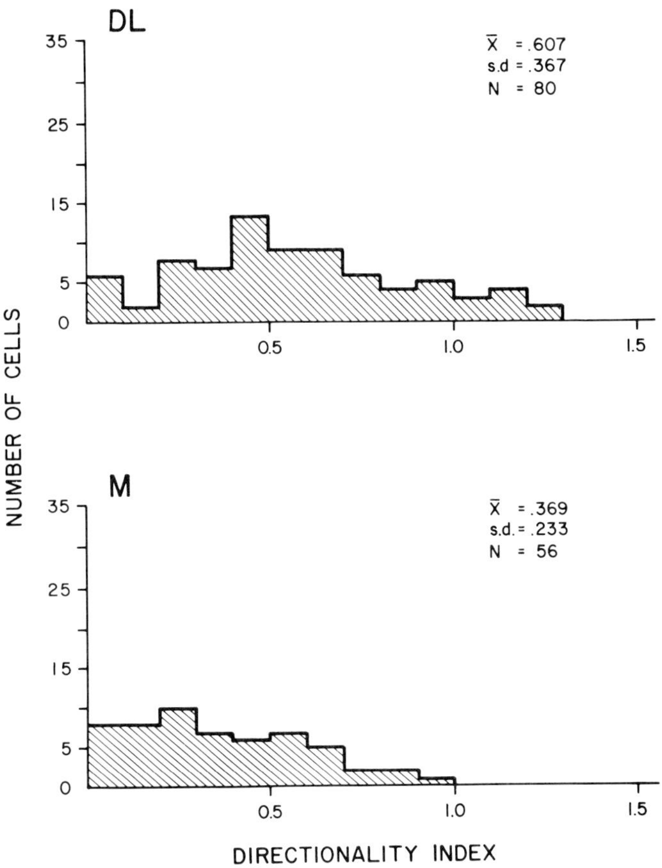
DL
X̄ = .607
s.d = .367
N = 80
M
X̄ = .369
s.d. = .233
N = 56
NUMBER OF CELLS
DIRECTIONALITY INDEX
0
5
15
25
35
0.5
1.0
1.5

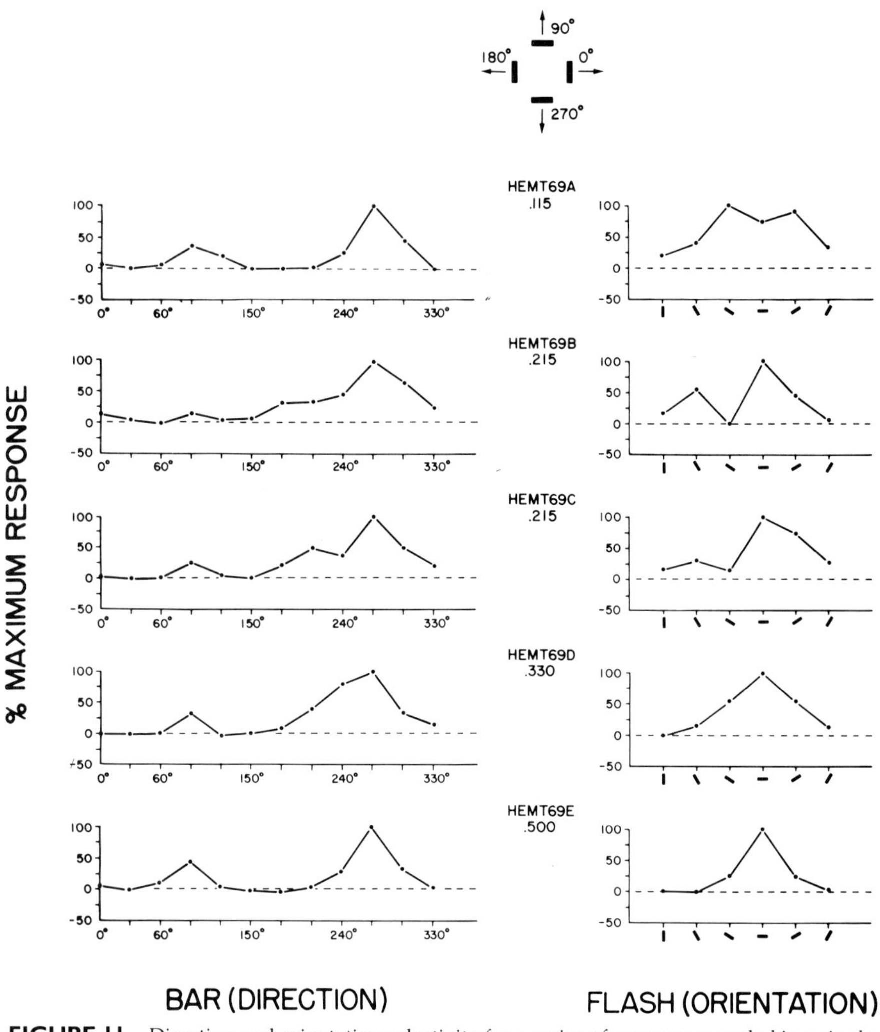

FIGURE 11 Direction and orientation selectivity for a series of neurons recorded in a single penetration nearly perpendicular to the surface of MT. A pair of graphs is illustrated for each unit (HEMT69A through L). The depth beneath the surface at which each cell was recorded is given beneath each identifying number. An electrolytic lesion was made at the bottom of the microelectrode track. Cells A through J were located in layers II and III: cells K and L were located in layer IV. Graphs on the left illustrate the average response of each cell to five presentations of a 20 × 1 degree light bar at 12 different angles. HEMT69A through F preferred 270 degrees; HEMT69G through K preferred 90 degrees. Graphs on the right illustrate the average response of the cells to 10 presentations of a flashed bar at the orientations shown. All cells except the last preferred the horizontal orientation; the last was inhibited by horizontal bars. The direction of movement was 90 degrees to the bar orientation, thus the preferred directions 270 degrees (down) and 90 degrees (up) are consistent with the preferred horizontal orientation. Reproduced from Baker *et al.* (1981) with permission.

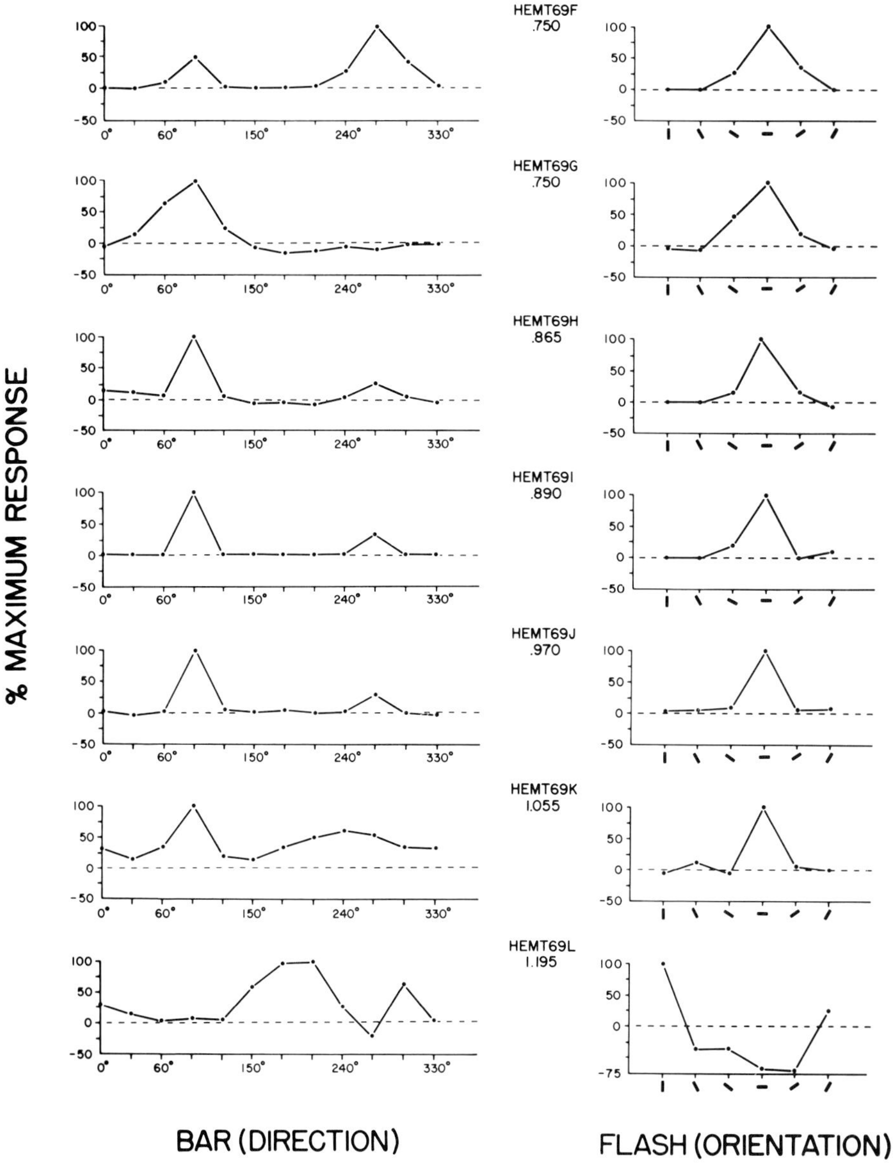
% MAXIMUM RESPONSE
HEMT69F
.750
HEMT69G
.750
HEMT69H
.865
HEMT69I
.890
HEMT69J
.970
HEMT69K
1.055
HEMT69L
1.195
BAR (DIRECTION)
FLASH (ORIENTATION)

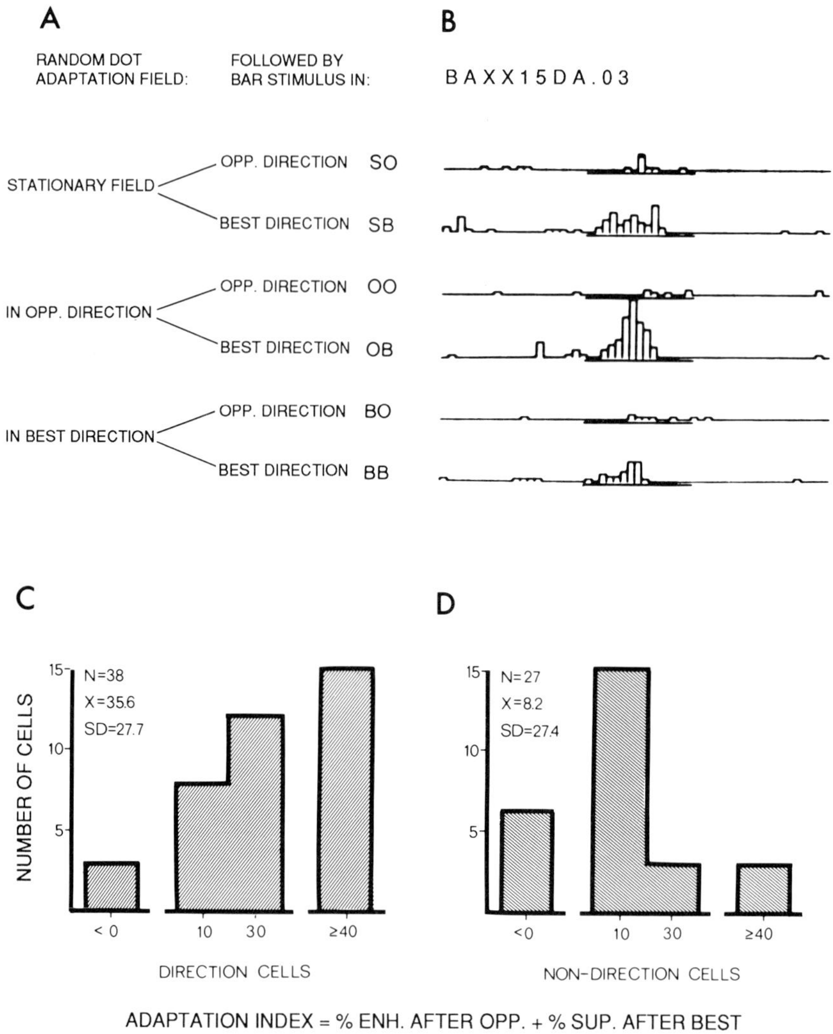

FIGURE 12 (A) Diagrammatic representation of an adaptation series. A 20-sec period of adaptation to stationary noise or noise moving in either the best or opposite directions was followed by a bar stimulus moving in the best direction or opposite direction. These six conditions were pseudorandomly interleaved and presented five times each. (B) An example of an adaptation series. Note that the response is strongest after adaptation in the opposite direction, and weakest after adaptation in the best direction. (C, D) The distribution of adaptation indices for directional and nondirectional cells. The AI was calculated by adding the enhancement after adaptation in the opposite direction to the suppression after adaptation in the best direction. The AI's for directional cells are generally high, whereas those for the nondirectional cells tend to congregate near zero. The difference between these distributions is statistically significant at the 0.001 level. Reproduced from Petersen *et al.* (1985) with permission.

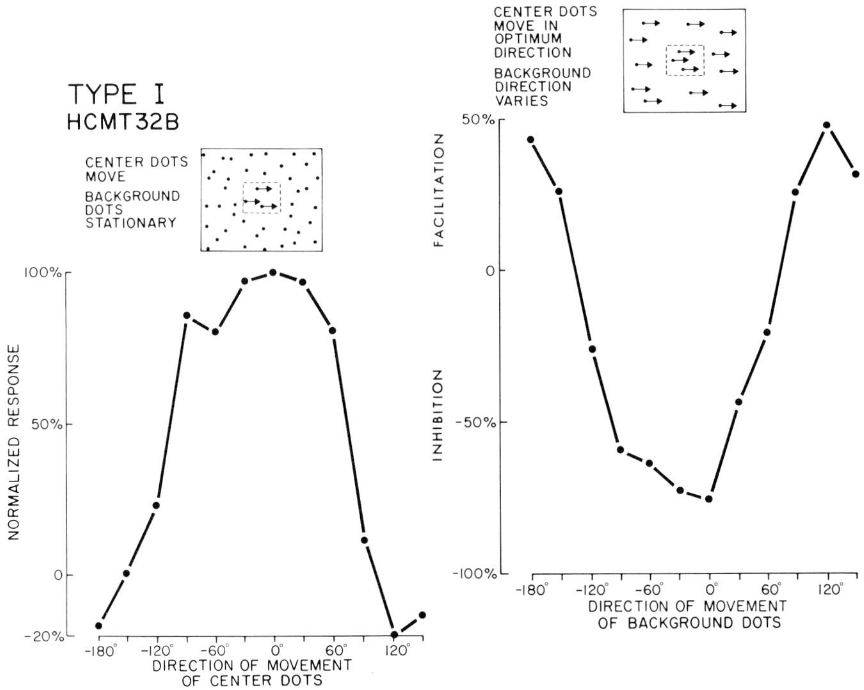

FIGURE 13 Direction-selective neuron with an antagonistic direction-selective surround recorded from area MT in an owl monkey. The left graph depicts the response of the cell to 12 directions of movement in an array of random dots within an area coextensive with its CRF. The response is normalized so that 0% is equal to the average level of spontaneous activity sampled for 2-sec periods before each presentation. Negative percentages in the left graph indicate inhibition relative to the level of spontaneous activity. In the left graph, the response to the optimum direction is 100%. The right graph depicts the response of the cell to different directions of background movement while the classical receptive field was stimulated by the array moving in the optimum direction during the 2-sec sample periods preceding background movement, and thus a response of -100% in the left graph is equivalent to 0% in the right graph. The stimulus conditions are depicted schematically above each graph. In the experiment, the dots were 50% dark and 50% light and the background was much larger relative to the center than is depicted schematically. Reproduced from Allman *et al.* (1985) with permission of Annual Reviews.

perception. This type of neuron constituted 44% of the MT population; 30% were suppressed equally by all directions of background movement; another 18% had more complex surround responses; and only 8% lacked a response from the surround.

Figure 14 is the combined histogram for 42 MT neurons that showed surround responses. In the lower histogram there were continuously moving random dots presented in the classical receptive field and background

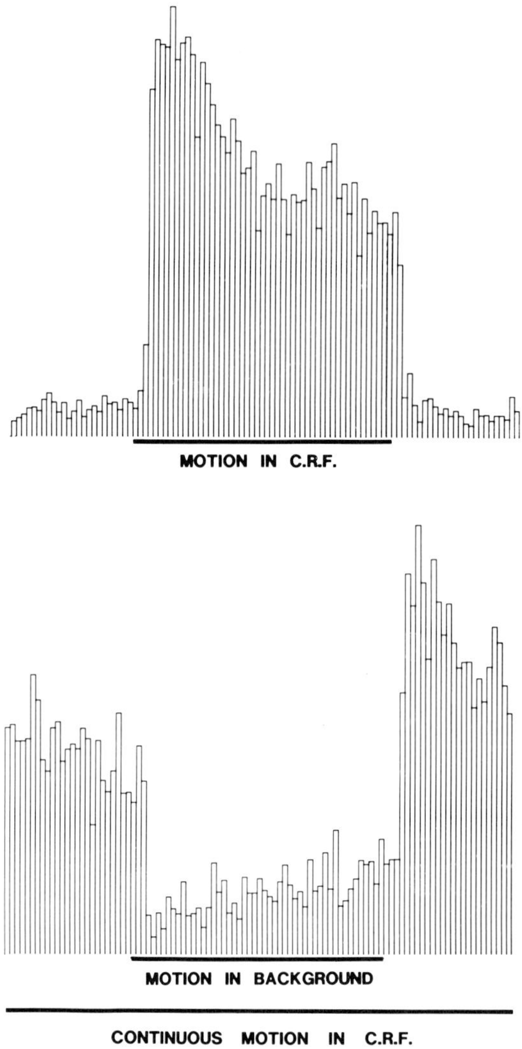

FIGURE 14 The top histogram illustrates the combined responses of 42 MT neurons to random dots moving for a 2-sec period in the preferred direction within their classical receptive fields, with the background stationary. In the lower histogram, the same MT neurons were stimulated continuously with random dots moving in the preferred direction within their classical receptive fields and then tested for a 2-sec period in which the random dots in the surrounding regions also moved in the same direction. Each bin represents 40 msec. The histograms were constructed by normalizing with the largest 40-msec bin in the histogram for each cell and then combining the histograms. Reproduced from Allman *et al.* (1985) with permission of Annual Reviews.

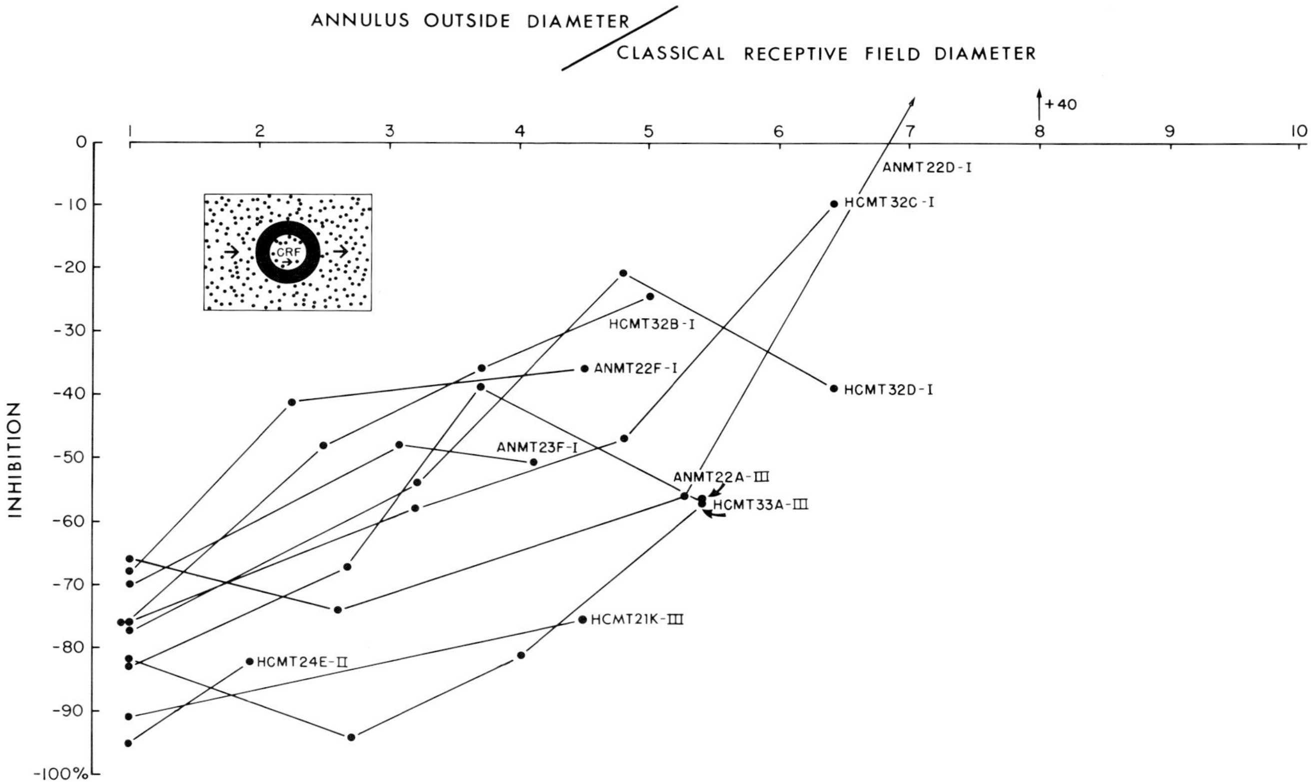

FIGURE 15 The effect of varying the outside diameter of masking annuli of the surround on the response from the classical receptive field in 10 neurons recorded from area MT in the owl monkey. The stimulus conditions are depicted schematically, but in the experiments the dots were 50% dark and 50% light and the surround was much larger. The dots in the center and the surround moved in the optimum direction for the classical receptive field. The inside diameter of the masking annulus corresponds approximately to the diameter of the classical receptive field. The abscissa corresponds to the ratio of the outside diameter of the masking annulus to the diameter of the classical receptive field. Reproduced from Allman *et al.* (1985) with permission of Annual Reviews.

motion in the same direction was presented for 2 sec. There was immediate strong suppression due to the surround stimulation followed by a strong rebound in activity following cessation of the surround movement.

Figure 15 illustrates the extent of the surround in 10 MT neurons. An annulus of varying diameter was used to mask parts of the surround. The annuli had to be very large, more than 7 times the diameter of the classical receptive field, in order to block the surround inhibition. Thus the area of the true receptive field for MT neurons is on the order of 50 times greater than that of the classical receptive field.

The antagonistic organization of MT neurons extends to the velocity of motion. Figure 16 illustrates a velocity tuning curve for a MT neuron with an optimal velocity of 16 degrees per second. When this neuron was stimulated with a bar sweeping at 16 degrees per second and simultaneously the velocity of background motion was varied, there was maximal suppression of background motion at 16 degrees per second. This effect was even more pronounced in V1 and V2 (Fig. 17). Receptive fields with this antagonistic velocity processing were predicted in a theoretical paper by Nakayama and Loomis (1974) as a possible neural basis for kineopsis, or depth perception through the analysis of the differential motion.

E. Sensitivity to Stimulus Shape in Area DL

The responses of DL neurons stand in marked contrast to those of MT. They are slower and more sustained, and can give continuous responses to stationary stimuli. DL neurons are very sensitive to the shape of stimuli, irrespective of their exact location within their classical receptive fields. Figure 18 illustrates the dimensionally selective properties of DL neurons in contrast to responses obtained from other areas. For example, in the upper left graph in Fig. 18, the responses of a DL and a DM neuron are shown for stimuli of different length. The DM neuron's response reached a plateau but did not diminish with increasing bar length; the DL neuron's response after reaching a maximum decreased with increasing bar length. When the best bar length for each DL neuron was compared with the classical receptive field length in the same dimension, it was clear that many DL neurons preferred lengths that were much shorter than the length of the classical receptive field. This was not the case for neurons in MT, DM, and M. Similar results were obtained for bar width and spot diameter. Figure 19 illustrates the preferred shapes observed in DL neurons.

F. Stereopsis

Stereopsis has been relatively little studied in the owl monkey. However, Fig. 20 illustrates data that indicate that tuned excitatory and near-far cells exist in the owl monkey as has been reported in macaque monkeys (Poggio

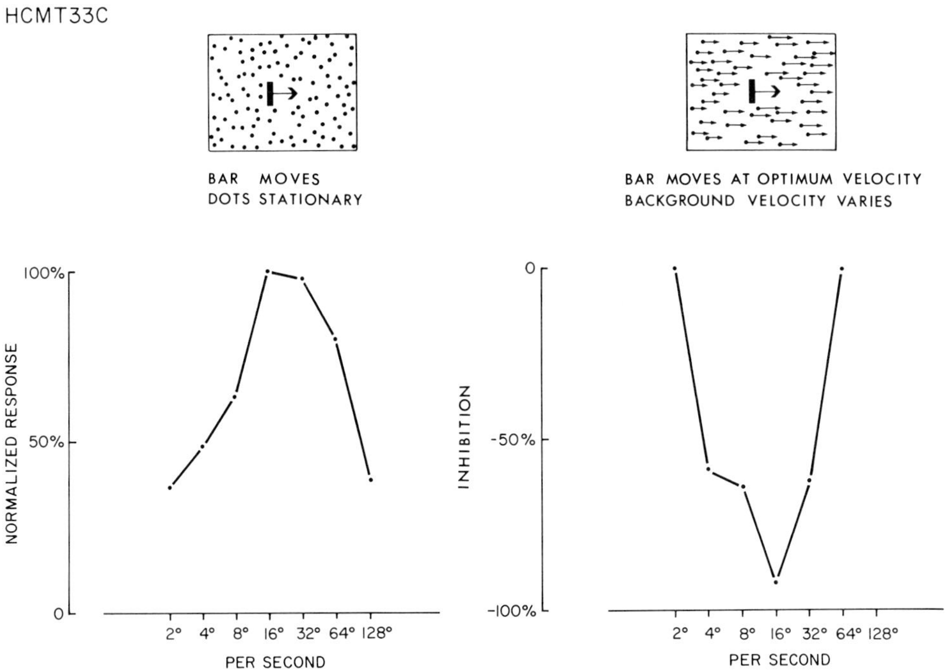

FIGURE 16 The effect of center and background velocity on a neuron recorded from area MT in the owl monkey. The left graph is a velocity-tuning curve for a bar moving in the preferred direction with the background stationary. The right graph is a velocity-tuning curve for background movement while simultaneously presenting the bar moving at the optimum velocity (16 degrees per second). In the background velocity experiments, the background dots stimulated both the classical receptive field and the surround; however, in these experiments, covering the surround eliminated the inhibitory effect. Reproduced from Allman *et al.* (1985) with permission of Annual Reviews.

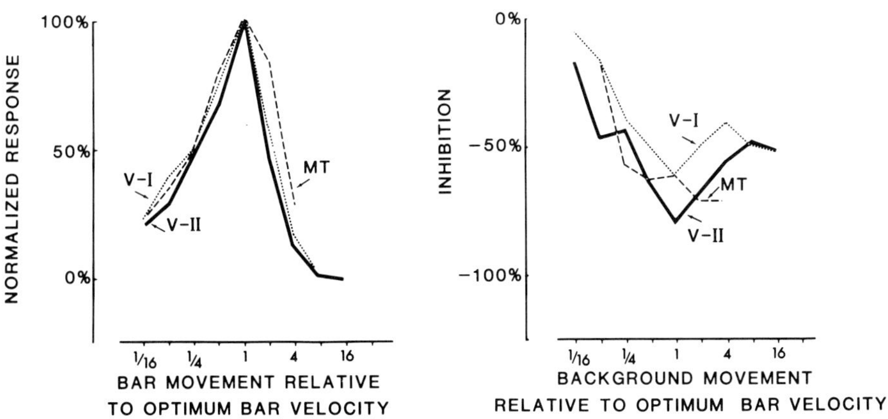

FIGURE 17 Median response curves for bar and background velocity experiments in V1, V2, and MT. Reproduced from Allman *et al.* (1990) with permission.

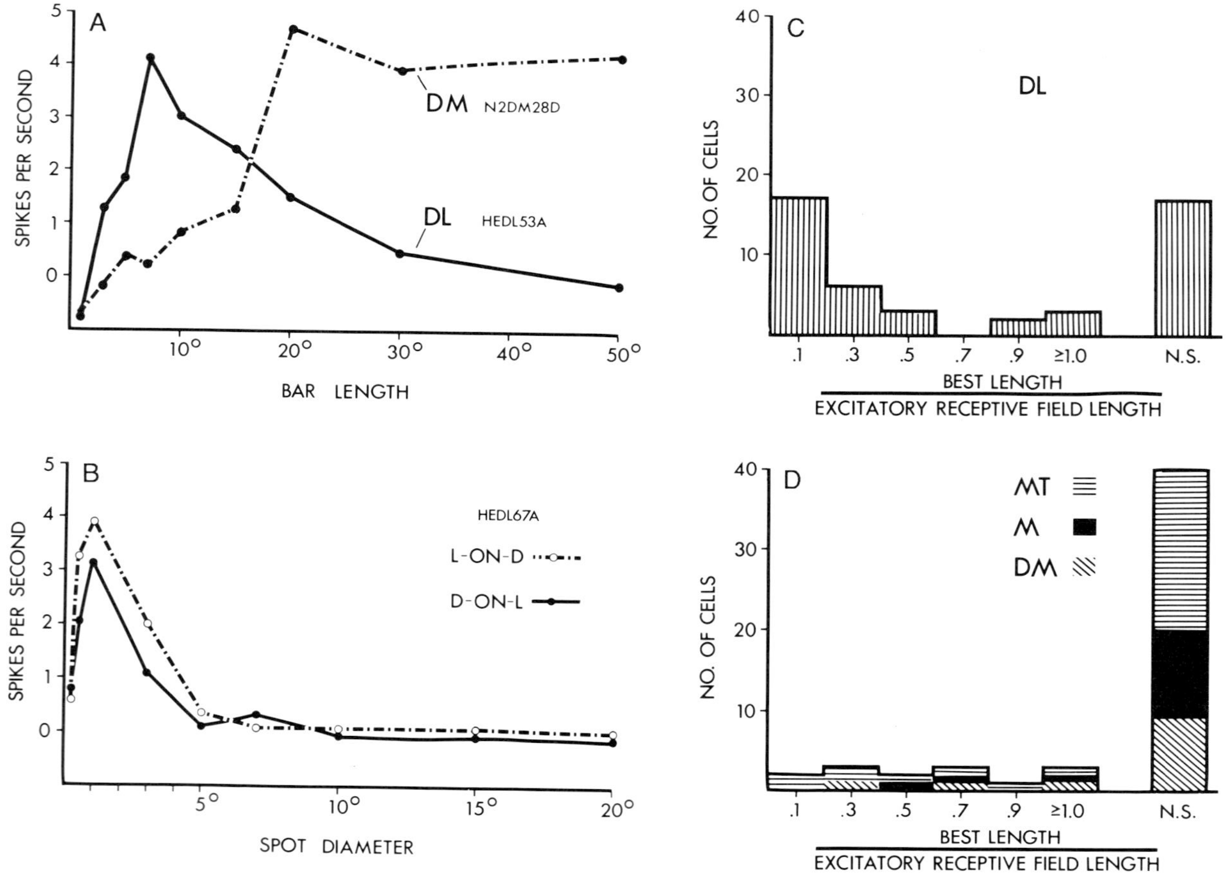
A
SPIKES PER SECOND
5
4
3
2
1
0
10°
20°
30°
40°
50°
BAR LENGTH
DM N2DM28D
DL HEDL53A
B
SPIKES PER SECOND
5
4
3
2
1
0
5°
10°
15°
20°
SPOT DIAMETER
HEDL67A
L-ON-D
D-ON-L
C
DL
NO. OF CELLS
40
30
20
10
.1
.3
.5
.7
.9
≥1.0
N.S.
BEST LENGTH
EXCITATORY RECEPTIVE FIELD LENGTH
D
NO. OF CELLS
40
30
20
10
MT
M
DM
.1
.3
.5
.7
.9
≥1.0
N.S.
BEST LENGTH
EXCITATORY RECEPTIVE FIELD LENGTH

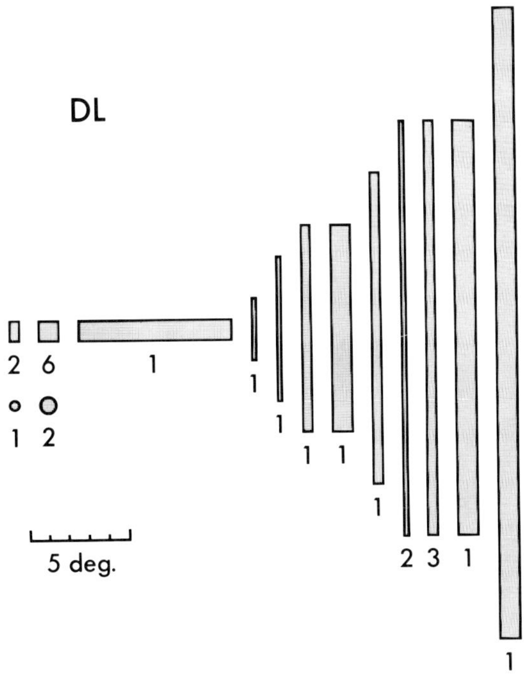

FIGURE 19 Optimal stimulus dimensions for cells recorded from area DL in the owl monkey. From S. E. Petersen, J. F. Baker, and J. M. Allman, unpublished data.

and Fischer, 1977). We trained an owl monkey to report the orientation of dynamic random dot stereograms (J. Allman and F. Miezin, unpublished data) and thus we believe that owl monkeys see in stereoscopic depth.

IV. THE BEHAVING OWL MONKEY PREPARATION

The owl monkeys offers important advantages for use in the study of visual cortex function in the behaving animal. The cortical areas have been

---

FIGURE 18 (A) Responses of two units to different bar lengths. Each data point is the average of five stimuli presented in pseudorandom order. HEDL53A (solid line) was recorded from DL and shows marked selectivity. N2DM28D (dashed line) illustrates the typical response profile for cells outside of DL, in which the cell summates up to a certain value, whereupon the response levels off. The length of the classical receptive fields for both cells was 20 degrees. (B) Responses of a DL neuron to light (open circles) and dark (solid circles) spots of different diameters. (C, D) Optimal bar length is expressed as a percentage of the comparable dimension of the classical receptive fields. Cells with a length selectivity index of less than 0.5 were considered to be nonselective and are represented by bins at the right. Reproduced from Petersen *et al.* (1980) with permission.

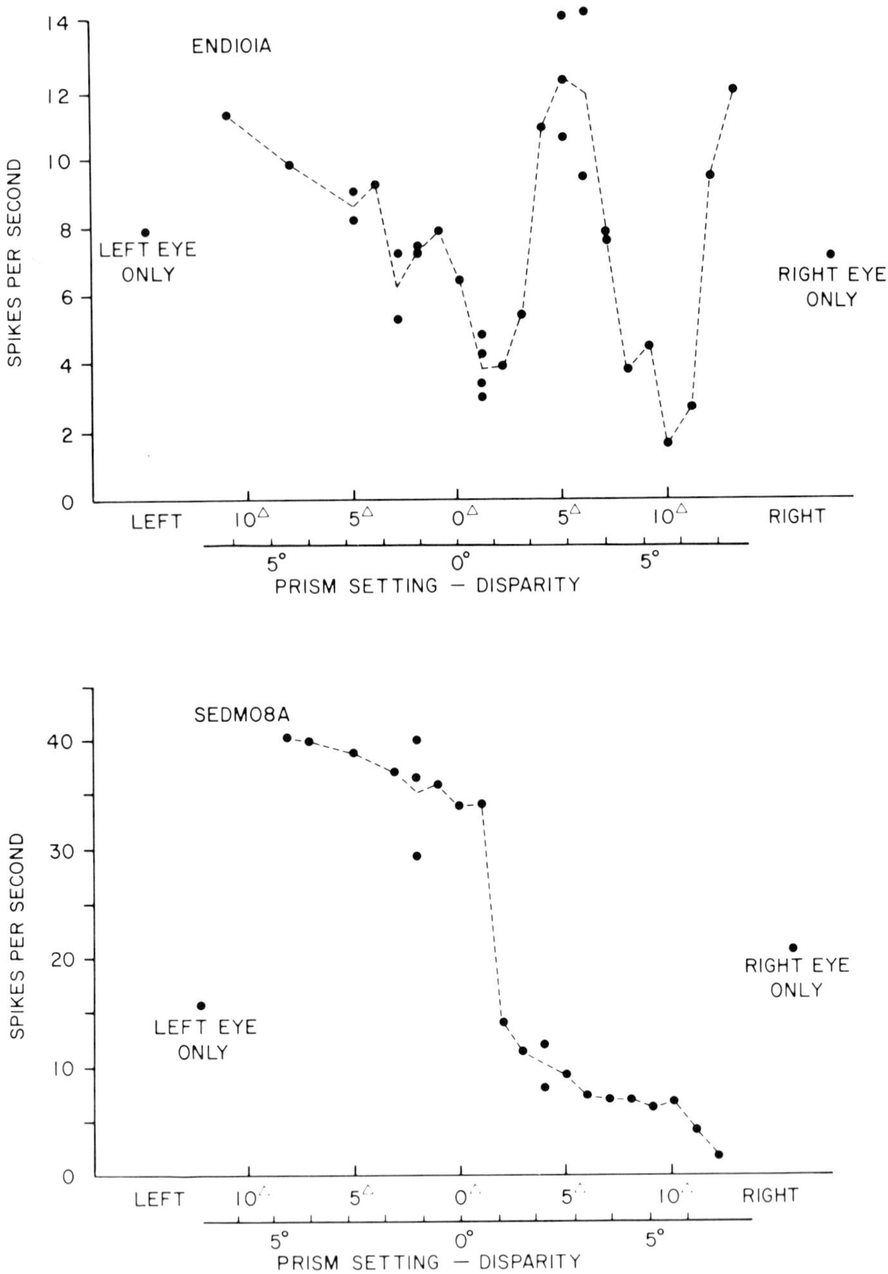

FIGURE 20 An example of a tuned excitatory binocular disparity neuron recorded from area DI (upper graph) and a near-far disparity neuron recorded from area DM (lower graph) in the owl monkey. The disparity values are not corrected for eye position. From J. M. Allman, S. E. Petersen, and J. F. Baker, unpublished data.

extensively mapped and many are located on the smooth dorsolateral surface of the brain. Thus it is relatively easy to position recording electrodes in the desired cortical location. This would also facilitate the injection of agents to block temporarily any neural activity, such as muscimol, during recording and perceptual experiments. The large array of cortical visual areas on the smooth dorsolateral surface would also greatly facilitate optical recording experiments.

We have developed new training procedures and built equipment specially adapted for owl monkey training and electrophysiology, because owl monkeys do not respond well to standard procedures used for training macaque monkeys. Owl monkeys, and all other New World monkeys, cannot sit comfortably in training chairs designed for macaque monkeys because they lack ischial callosities. We designed a special training apparatus consisting of a lexan alcove with two keys, which attaches to the cage in which the monkey lives. This alcove is designed to fit the owl monkey's natural squatting position. The monkey learns to enter the alcove and report discrimination of various images by pressing one of the two keys for a fruit juice reward. We monitor fluid intake to ensure that the monkey receives adequate amounts, and free water is offered for several hours after each training session. Owl monkeys also require much more gentle habituation to restraint than do macaque monkeys. To perform microelectrode or optical recording while the monkey performs a visual task, the monkey's head must be held in a stationary position during recording. Stability is accomplished by cementing a small stainless-steel triangular peg to the frontal bone of the skull under general anesthesia. The monkey is then trained to position the peg in a V-shaped receptable equipped with a microswitch. When the switch is activated, the computer then begins to present visual stimulation for the monkey to discriminate and thus gain fruit juice rewards. The monkey's head is fixed by securing the peg once it is placed into the V-shaped receptacle.

V. THE STUDY OF PERCEPTUAL MEMORY IN BEHAVING OWL MONKEYS

In our laboratory, owl monkeys are being used to explore the mechanisms of perceptual memory. Viewed from an ecological perspective, the main task of the visual cortex is to extract behaviorally meaningful patterns from noisy and ambiguous images of the natural world. Because of the inherent ambiguity of natural images, multiple and conflicting interpretations may be possible. The task of the visual cortex is to select the most appropriate interpretation quickly and reliably. This task must often be performed on the basis of incomplete information. This function requires a kind of perceptual learning that, we hypothesize, takes place in the visual cortex.

We are exploring this putative learning function of visual cortex through the use of cognitively defined contours, which are the borders of patterns embedded in a noisy background (Fig. 21). These patterns are difficult to see until revealed by special cues such as the brief movement of the pattern with respect to the background. Our perceptual experience with these cognitive contours is that, once they have been revealed, they are very readily seen again and that this capacity persists for months or years for a particular stimulus array. This perceptual learning is reminiscent of a type of long-term learning described by Warrington and Weiskrantz (1968) in amnesic patients. These patients, who had severe deficits in their capacity to remember recent events, showed a remarkable ability to identify visual patterns when they viewed fragmented residues of whole patterns. The amnesic patient, HM, whose medial temporal lobe including hippocampus had been surgically removed bilaterally, showed retention of his capacity to identify fragmented images in these tests (Milner *et al.*, 1968). This capacity has been termed "priming" and has been postulated to be performed by a "perceptual representation system" thought to reside in extrastriate visual cortex (Tulving and Schacter, 1990). Retention due to priming is long-lasting with no decay after 1 week (McAndrews *et al.*, 1987; Musen and Treisman, 1990), and some retention after 3 months (Warrington and Weiskrantz, 1968).

Position emission tomography studies have revealed that priming with visual images produces reduced blood flow upon retesting in a region of

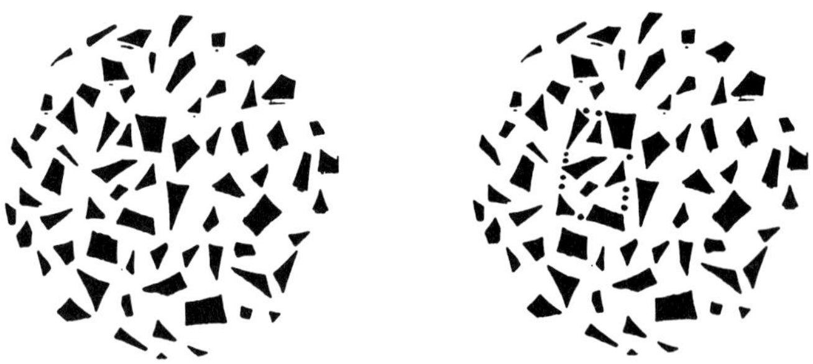

FIGURE 21 Sample stimulus pattern to test visual perceptual memory. A rectangular figure is embedded in the random polygon array. The right figure is not obvious until it is cued (by closure in this example), shown on the right. The subject's task is to report the orientation of the embedded figure before, during, and after cuing. The stimulus was generated by cutting up small slips of black paper and sprinkling them on a white background. The image was then captured as a computer bit map, using a CCD camera with a frame grabber. We can generate an unlimited number of novel stimuli with this simple method. From Jeo and Allman, unpublished stimulus pattern.

right occipital extrastriate visual cortex in human subjects (Squire *et al.*, 1991). This result suggests that priming facilitates the neural solution to perceiving images with consequently less energy expended and presumably a higher signal-to-noise ratio in the neural populations involved. Access to priming memory is hyperspecific in that it depends among other things on the exact geometrical configuration of the priming stimulus (Tulving and Schacter, 1990). Hyperspecificity of access suggests that priming memory may be stored within the visuotopically mapped cortical visual areas.

These experiments can be regarded as a further exploration of the nonclassical responses of visual cortical neurons described earlier in this chapter. These results have been extended to "illusory contours" that were implied by stimuli that were entirely beyond the classical receptive field for neurons in the second visual area (Peterhans and von der Heydt, 1989).

The functional properties of visual cortical neurons have been viewed as very highly specialized filters set to detect different aspects of visual stimuli such as the direction of motion, binocular disparity, orientation, or color. We suggest that in addition to these classical features of visual cortical neurons, there is the capacity to respond selectively to patterns on the basis of prior experience. In other words, visual cortex can "learn to see" ambiguous patterns embedded in a noisy background.

We further suggest that these perceptual learning capacities may be localized in a manner analogous to the evident perceptual specializations present in the various areas. For example, perceptual learning related to differential motion might be preferentially related to area MT. To test this idea we can use different types of cues (differential motion, closure, shading, and stereoscopic depth) to determine whether the nature of the cue might affect the site of storage within the visual cortical areas.

We use stimuli with embedded rectangles that are either horizontal or vertical. The monkeys are trained in an apparatus with two paddle keys and respond to horizontal stimuli by pressing the right key and to vertical stimuli by pressing the left key. The monkeys are then tested to determine whether they respond correctly to the orientation of a series of embedded figures before and after cuing; response time is also measured. Owl monkeys are easily trained to perform visual discrimination tasks and are very good psychophysical subjects. In a typical experiment, a series of different embedded figures are presented in pseudorandom order. Initially, we display each stimulus array without any cues for the embedded figure. We allow the subject to respond to the novel stimulus, then immediately activate the cue for the imbedded rectangle, and the subject is allowed to respond again. After several such presentations, the stimuli is presented without any cues.

To ensure that the subjects are performing the figure detection and not simply memorizing the entire figure, we plan to perform two control experiments. The first will be to measure response to a random polygon array without any embedded figures, where only one orientation response will be reinforced. The second control experiment will be to compare the effects of rotating elements in the surround with rotating elements forming the contour boundary in the embedded figure. If the subject is responding to the embedded figure and not simply memorizing the entire array, disturbing elements within the embedded figure should disrupt performance more than disturbing surround elements.

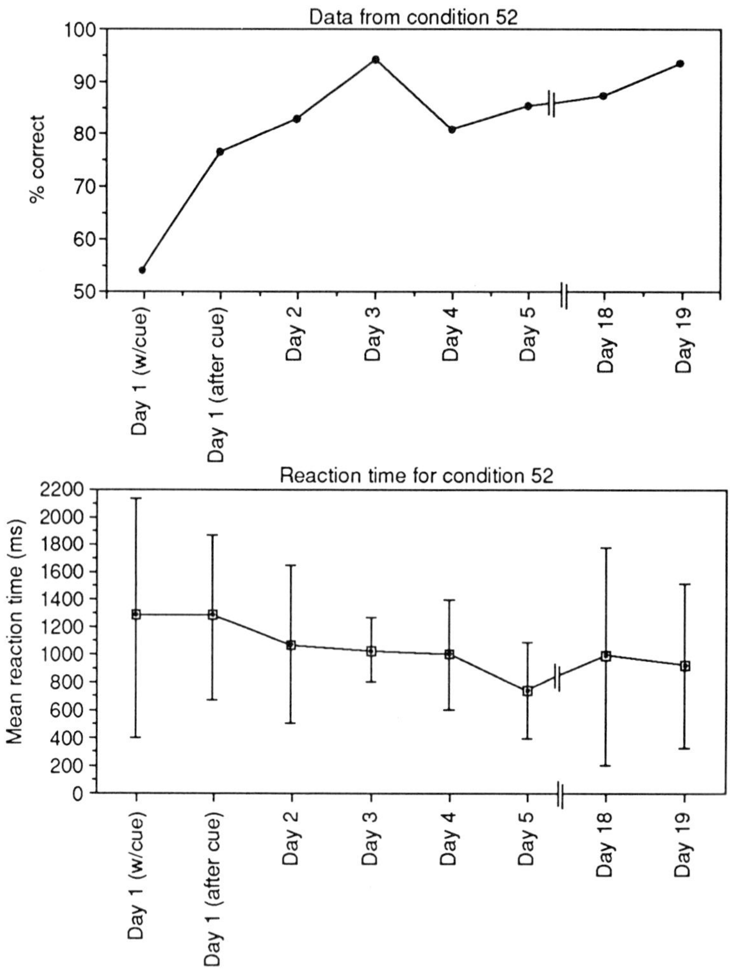

FIGURE 22 Typical owl monkey performance for a new stimulus pattern. Top: Performance goes from chance levels (54%, day 1 without cue) to 76% (Day 1 after cue) and stays at steady levels even after a 13-day break (Day 5–Day 18). Bottom: Reaction time improves slightly from Day 1 to Day 5 and remains constant after a 13-day break. Standard deviation of reaction time decreases from Day 1 to Day 5, but increases after the break. From Jeo and Allman, unpublished data.

We have trained two owl monkeys to perform the perceptual memory task. The monkeys report the orientation of the embedded horizontal or vertical rectangle in a random array of polygons, described earlier. Initially, we displayed stimulus array without any cues for the embedded figure. We allowed the monkey to respond to the novel stimuli and rewarded for a correct answer, then immediately activated the cue for the embedded rectangle, and the monkey was allowed to respond again. After several such presentations, the stimuli were presented without any cues.

Both owl monkeys performed this task very well; one monkey learned over 100 different stimulus patterns. The principle results are: performance is not rotation invariant, that is, when we rotate the entire stimulus 90 degrees, accuracy degrades to chance levels; displaying the stimuli as a mirror image similarly degrades performance; however, performance is contrast sign invariant, such that displaying the images in reverse contrast does not affect accuracy.

The monkeys work daily, can perform up to 1500 trials per day, and will work steadily for 2 to 3 hr. The time that it takes for the monkey to learn a new stimulus varies from one cued presentation for easy figures to many exposures over several training sessions for more difficult ones. Reaction time and its standard deviation both decrease as the stimuli are learned. This knowledge is retained even after several weeks without exposure to the same stimuli (Fig. 22).

REFERENCES

Allman, J., and Kaas, J. (1971a). A representation of the visual field in the caudal third of the middle temporal gyrus in the owl monkey (*Aotus trivirgatus*). *Brain Res.* **31,** 85–105.

Allman, J., and Kaas, J. (1971b). Representation of the visual field in striate and adjoining cortex of the owl monkey (*Aotus trivirgatus*). *Brain Res.* **35,** 89–106.

Allman, J., and Kaas, J. (1974a). The organization of the second visual area (VII) in the owl monkey: A second order transformation of the visual hemifield. *Brain Res.* **76,** 247–265.

Allman, J., and Kaas, J. (1974b). A crescent-shaped area surrounding the middle temporal area (MT) in the owl monkey (*Aotus trivirgatus*). *Brain Res.* **81,** 199–213.

Allman, J., and Kaas, J. (1975). The dorsomedial cortical visual area: A third tie in the occipital lobe of the owl monkey (*Aotus trivirgatus*). *Brain Res.* **100,** 473–487.

Allman, J., and Kaas, J. (1976). Representation of the visual field on the media wall of occipital–parietal cortex in the owl monkey. *Science* **191,** 572–575.

Allman, J., Miezin, F., and McGuinness, E. (1985). Stimulus specific response from beyond the classical receptive field: Neurophysiological mechanisms for local–global comparisons in visual neurons. *Annu. Rev. Neurosci.* **8,** 407–430.

Allman, J., Miezin, F., and McGuinness, E. (1990). The effects of background motion on the responses of neurons in the first and second visual areas (V-I and V-II). In "Local and Global Order in Perceptual Maps" (G. M. Edelman, W. E. Gall, W. M. Cowan, eds.), pp. 131–142, Wiley-Liss, New York.

Allman, J., and Zucker, S. (1991). Cytochrome oxidase and functional coding in primate striate cortex: A hypothesis. *Cold Spring Harbor Symposia on Quantitative Biology* **55,** 979–982.

Baker, J. F., Petersen, S. E., Newsome, W. T., and Allman, J. M. (1981). Visual response properties of neurons in four extrastriate visual areas of the owl monkey (*Aotus trivirgatus*):

A quantitative comparison of medial, dorsomedial, dorsolateral and middle temporal areas. *J. Neurophysiol.* **45,** 397–416.
Born, R. T., and Tootell, R. B. H. (1992). Segregation of global vs. local processin in primate area MT. *Nature (London)* **357,** 497–499.
Horton, J. (1984). Cytochrome oxidase patches: A new cytoarchitectronic feature of monkey cortex. *Philos. Trans. R. Soc. London, B Ser.* **304,** 199–253.
Livingstone, M., and Hubel, D. (1984). Anatomy and physiology of a color system in primate visual cortex. *J. Neurosci.* **4,** 309–356.
McAndrews, M., Glisky, E., and Schacter, D. (1987). When priming persists: Long-lasting implicit memory for a single episode in amnesic patients. *Neuropsychologia* **25,** 497–506.
Milner, B., Corkin, S., and Teuber, H.-L. (1968). Further analysis of the hippocampal amnesic syndrome: A 14-year follow-up study of H.M. *Neuropsychologia* **6,** 215–234.
Musen, G., and Treisman, A. (1990). Implicit and explicit memory for visual patterns. *J. Exp. Psychol.: Learning Memory, Cognition* **16,** 127–137.
Nakayama, K., and Loomis, J. (1974). Optical velocity patterns, velocity-sensitive neurons, and space perception: A hypothesis. *Perception* **3,** 63–80.
Newsome, W. T., and Allman, J. M. (1980). The interhemispheric connections of visual cortex in the owl monkey. *Aotus trivirgatus,* and the bushbaby, *Galago senegalensis. J. Comp. Neurol.* **19,** 209–233.
Peterhans, E., and von der Heydt, R. (1989). Mechanisms of contour perception in monkey visual cortex. I. Contours bridging gaps. *J. Neurosci.* **9,** 1749–1763.
Petersen, S. E., Baker, J. F., and Allman, J. M. (1980). Dimensional selectivity of neurons in the dorsolateral visual area of the owl monkey. *Brain Res.* **197,** 507–511.
Petersen, S. E., Baker, J. F., and Allman, J. M. (1985). Direction-specific adaptation in area MT of the owl monkey. *Brain Res.* **346,** 146–150.
Petersen, S. E., Miezin, F. M., and Allman, J. M. (1988). Transient and sustained responses in four extrastriate visual areas of the owl monkey. *Exp. Brain Res.* **70,** 55–60.
Poggio, G. F., and Fischer, B. (1977). Binocular interactions and depth sensitivity in striate and prestriate cortex of behaving rhesus monkeys. *J. Neurophysiol.* **40,** 1392–1405.
Sereno, M., and Allman, J. (1991). Cortical visual areas in mammals. In "The Neural Basis of Visual Function" (A. Levinthal, ed.), pp. 160–172. Macmillan, London.
Sereno, M., MacDonald, C., and Allman, J. (1987). Multiple visual areas between V2 and MT in the owl monkey. *Soc. Neurosci. Abstr.* **13,** 625.
Squire, L., Ojemann, J. Miezin, F., Petersen, S., Videen, T., and Raichle, M. (1991). Activation of the hippocampus in normal humans: A functional anatomical study of human memory. *Proc. Natl. Acad. Sci. U.S.A.* **89,** 1837–1841.
Tootell, R. B. H., Hamilton, S. L., and Silverman, M. S. (1985). Topography of cytochrome oxidase activity in owl monkey cortex. *J. Neurosci.* **5,** 2786–2800.
Tootell, R. B. H., Silverman, M., Hamilton, S., DeValois, R., and Switkes, E. (1988). Functional anatomy of macaque striate cortex. III. Color. *J. Neurosci.* **8,** 1569–1593.
T'so D., and Gilbert, C. (1988). The organization of chromatic and spatial interactions in primate striate cortex. *J. Neurosci.* **8,** 1712–1727.
Tulving, E., and Schacter, D. (1990). Priming and human memory systems. *Science* **247,** 301–306.
Warrington, E., and Weiskrantz, L. (1968). New method of testing long-term retention with special reference to amnesic patients. *Nature (London)* **217,** 972–974.
Weller, R., and Kaas, J. (1985). Cortical projections of the dorsolateral visual area in owl monkeys: The prestriate relay to inferior temporal cortex. *J. Comp. Neurol.* **234,** 35–59.
Weller, R., and Kaas, J. (1987). Subdivisions and connections of inferior temporal cortex in owl monkeys. *J. Comp. Neurol.* **256,** 137–172.
Wong-Riley, M., and Carroll, E. (1984). The effect of impulse blockage on cytochrome oxidase activity in the monkey visual system. *Nature (London)* **307,** 262–264.

12

The Organization of Sensory and Motor Cortex in Owl Monkeys

Jon H. Kaas
Department of Psychology
Vanderbilt University
Nashville, Tennessee 37240

I. INTRODUCTION

The neocortex makes up the bulk of the human brain and is responsible for most of the higher-order processing of sensory information, for making decisions, and for effecting behavioral responses. Recently, interest in the organization of neocortex has greatly increased because advances in methods of noninvasively exploring the human brain promise major increases in understanding. Nevertheless, many of our concepts of how the human brain is organized come from studies of other mammals, especially primates, and reasearch on other mammals will continue to be a major source of information and theories about brain organization and function. In addition, studies on other mammals illustrate how different types of brain organization relate to specific environmental adaptations, and how general problems are addressed by neural systems.

Several different primate taxa have been used in studies of brain organization, but owl monkeys have been one of the most useful. Old World monkeys have been commonly used as experimental mammals because of their close phyletic relationship to humans, but their large, highly fissured, and presumably complex brains are difficult to investigate. New World owl monkeys are somewhat less closely related to humans, but they offer the technical advantages of a smaller and less fissured brain, with more cortex accessible on the surface, and brains that are more easily processed histologically. In addition, cortical organization may be less complex than in Old World monkeys or humans, and it can be advantageous to investigate a simpler plan to help understand more complex plans of cortical organization. As the only nocturnal monkey, owl monkeys are especially useful in studies of the visual system in that they are less specialized for central or foveal vision than diurnal monkeys. This allows patterns of retinotopic change in visual structures to be more easily detected.

This chapter provides an outline of our current understanding of cortical organization in owl monkeys. Such an outline can usefully guide further studies, as well as allow comparisons with other mammals. It is important to remember, however, that this current portrayal of cortical organization constitutes a broad theory of cortical organization that is subject to modifications and change as further information is incorporated.

The basic unit of cortical organization is the cortical area or field. The classical neuroanatomist, Brodmann (1909), regarded cortical areas as the organs of the brain on the assumption that the cortex is composed of a patchwork of tilelike fields that have unique functional roles. Areas can be distinguished because their unique functional roles require structural specializations (architectonic distinctiveness), unique patterns of inputs and outputs (connections with other areas and subcortical structures), and neurons with specific ways of responding to stimulus events (see Kaas, 1989). Evidence for how cortex is subdivided into areas and how these areas are interconnected is often incomplete and ambiguous, but great progress has been made. Although the evidence validates the basic premise of Brodmann (1909) that cortex consists of a patchwork of areas, the cortical plan derived from recent research differs considerably from that proposed by Brodmann. Only a few cortical areas of Brodmann are retained in current proposals.

Our current theory of how neocortex in owl monkeys is subdivided into areas is presented in Fig. 1. Part of the figure illustrates a dorsolateral view of an owl monkey brain with several exposed cortical areas superimposed. Though the owl monkey brain has few fissures, some cortex remains hidden in fissures, on the medial wall between the two cerebral hemispheres, and on the ventral surface. Thus, it is also useful to portray the areas on a flattened view of cortex (Fig. 1, upper right), as cortex is a thin and rather uniform sheet that can be separated from the underlying connection fibers, unfolded by opening fissures, and flattened so that all the cortical surface is visible. In such a preparation, the otherwise hidden areas are visible. The outlined fields are the areas related to the somatosensory, visual, auditory, and motor systems, because these systems have been well studied in owl monkeys and other primates. Though other fields related to these systems will probably be discovered, it is obvious from Fig. 1 that most of the cortex in owl monkeys is "sensory" or "motor." This cortex is devoted to analyzing and storing sensory information and planning and eliciting motor behavior. Other areas that are not shown and are poorly understood relate to taste, vestibular, and probably olfactory information. These and other fields, such as those related to motivational and emotional systems and short-term memory, would fill in much of the unlabeled cortex in the summary map (Fig. 1).

II. VISUAL CORTEX

The major features of visual cortex have been extensively studied in owl monkeys, and a number of reviews are available (e.g., Kaas and Krubitzer, 1991; Kaas, 1986; Allman, Jeo, and Sereno, Chapter 11, this volume). In brief, visual cortex consists of a complex of 20 or more interconnected visual areas (Fig. 1), each also connected with a number of subcortical structures, including the pulvinar complex, claustrum, basal ganglia, superior colliculus, and, for some areas, the lateral geniculate nucleus (see Kaas and Huerta, 1988, for review). Some of the visual areas have been well defined by multiple criteria (see Kaas, 1989) and correspond to subdivisions also known in other primates. The evidence for other areas is more equivocal, and thus schemes may change with more evidence, and current proposals by various investigators may differ.

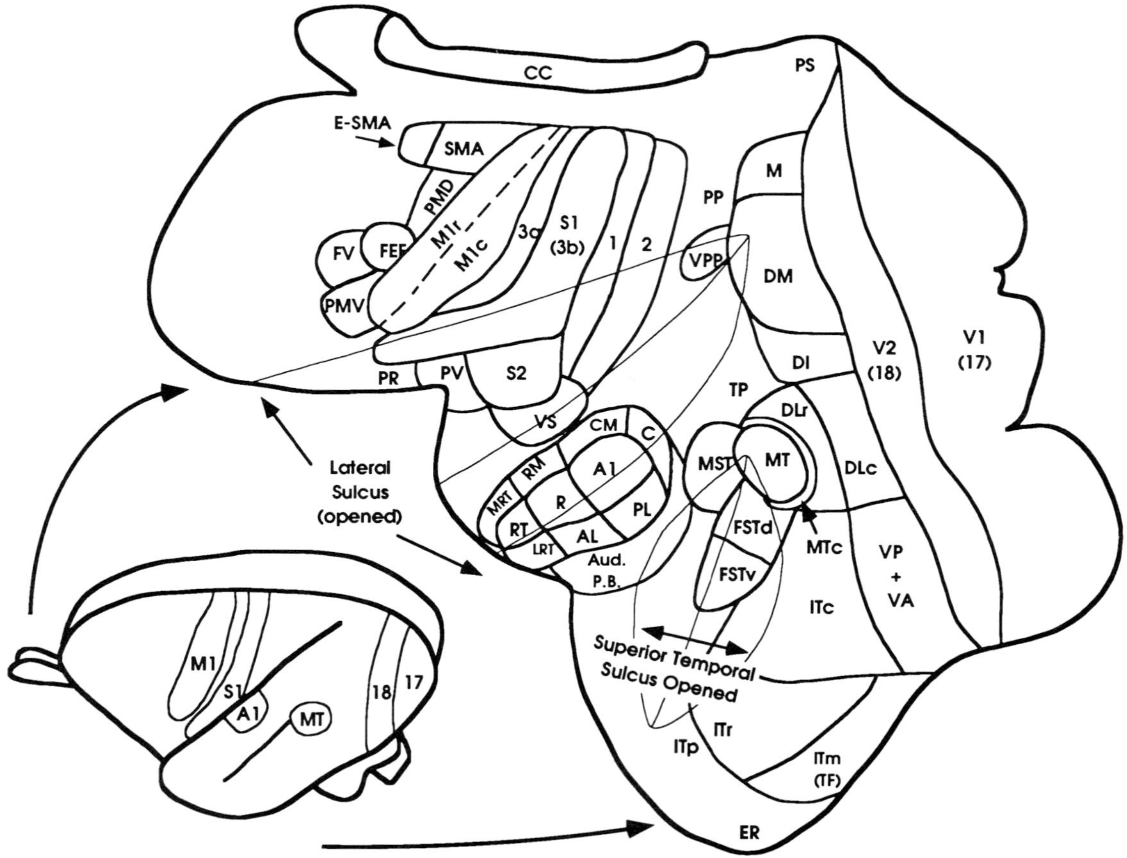
CC
PS
E-SMA
SMA
PMD
M1r
M1c
3a
S1
(3b)
1
2
PP
M
VPP
DM
FV
FEF
PMV
V2
(18)
V1
(17)
DI
PR
PV
S2
TP
DLr
VS
CM
C
MST
MT
DLc
RM
A1
MRT
R
PL
RT
AL
LRT
Aud.
P.B.
FSTd
MTc
FSTv
VP
+
VA
ITc
Lateral
Sulcus
(opened)
Superior Temporal
Sulcus Opened
M1
S1
A1
MT
18
17
ITr
ITp
ITm
(TF)
ER

A. Area 17 or V1

As in all or most mammals (Kaas and Krubitzer, 1991), the primary area is easily defined in owl monkeys by its distinct histological appearance (see Allman and Kaas, 1971a; Tootell *et al.*, 1985; Krubitzer and Kaas, 1990b), the presence of a systematic map or representation of the contralateral visual hemifield (Allman and Kaas, 1971b), and major inputs from the lateral geniculate nucleus or LGN (Kaas *et al.*, 1976; Norden and Kaas, 1978; Diamond *et al.*, 1985). Major output targets (Fig. 2) include visual areas V2, MT, DM, and the DL complex, as well as feedback to the LGN (Lin and Kaas, 1977) and projections to the central nucleus of the inferior pulvinar complex (Lin and Kaas, 1979), the superior pulvinar, the pretectum, and the superior colliculus (Graham *et al.*, 1979). Area 17 constitutes nearly 15% of neocortex and provides the activating input, directly or indirectly, for most of the visual cortex (Kaas and Krubitzer, 1992).

Information relayed from the retina to the LGN and then to area 17 is segregated into three functionally distinct channels sometimes referred to as the X, Y, and W streams (Sherman, *et al.*, 1976), but more commonly the parvocellular (P), magnocellular (M), and interlaminar (I) streams (for the names of the LGN layers where X, Y, and W inputs relay to area 17,

---

FIGURE 1 Subdivisions of neocortex in owl monkeys. A dorsolateral view of an owl monkey brain on the lower left shows fissures and several cortical areas. The cortex can be separated from the rest of the brain, fissures opened, and cortex flattened into a single sheet, as shown on the upper right. Subdivisions of cortex that have been proposed for owl monkeys are outlined on this view of flattened cortex so that areas normally hidden or partially hidden in the lateral and superior temporal fissures or the medial wall of the hemisphere can be seen. Visual areas include primary (V1) and secondary (V2) areas in architectonic areas 17 and 18, respectively; dorsomedial (DM), dorsointermediate (DI), and dorsolateral areas with rostral (DLr) and caudal (DLc) divisions; caudal (ITc), rostral (ITr), medial (ITm), and polar (ITp) divisions of inferotemporal cortex; the middle temporal area (MT) and a crescentlike area surrounding MT (MTc for crescent); dorsal (FSTd) and ventral (FSTv) divisions of the fundal area of the superior temporal sulcus; a medial area (M); a temporal-parietal region (TP); and a ventral posterior parietal area (VPP). Area prostriata (PS) is a limbic area with probable visual functions, whereas entorhinal cortex (ER) connects visual with hippocampal cortex. Auditory cortex includes a core of the primary area (A1) and primarylike rostral (R) and rostrotemporal (RT) areas. An auditory belt of anterior lateral (AL), posterior lateral (PL), caudal (C); caudomedial (CM), rostromedial (RM), medial rostrotemporal (MRT), and lateral rostrotemporal areas (LRT). An auditory parabelt region (Aud. p.b.) receives auditory projections and relays to the adjoining temporal lobe. Somatosensory cortex includes areas 3a, 3b, and 1, and 2 of anterior parietal cortex; the second area (S2); the parietal ventral area (PV), the ventral somatic area (VS); the parietal rostral region (PR), and posterior parietal cortex (PP). Motor and visuomotor areas of the frontal lobe include rostral and caudal divisions of primary motor cortex (MIr and MIc); the supplementary motor area (SMA) with a rostral part devoted to eye movements (E-SMA); dorsal and ventral premotor fields (PMD and PMV); the frontal eye field (FEF); and the frontal visual area (FV). The corpus callosum (cc) connects the two hemispheres. See text for further details.

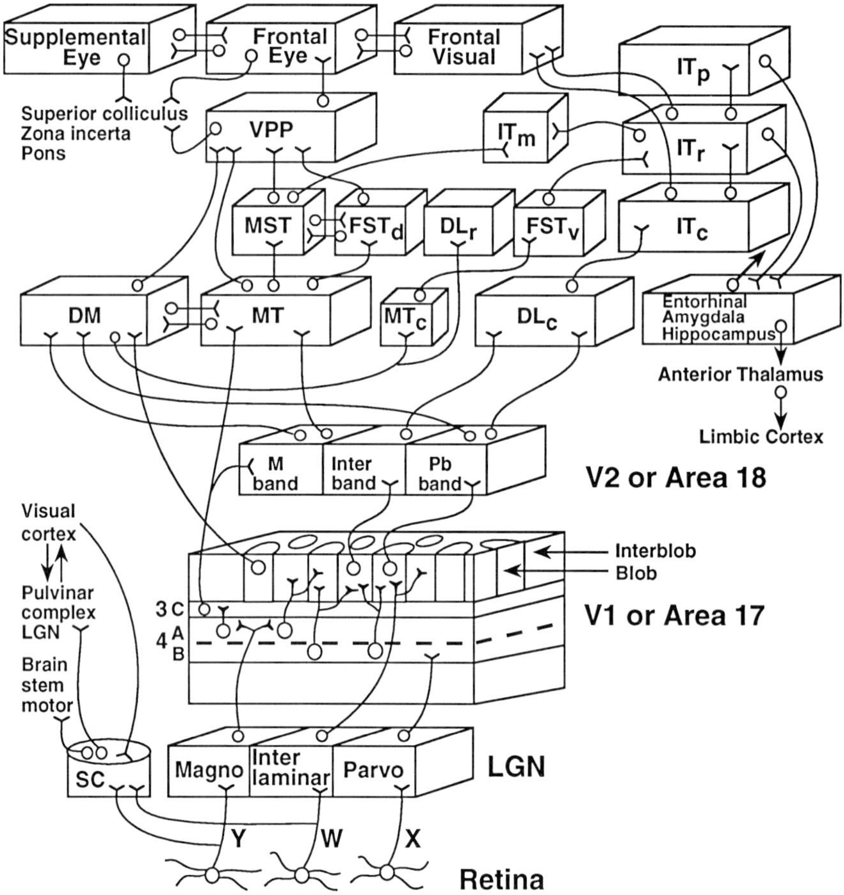

FIGURE 2 An outline of the functional organization of the visual system in owl monkeys emphasizing the major components of serial and parallel processing streams. Visual information is relayed from the retina via three classes of ganglion cells, called here Y, W, and X after names given originally in cats. These classes are also known as alpha, gamma, and beta cells when characterized morphologically. Y cells project to the magnocellular layers of the lateral geniculate nucleus (LGN) and the superior colliculus (SC). W cells project to cells between layers (the interlaminar zones) and the S layers of the LGN (see Kaas *et al.*, 1972; 1978) as well as the SC. X cells project to the parvocellular layers. Magnocellular layers, as part of the magnocellular or M stream, project to the upper half of layer 4 (4A), where information is relayed to output cells in layers 3C and 5. These M-stream neurons project to M bands in the second visual area, V2, and to the middle temporal visual area (MT). Parvocellular layers project to lower layer 4 (4B), where information is relayed to subregions of layer 3 called blob and interblob modules. Blob neurons project Pb (Parvo-blob) bands in V2 and interblob neurons project to interbands (or PI for parvo-interblob) of V2. Subsequent processing involving the M stream includes relays from MT to the medial superior temporal area (MST) and the dorsal fundal superior temporal area (FSTd), both of which project with MT to the ventral posterior parietal area (VPP), which access visuomotor centers, including the frontal eye field and the superior colliculus. These areas and connections form the backbone of the "dorsal" stream of visual processing for visual attention and tracking and spatial vision (Ungerleider and Mishkin, 1982). The PI and Pb bands of V2 access the caudal dorsolateral area (Dlc), which provides inputs to a series of processing steps over caudal, rostral, polar,

see Kaas and Huerta, 1988). Neurons in the X or P stream have small receptive fields, respond with a maintained discharge to a steady stimulus, may be color selective, and constitute the majority of the retinal ganglion cells and LGN relay cells. The major functional role of the P stream is thought to be in detailed object vision. Consistent with this proposed role, the P stream is proportionately reduced in nocturnal owl monkeys as compared to other monkeys. Neurons in the Y or M stream have large receptive fields, respond to stimulus change, and are nonselective for color. They constitute roughly 10% of the retinal ganglion cells, and these cells relay to the superior colliculus for visuomotor functions, as well as the LGN. The M stream is thought to be important in detection and visual tracking. Finally, the W or I stream consists of a poorly characterized group of small neurons with medium to large receptive fields and often weak responses to visual stimuli. They also relay to both the LGN, terminating in the interlaminar zones, and the superior colliculus. The functions of the I stream are uncertain, but a role in modulating the other streams has been suggested (see Kaas and Huerta, 1988).

Area 17 has a laminar arrangement of neurons that is more distinct than in other cortical areas. As in other primates, P inputs terminate in sublayer 4B and M inputs terminate in sublayer 4A (Diamond *et al.*, 1985; Pospichal *et al.*, 1994). In owl monkeys, the inputs related to one eye or the other are not sharply segregated into the "ocular dominance bands" that are apparent in many species of monkeys (see Florence and Kaas, 1992), but instead these inputs overlap extensively in cortex (Kaas *et al.*, 1976). Inputs from the W or I stream terminate in a dotlike array of cell clusters or "CO blobs" that express high amounts of the metabolic enzyme cytochrome oxidase (CO). This input also terminates superficially in layer I on the apical dendrites of deeper pyramidal neurons. Neurons in the CO blobs appear to be dominated by P stream inputs relayed from outer layer 4, but M-stream influences exist as well (see Lachica *et al.*, 1992). Neurons in blobs do not obviously reflect W-stream properties. Thus, W inputs are thought to modulate ongoing activity rather than activate neurons.

As in other primates, the intrinsic circuitry of area 17 transforms LGN inputs so that neurons beyond layer 4 (Felleman and Kaas, 1982) are binocular, largely selective for elongated stimuli (orientation selective), and often preferentially respond to stimuli moving in particular directions (direction

---

and medial subdivisions of inferior temporal cortex (ITc, ITr, ITp, and ITm). These fields form the backbone of the ventral stream for object vision: IT cortex has interconnections with the hippocampus via entorhinal cortex and with frontal visual cortex. The dorsomedial area (DM) appears to integrate information from the M and P streams, but DM largely feeds the M stream via projections to VPP. Other subdivisions of visual cortex include the MT crescent (MTc), the rostral dorsolateral area (DLr), and areas such as the medial area (M) that are not included in the figure.

selective). Pyramidal cells dominated by M-stream inputs relay from layer 3 to one set of the dark CO bands in V2, and from layers 3 and 5 to MT (Diamond *et al.*, 1985). More superficial pyramidal neurons in layer 3 project to the other set of CO-dense bands as well as the CO-light interbands in V2 (see Kaas and Krubitzer, 1991). Finally, neurons largely within the CO blobs project to DM (Lin *et al.*, 1982; Krubitzer and Kaas, 1990c, 1993). Thus, the circuitry in area 17 transforms inputs, provides some interaction between P, M, and I streams, and creates a new channel by subdividing the P stream into two parts.

B. Area 18 or V2

All or most mammals (see Kaas and Krubitzer, 1991) appear to have a second representation of the visual hemifield (V2) that is located along the border of V1 that corresponds to the line of decussation of the retina (the zero vertical meridian through fixation). The first complete map of V2 in a primate was achieved in owl monkeys (Allman and Kaas, 1974a). A distinctive feature of V2 is that its outer (rostral) border corresponds to the zero horizontal meridian, so that the striplike V2 is "split," with the lower visual quadrant represented medial to the upper quadrant. V2 is about half the size of V1 in owl monkeys, constituting about 7% of neocortex (Krubitzer and Kaas, 1990b).

Histologically, V2 is most clearly defined in surface-view preparations (Fig. 1) from which sections have been stained for CO (Tootell *et al.*, 1985; Krubitzer and Kaas, 1989, 1990b) or myelin (Krubitzer and Kaas, 1989, 1990b). In the CO preparations, V2 is characterized by an array of CO-dense bands, separated by CO-light strips, extending across the width of the field. In sections stained for myelin, the CO-light strips appear as myelin-dense bands. The CO bands are of two types, one receiving input from the M stream of V1, and the other from blobs of the P stream (Pb stream); the interbands receive inputs from the interblob P stream (PI stream). The M bands project to MT, as well as to other targets, and the PI and Pb bands project to different modules in DL_c, as well as other targets. Thus, V2 consists of a series of three types of alternating bands, each with different inputs and outputs, and each with largely segregated intrinsic circuits. These three sets of bands or modules constitute three distinct, disjunctive, and interdigitated maps of the contralateral visual hemifield.

C. Early Areas of the Dorsal Stream: The Middle Temporal Visual Area and Associated Fields

Ungerleider and Mishkin (1982) have argued that after V1 and V2, visual information is processed in stations and pathways that can be grouped into a "dorsal" stream for visual attention and spatial vision and a "ventral" stream for object vision. Several areas of visual cortex in owl monkeys appear to be part of the dorsal stream. One of these areas is the middle temporal visual area, MT, a small oval of densely myelinated cortex in the upper temporal lobe of primates. MT was first defined in owl monkeys (Allman and Kaas, 1971a) and also was the first systematic representation of the visual field described outside of V1 and V2 in primates. MT can be easily identified in all primates by its myelination, direct projections from area 17, and characteristic map of the visual hemifield with central vision caudal and lower vision medial (see Krubitzer and Kaas, 1990b). MT is approximately one-fifteenth the size of V1 and one-sixth the size of V2, and yet it receives a substantial output from V1 and roughly one-third of the forward cortical projections of V2 (the M band output). These inputs from areas 17 and 18 appear to be critical for the functions of MT in owl monkeys, since lesions of area 17 deactivate neurons in MT (and area 18) so that they become unresponsive to visual stimuli (Kaas and Krubitzer, 1992). The area 17 neurons that project to MT are located in layers 3C and 5 (layers named after the scheme of Hassler, 1966). These neurons are selective for stimulus orientation, width, and usually direction of movement, and MT neurons themselves are highly selective for specific orientation and direction of moving bars, as well as selective for narrow bar widths (Felleman and Kaas, 1984). Receptive fields for neurons in MT are 5–10 times larger than those for neurons in V1, and MT neurons have large, antagonistic "surrounds" (Allman *et al.*, 1985; Petersen *et al.*, 1985) so that movements within the receptive field center are contrasted with more global movement of the visual world within the surround. There is some evidence that classes of neurons exist in MT and that they are segregated into arrays of interdigitating modules (Tootell and Born, 1991). MT closely interacts with two adjoining visual areas, the dorsal division of "the area of the fundus of the superior temporal sulcus," or FST_d (Kaas and Morel, 1993), and the medial superior temporal area, MST (Weller *et al.*, 1984; Krubitzer and Kaas, 1990b). These fields appear to process inputs from MT and have neurons with response features related to visual tracking. MT, FST_d, and MST all project to the ventral posterior parietal area, VPP, and adjoining posterior parietal cortex, regions highly concerned with visuomotor behavior and spatial aspects of vision (see Ungerleider and Mishkin, 1982).

D. Other Areas of the Dorsal Stream

A small wedge of cortex between V2 and the end of the lateral sulcus contains a systematic representation of the contralateral visual hemifield, with upper vision medial and lower vision lateral (Allman and Kaas, 1975). This dorsomedial area, DM, is coextensive with a region of cortex that is densely myelinated (Allman and Kaas, 1975; Krubitzer and Kaas, 1990c, 1993), and it is bordered medially by the medial area, M (Allman and Kaas,

1976), which projects widely to other areas of cortex (Graham *et al.*, 1978), and laterally by the dorsointermediate area, a less understood region of cortex. DM, approximately the size of MT (Krubitzer and Kaas, 1990b), receives a sparse but direct input from V1 (Lin *et al.*, 1982), which originates largely from neurons in the CO-dense blobs (Krubitzer and Kaas, 1990c). Both M bands and Pb bands of V2 also project to DM (Krubitzer and Kaas, 1990c). Most neurons in DM are selective for stimulus orientation and direction of movement (Baker *et al.*, 1981). They appear to be dominated by the M stream, but they also have sustained responses to maintained stimuli that indicate P-stream influences (Petersen *et al.*, 1988). The major outputs are to a moderately myelinated oval of cortex, the ventral posterior parietal area (Krubitzer and Kaas, 1990c), which relays to visuomotor areas of the frontal lobe (Kaas *et al.*, 1977), the frontal eye field, and the supplementary eye field (Huerta *et al.*, 1987; Stepniewska *et al.*, 1991). Other connections are with MT, MTc, and DLr (Wagor *et al.*, 1975).

The MT_c or "MT crescent" and rostral DL (DLr) are subdivisions of the dorsolateral complex, once thought to be a single field (Allman and Kaas, 1974b). MT_c is characterized by a string of small patches of CO-dense neurons within a narrow band of tissue that surrounds much of MT (Tootell *et al.*, 1985). MT_c has major connections with FST_v, the ventral division of the fundal superior temporal complex (Kaas and Morel, 1993). DLr is less densely myelinated than DLc, and DLr has major connections with DM (see Weller *et al.*, 1991). Overall, these fields process information in the M stream that is critical for visual tracking and locating objects in space, but P-stream inputs are significant, and these areas also help integrate M and P streams.

The medial area, M, is a representation of the visual hemifield on the medial wall of the cerebral hemisphere of owl monkeys that magnifies paracentral and peripheral rather than central vision (Allman and Kaas, 1976), and thus would seem to be involved in visual attention or spatial vision rather than detailed object vision. M projects to posterior parietal cortex, VPP, DM, and DI in a laminar pattern that suggests a modulatory role (Graham *et al.*, 1978). As with DM and PP, M also projects to the deeper portions of the superficial superior colliculus (Graham *et al.*, 1979), a projection suggesting an important role in directing eye movements.

E. Subdivisions of the Ventral Stream

The ventral stream of processing, which is specialized for object or form vision (Ungerleider and Mishkin, 1982), starts with a relay from both Pb bands and PI interbands of V2 to DLc, the caudal division of DL (see Cusick and Kaas, 1988). The bandlike nature of the projection pattern of V2 to dorsolateral visual cortex was first demonstrated in owl monkeys (Weller and Kaas, 1985). Consistent with its proposed role in object vision, DLc is characterized by neurons that are highly selective for the sizes (lengths and widths) of visual stimuli (Petersen *et al.*, 1988) and by neurons with sustained responses to maintained visual stimuli (Petersen *et al.*, 1988). DL represents the contralateral hemifield with the upper field medial, the zero horizontal meridian along with the V2 border, and the zero vertical meridian rostral (Allman and Kaas, 1974a; Sereno *et al.*, 1986).

The major output of DLc is to the upper or caudal division of inferior temporal cortex, ITc (Weller and Kaas, 1985). Other outputs include the ventral posterior (VP) region (Fig. 1), a poorly understood region that is implicated by connections in the ventral stream. DLc also has weak connections with FST_d of the dorsal stream (Kaas and Morel, 1993). The major output of ITc is to more rostral inferior temporal cortex, ITr (Weller and Kaas, 1987), which also receives major outputs of FST_v (Kaas and Morel, 1993). ITr has major projections to the pole of inferior temporal cortex, ITp, and minor connections link DL, ITc, and ITr to medial inferior temporal cortex, ITm, which corresponds to an architectonically distinct (myelin light) region of cortex (Weller and Kaas, 1987). However, other connections of ITm with MST (Weller and Kaas, 1987) suggest that ITm is largely part of the "dorsal" stream of visual processing. ITc, ITr, and probably ITp and ITm provide access to the entorhinal cortex, the hippocampus, and amygdala, structures that are critical in the formation of visual memories and in assigning emotional significance to visual stimuli (see Kaas and Krubitzer, 1991).

F. Visual Connections between the Hemispheres

Most studies of the connections of visual areas in owl monkeys have included descriptions of the interhemispheric as well as the ipsilateral connections, and thus much is known about how areas on one side of the brain connect with those on the other side (see Cusick and Kaas, 1986, for review). In addition, the total pattern of interhemispheric connections has been described for owl monkeys in two independent studies, one based on stains for degenerating axons after section of the corpus callosum (Newsome and Allman, 1980) and one on the bidirectional transport of the tracer horseradish peroxidase (Cusick *et al.*, 1984). The results indicate that almost all the visual cortex is subserved by the corpus callosum, and only the pole of the temporal lobe (including ITr) has anterior commissure connections (Weller and Kaas, 1987). V1 is almost devoid of callosal connections, although a few callosally projecting cells are located in area 17 within 1–2 mm of the border. Other visual areas have more extensive, but unevenly distributed callosal connections, so that neurons in different parts of these areas differ in the extent of interhemispheric influences on receptive fields. Most visual areas have neurons with excitatory receptive fields only in the contralateral hemifield and callosal connections appear to contribute only to the larger, antagonistic surrounds that include parts of the ipsilateral hemifield (Allman *et al.*, 1985). However, neurons in IT cortex have large excitatory receptive fields extending into the ipsilateral visual hemifield (Pollen *et al.*, 1984), and thus a more extensive role for interhemispheric connections is expressed.

The callosal connections of restricted locations within visual areas are highly similar across most higher-order visual areas. Each location in a visual area typically has dense callosal connections with a mirror image or matched location in the corresponding visual area of the other hemisphere, even though such matched locations may often be mismatched for the representation of visual space by relating to mirror image locations in the right and left visual hemifields (see Cusick and Kaas, 1986). Sparse additional connections may be more scattered in the corresponding area, and they may favor locations representing space near the zero vertical meridian. Finally, each visual area typically has callosal connections, not only with its counterpart, but with two to three other visual areas that are also the main targets of ipsilateral projections, that is, the visual areas immediately above and below the area in the processing hierarchy. Thus, callosal connections subserve the hierarchical processing of information, as well as the transfer of information between matched areas at the same level of processing.

III. AUDITORY CORTEX

The organization of auditory cortex has not been extensively investigated in any primate, but more is known about the auditory cortex in owl monkeys than in any other primate as a result of an early landmark microelectrode mapping and architectonic study of Imig *et al.* (1977), and subsequent investigations of the connections of two auditory areas, A1 and R, with cortical and subcortical targets (FitzPatrick and Imig, 1978, 1980). Auditory cortex organization has been further investigated in owl monkeys in a study combining architectonic methods, microelectrode mapping, and the use of tracers to reveal connections (Morel and Kaas, 1992). The results led to a theory of parcellation (Fig. 1) that includes a core of three primarylike areas, A1 (corresponding to the traditional primary area), the rostral area, R (first described by Merzenich and Brugge, 1973, in macaque monkeys), and RT (a newly discovered field). The primarylike fields are distinguished from other auditory fields by having neurons that are highly responsive to pure tones and highly selective for specific frequencies (frequency selective or tuned). Neurons within these fields are systematically arranged so that frequencies are represented from high to low across each field (the fields are tonotopically organized). Neurons in fields surrounding the core (the belt fields) are less responsive to tones, less sharply tuned, and thus they are difficult to characterize. However, the separate higher-order auditory fields can be distinguished from each other by differences in architecture (histological appearance) and by patterns of connections. Cortex immediately lateral to the belt (the auditory parabelt) is less responsive to auditory stimuli in anesthetized animals, but it has dense connections with the auditory belt. The parabelt relays auditory information to the frontal lobe.

A. The Primarylike Core of A1, R, and RT

The core areas have a number of features that make them primarylike (Morel and Kaas, 1992). First, the neurons in these fields are highly responsive to tones and other auditory stimuli, even in anesthetized owl monkeys. In addition, the neurons are narrowly tuned, responding best to one frequency, and responding progressively less for the same sound intensity for frequencies displaced from the best frequency. Thus, by recording at many locations in these fields, the ways that tones (best frequencies) are represented within the fields can be determined.

A1, the rostral area, R, and the rostral temporal area, RT, have characteristic tonotopic organizations that help identify and distinguish these fields (Fig. 3). A1 represents low to high frequencies in a sequence from the

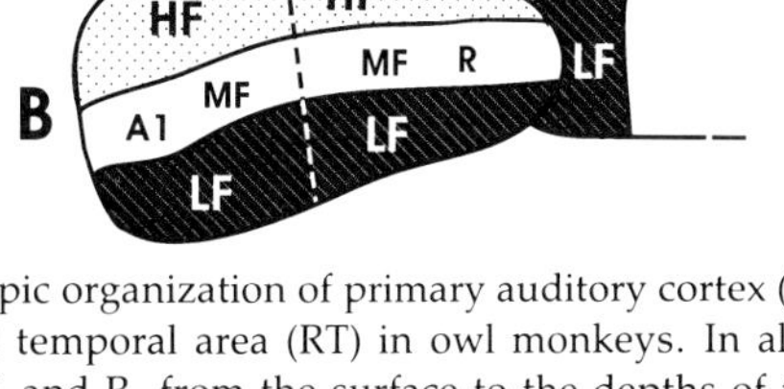

FIGURE 3 The tonotopic organization of primary auditory cortex (AI), the rostral auditory area (R), and the rostral temporal area (RT) in owl monkeys. In all cases, progressions of recording sites across A1 and R, from the surface to the depths of the lateral sulcus (lower to upper in parts A and B), encounter neurons first best activated by lower and then successively higher tones. However, for different owl monkeys, two different patterns are seen. In some cases, the lines of isofrequency (directions across A1 and R where neurons all have the same best frequency) course at a lateromedial angle so that neurons with middle-range frequencies reach the caudolateral edge of AI, and disjunctions in the frequency representations occur at the A1–R border (pattern A). In other cases, the lines of isofrequency continue uninterrupted across the two areas (pattern B). In A and B, the continuous change of representation of frequency from high to low is portrayed as three steps. HF, high-frequency neurons; MF, middle frequency, LF, lower frequency. Modified from Morel and Kaas (1992).

exposed lip of the lower bank of the lateral sulcus to cortex in the depth of the sulcus. In the opposite direction, neurons with the same or similar best frequencies are aligned in rows or columns across the length of the field in the so-called isorepresentation lines or bands. From case to case, the orientations of the lines of isorepresentation vary from coursing along the rostrocaudal length of the area to being at somewhat of an angle. A1 and R have similar tonotopic organizations, with neurons deeper in the fissure tending to respond to progressively higher frequencies for both areas. In cases where the lines of isorepresentation are angled, mismatches along the junctions in the maps in A1 and R help distinguish the two fields (Fig. 3A). In other cases, the lines of isorepresentation merge so that the maps provide little information about the location of the junction of the two fields (Fig. 3B). R has not been completely mapped. The caudal part of R represents low frequencies. Higher frequencies may be represented more rostrally in the temporal lobe or deeper in the fissure.

A1, R, and RT are also primarylike in their architecture or histological appearance. They all have a well-developed layer 4 that is densely packed with small neurons, thus producing the "koniocortical" appearance of primary cortex. These three fields are densely myelinated and react densely for cytocrome oxidase, also features of primary areas of cortex. Nevertheless, these primarylike features are graded across the three fields, being most pronounced in A1, slightly less in R, and even less in RT. Thus, it is possible to architectonically distinguish AI, R, and RT.

Finally, A1, R, and RT are all primarylike in that they all receive inputs from the principal (ventral) relay nucleus of the medial geniculate complex of the auditory thalamus (Fig. 4A). However, RT receives inputs from only the caudal portion of the nucleus, and both R and MT receive more input than A1 from medial (or magnocellular) and dorsal nuclei of the complex. Thus, differences in thalamic connections also distinguish the three fields.

The sequence of processing in auditory cortex of primates is not well understood. However, A1 projects strongly to R, and R to RT. Thus, these fields appear to form a series of processing steps (Fig. 4B), yet they are also activated in parallel from the ventral nucleus. Intrinsic connections of A1 spread laterally, largely along the isofrequency lines, suggesting mutual facilitation. Callosal projections largely connect parts of A1 of each hemisphere that represent the same frequency.

B. The Auditory Belt

A narrow 2-mm-wide belt of cortex around the auditory core has neurons that respond to auditory stimuli and direct connections with the core (Fig. 4B). It is not completely clear how this belt is divided into processing areas, but connection and myelination patterns suggest the existence of seven

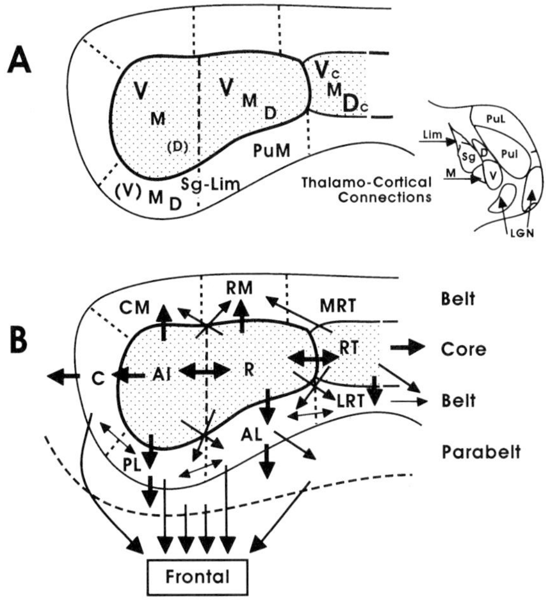

FIGURE 4 Cortical and thalamic connections of A1, R, and RT in owl monkeys. (A) Thalamic inputs to fields are indicated by superimposed letters, with larger letters denoting more input. AI, for example, gets major inputs from the ventral (V) nucleus of the medial geniculate complex, less input from the medial or magnocellular nucleus (M), and sparse input from the dorsal nucleus (D). Parts of the auditory belt also receive inputs from the medial pulvinar (PuM) and the supergeniculate–limitams complex (5g–Lim). A frontal brain section through the auditory thalamus is drawn on the lower right to show the spatial arrangement of the auditory nuclei relative to the lateral geniculate nucleus (LGN), the inferior pulvinar (PuI), and the lateral pulvinar (PuL). (B) The cortical connections of subdivisions of auditory cortex. Arrow sizes indicate the magnitudes of connections. Auditory areas of the belt include caudal (C), caudomedial (CM), rostromedial (RM), medial rostral temporal (MRT), lateral rostral temporal (LRT), anterior lateral (AL), and posterior lateral (PL) fields. Frontal cortex is in the frontal lobe. Modified from Morel and Kaas (1992).

areas (Morel and Kaas, 1992). Four of these fields (posterolateral or PL, anterolateral or AL, caudomedial or CM, and rostromedial or RM) correspond to those described by Imig *et al.* (1977), except that the extents of PL and CM have been reduced to accommodate a caudal area, C, just caudal to A1. In addition, AL has been reduced to allow for the newly defined core area, RT, and bordering belt areas, medial rostrotemporal (MRT) and lateral rostrotemporal (LRT), have been added.

Area CM is distinguished by moderately dense myelination and less densely packed neurons in granular and supragranular layers than A1, and neurons responsive to tones, but with more variable responses and broader frequency tuning than A1. C is clearly different from CM in having much less dense myelination. In addition, single injections of tracers in A1 reveal separate projections to CM and C. RM is also distinguished from CM by a marked reduction in myelination. RM is less koniocortical in appearance than A1 and R, and its major input is from R rather than A1. Neurons in RM respond to auditory stimuli, but response characteristics have not been determined. Areas PL and AL are lightly myelinated fields bordering A1 and R, respectively. Neurons in these fields have reduced responsiveness to tones and broad-frequency tuning. Best frequencies, when determined, tend to match those of neurons in adjacent parts of A1 and R so that two crude tonotopic patterns might also distinguish PL and AL. PL is more densely connected with A1, and AL with R. Areas MRT and LRT are distinguished by position and connections. LRT has major connections with RT rather than R. MRT seems to have few connections with A1 or R, and connections with RT seem likely, but have not been clearly demonstrated. Both MRT and LRT are lightly myelinated.

The major cortical connection patterns of the belt are summarized in Fig. 4B. Injections of tracers in the core areas indicate that CM, C, and PL all receive major feedforward projections from A1. RM and AL receive similar inputs from R, and RT probably provides the activating inputs for LRT. Only the connections of PL and AL have been further investigated, and these fields project into the auditory parabelt. PL and AL receive little or no input from the ventral nucleus of the medial geniculate complex, and thus are not primarylike, but they are interconnected with other parts of the auditory thalamus and the more widely projecting medial pulvinar (Fig. 4A).

C. The Auditory Parabelt

A strip of cortex lateral and caudal to the auditory belt has connections with the auditory belt (Fig. 4B) and parts of the auditory thalamus. The full extent of the parabelt cortex has not been determined, but the region with connections with the belt extends onto the upper bank of the superior temporal sulcus. Injections in the parabelt region reveal inputs from the belt, connections with nearby locations in the temporal lobe, and projections to locations rostral and medial to the frontal eye field in the frontal lobe. The connections with the frontal lobe have been further demonstrated by injections in frontal cortex that label neurons in the parabelt (Morel and Kaas, 1992). Thus, as for the visual system, the third level of auditory fields relays information to regions of prefrontal cortex that are presumably important in using auditory information to guide and formulate motor plans. How auditory information is relayed to limbic centers that are important for memory and evaluation is not yet certain. However, there is evidence from other primates that the auditory thalamus projects directly to the amygdala.

IV. SOMATOSENSORY CORTEX

Much of our understanding of the organization and connections of somatosensory cortex in primates is based on studies in owl monkeys. For example, the first compelling evidence that the "primary area" or "S1" of monkeys actually consists of four separate areas or representations of the body was obtained from owl monkeys (Merzenich *et al.*, 1978; Kaas *et al.*, 1979). In addition, the ventral somatosensory area, VS, is a recently discovered subdivision of somatosensory cortex that so far has been described only in owl monkeys (Cusick *et al.*, 1989). Finally, the subdivision of "S1 proper" or area 3b (see the following) into modules of neurons related either to slowly or to rapidly adapting cutaneous receptors was first described in owl monkeys (Sur *et al.*, 1981).

The basic features of the somatosensory system in primates (Figs. 5 and 6) have been reviewed elsewhere (Kaas and Pons, 1988). The physiological characteristics of peripheral nerve afferents have not been investigated in owl monkeys, but they probably correspond to those in other primates. In order to identify objects by touch, it is important to know the positions of skin surfaces. Muscle spindle receptors (MS) provide information about limb, hand, and digit positions; slowly adapting type II receptors (SA-II) respond to skin or tendon stretch and they may help indicate limb position. Slowly adapting type I (SA-I) receptors are activated by light pressure, and they signal static pressure and the location of edges of objects pressed on the skin. Rapidly adapting type I afferents (RA-I) respond to changes in skin indentation and they are important in the discrimination of textures moved across the skin and in the sensation of flutter. The rapidly adapting type II (RA-II) or Pacinian (PC) afferents subserve the sensations produced by high-frequency vibrations. Other afferents respond to tactile stimuli, with increasing response magnitude as the intensity of the stimulus increases into the painful range. These wide dynamic range (WDR) afferents may help signal stimulus intensity and noxious stimuli.

All of these afferents enter the brain stem or spinal cord where they terminate in patterns that preserve the spatial topography (somatotopy) of the receptor array (see Florence *et al.*, 1991), while the submodalities remain segregated. Thus, inputs from digits 1–5 of the hand terminate in a 1–5 order in the dorsal horn of the spinal cord and in the cuneate nucleus of the dorsal column–trigeminal complex in the lower medulla. More specifically, MS afferents for the hand muscles terminate in the lower brain stem, where they relay to the cerebellum and to the ventroposterior superior nucleus of the thalamus. Classes of SA and RA afferents from the skin of the hand project to cell clusters in the cuneate nucleus, where they apparently remain segregated from each other, and they relay to separate SA or RA cell groups in the ventroposterior nucleus. The inputs to the ventroposterior inferior nucleus, VPI, have not been clearly identified, but they are largely spinothalamic projections that may be dominated by wide dynamic range neurons.

A. Areas 3a, 3b, 1, and 2

Anterior parietal cortex of monkeys and higher primates includes four primarylike fields corresponding to the four classical architectonic areas "3a, 3b, 1, and 2" (Kaas and Pons, 1988). From the late 1930s, these four architectonic subdivisions were considered parts of a single somatotopic representation of the body surface, but research on owl monkeys demonstrated instead that each of the four fields contains a systematic representation of body receptors (Fig. 5; Merzenich *et al.*, 1978, 1987; Kaas *et al.*, 1979). Area 3b and area 1 are both highly responsive to light tactile stimuli, and both contain systematic maps of body surface receptors. The maps are organized in parallel, so that they both represent the body from tail to tongue in a mediolateral progression and are largely mirror images of each other, so that the glabrous hand, for example, is represented palm to palm at the 3b/1 border with digit tips pointing rostrally in area 3b and caudally in area 1. Areas 3a and 2 respond to the stimulation of both cutaneous and deep (muscle spindle and probably joint and tendon) receptors, although the activation by deep receptors is typically more obvious, especially in area 3a. These fields contain maps organized in parallel to those in areas 3b and 1. Area 2, which has been more completely mapped, appears to be approximately a mirror image of area 1.

An early study in which the tracer horseradish peroxidase was injected in area 3b, area 1, or area 2 of owl monkeys (Lin *et al.*, 1979) demonstrated that VP projects to both area 3b and area 1, and that some neurons in VP are likely to project to both fields. A nucleus dorsal to VP, which we now call VPS, projects to area 2. We now have evidence from studies in squirrel monkeys (e.g., Cusick *et al.*, 1985) that VPS also projects to area 3a; as many as 40% of the neurons in VPS project to both area 3a and area 2, and about 20% of the neurons in VP project to both areas 3b and 1. Thus, MS information is provided to areas 3a and 2 by VPS, partly via the same neurons, and RA (and probably not SA, see the following) information is provided to areas 3b and 1, partly from the same neurons (Fig. 6).

A number of features of area 3b indicate that this field is homologous (the same field) with S1 as first described in cats and subsequently described in a range of nonprimate mammals (see Kaas, 1983, for review). Area 3b, as for other body representations, has disproportionally large amounts of cortex devoted to parts of the body with large numbers of cutaneous receptors, specifically the hand, lips, and tongue (Fig. 5). Neurons in these portions of area 3b with a large "magnification factor" have smaller receptive fields. Because of the inverse relation of "magnification" and receptive field size, similar amounts of tissue and neurons are devoted to

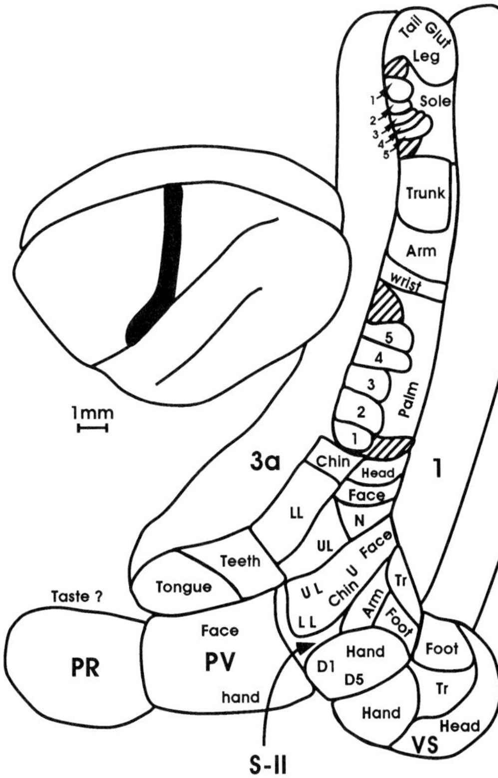

FIGURE 5 Subdivisions of somatosensory cortex in owl monkeys. On the left, the location of area 3b or S-I proper (see text) is indicated on a dorsolateral view of the brain. On the right, area 3b and areas adjoining area 3b on the dorsal surface of the brain and in the lateral fissure are magnified. Some somatotopic features of the systematic representation of body surface receptors in area 3b are indicated. In brief, the tail and gluteal region are represented medially in cortex, followed in successively more lateral locations by the foot and toes 5 through 1, the trunk, arm and wrist, the hand with digits 5 through 1, the chin, nose (N), lower lip (LP), upper lip (UL), teeth, and tongue. The representations of the dorsal parts of the foot and hand are shaded. Areas 3a, 1, and 2 form similar parallel representations (organizations not shown). Other systematic representations are coextensive with the second area, S2, the ventral somatosensory area (VS), the parietal ventral area (PV), and possibly the parietal rostral area (PR). Tr represents trunk. See text for more discussion.

processing information from similar receptor groups anywhere on the body surface (Sur *et al.*, 1980). Finally, area 3b actually contains two mosaiclike interdigitated maps, one for SA inputs and one for RA inputs (Sur *et al.*, 1981, 1984). Thus, both SA and RA groups of neurons relay to separate, distributed groups of cells in area 3b. However, recordings in area 1 detect only RA inputs. Therefore, SA neurons appear to relay only to 3b. The VP neurons terminate in area 3b with axon arbors that extend laterally over

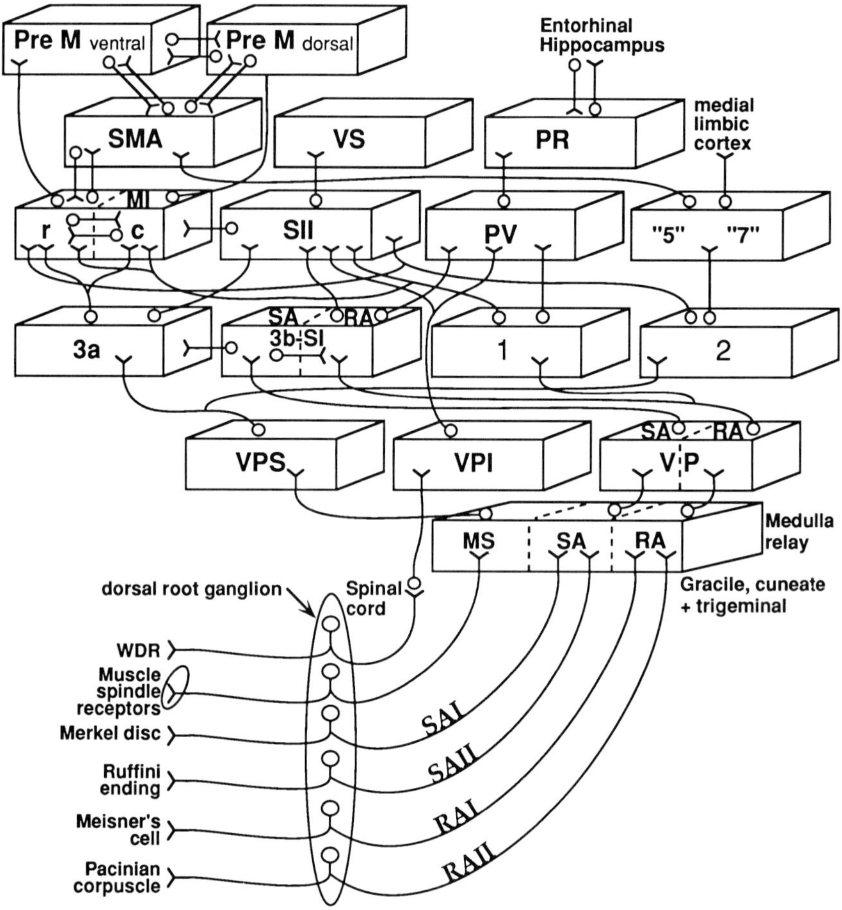

FIGURE 6 Processing hierarchy for somatosensory information related to the identification of objects by active touch. At the bottom, somatosensory afferents with cell bodies in the spinal root ganglion terminate in the dorsal horn of the spinal cord or the dorsal column–trigeminal complex of the lower medulla. Classes of afferents (slowly adapting type I, SA-I; slowly adapting type II, SA-II; rapidly adapting type I, RA-I; rapidly adapting type II, RA-II; muscle spindle afferents, MS; and wide dynamic range afferents, WDR) terminate on separate groups of cells that relay to the thalamus. The major thalamic nuclei are the ventroposterior (VP), the ventroposterior superior (VPS), and the ventroposterior inferior (VPI). Cortical areas include areas 3a, 3b, 1, and 2, the second area (S2), the parietal ventral area (PV), the ventral somatosensory area (VS), the parietal rostral area (PR), and posterior parietal cortex (areas "5" and "7"). Somatosensory areas project to motor areas, including rostral and caudal divisions of the primary field (M1r and M1c), dorsal and ventral premotor areas (PMD and PMV), and the supplementary motor area (SMA). See text for further details.

about 0.5 mm in layer 4 (Garraghty *et al.*, 1989), potentially activating a large number of cortical neurons.

Largely from studies on other monkeys (see Cusick *et al.*, 1985; Krubitzer and Kaas, 1990a), it has become clear that area 3b relays cutaneous information to area 3a, area 1, and to a lesser extent area 2. Area 1, in turn, projects to area 2. Thus, VPS provides the deep receptor information to areas 3a and 2, whereas relays from 3b and 1 provide the cutaneous receptor information. Furthermore, in both owl and squirrel monkeys, area 3b provides the dominant driving input to area 1; thus the direct thalamic input to area 1 has only a modulatory role (Garraghty *et al.*, 1990a). Corresponding to the higher processing level of area 1, receptive field sizes are larger and response properties are more complex for neurons in area 1 than area 3b (Sur, 1980). Area 2 constitutes a further step in processing where relays from area 1 and area 3b converge with MS information directly from VPS.

B. Somatosensory Areas of the Lateral Sulcus and Posterior Parietal Somatosensory Cortex

Cortex lateral to area 3b within the lateral nucleus contains three areas with systematic representations of the body (areas S2, PV, and VS) and one field of uncertain organization (area PR). In other monkeys (see Krubitzer and Kaas, 1992), and probably owl monkeys, VPI projects densely to S2 and to PV. S2 and PV receive additional inputs from all four areas of anterior parietal cortex, and these inputs are critical for the activation of S2 and PV (Garraghty *et al.*, 1990b). Thus, the VPI inputs to S2 and VP have the function of modulating the higher-order processing of convergent information from 3a, 3b, 1, and 2, perhaps by adding further information about stimulus intensity. S2 feeds forward to the ventral somatosensory area (VS), Pv, and the parietal rhinal region (PR). Thus all of these fields are higher processing stations. Further connections of these fields have not been adequately studied, but PR is the likely relay to entorhinal cortex and the hippocampus, structures that are critical for the long-term storage of tactile memories (Mishkin, 1979).

Area 2 (and to a lesser extent area 1) also projects to posterior parietal cortex, where further processing occurs. Posterior parietal cortex projects to the frontal lobe and to limbic cortex on the medial wall of the cerebral hemisphere (see Kaas and Pons, 1988). The organization of posterior parietal cortex, which includes presumptive areas "5" and "7," is not well understood.

V. MOTOR AND PREMOTOR CORTEX

Motor and premotor fields have been extensively studied in owl monkeys (see Gould *et al.*, 1986; Huerta *et al.*, 1987; Stepniewska *et al.*, 1991, 1993).

A primary area, M1, occupies a broad band of cortex just rostral to area 3a. We divide M1 into rostral and caudal subfields in owl monkeys because these two regions differ slightly in histological appearance, connections, and thresholds for evoked movements. Other fields include the supplementary motor field, the rostral part of which has been termed the supplementary eye field, the frontal eye field, and dorsal and ventral premotor areas.

A. Primary Motor Cortex

The primary field, M1, has a major role in the control of voluntary movements. Lesions of M1 in humans and other primates can greatly impair the ability to initiate movements, although considerable recovery can occur. The field has been called "agranular cortex" because a layer 4 of small granule cells is not clearly differentiated from layer 3, although it is probably incorrect to conclude that a layer 4 is missing. M1 is also characterized by large pyramidal cells that project to the brain stem and spinal cord (Nudo and Masterton, 1990), where they directly or indirectly activate motor neuron outputs to muscles. Electrical stimulation of M1 evokes movements of parts of the body, with medial locations in cortex tending to evoke movements of parts of the hindlimb, middle locations relating to the forelimb, and lateral cortex involved in face movements (see Gould *et al.*, 1986, for review).

Despite these distinctive features, M1 has not been adequately delineated in most primates. Part of the problem is that the pyramidal cells are larger medially, since they project further in the spinal cord to reach portions devoted to the hindlimb and relate to large muscles and gross movements, compared to more lateral pyramidal cells related to closer targets in the brain stem and upper spinal cord and fine movements of the tongue, lips, and digits. Thus, M1 has an uneven appearance, and mediocaudal parts have been termed area 4 and lateral parts as area 6. Since area 4 is widely considered to be M1, it seems reasonable to label the architectonic field that is coextensive with M1 as area 4.

The most precise and detailed maps of motor cortex result from the use of penetrating microelectrodes so that the effective stimulation zones can be restricted to those pyramidal cells near the tip of the electrode. The first extensive microelectrode map of M1 in a primate was that of Gould *et al.* (1986) in owl monkeys. This study confirmed earlier reports on the progression of somatopic order in M1 in that body movements were represented from hindlimb and tail to tongue in a mediolateral cortical sequence. Although several conflicting theories on the fine-grain organization of M1 exist, Gould *et al.* (1986) presented evidence that M1 is well described as a mosaic of small patches of cortex, with each patch devoted to a specific movement, and each movement encoded in several distributed patches.

Other research on M1 of owl monkeys (Stepniewska *et al.*, 1991, 1993) has led to modifications of the original description. First, M1 appears to include two subfields, each forming a mediolateral sequence of body movements as two parallel, mosaic maps, rostral M1 or $M1_r$ and caudal M1 or $M1_c$. The two maps are similar in that low levels of current evoke movements in each, but in some cases the thresholds for movements are higher in $M1_r$ than $M1_c$. The fields also differ slightly in architecture, with larger pyramidal cells in $M1_c$. But more importantly, the connections of the two fields differ somewhat. Inputs to M1 include somatosensory afferents from all four areas of anterior parietal cortex (Stepniewska *et al.*, 1993;), but very little input originates from area 3b. Area 3a appears to project more densely to $M1_c$, area 1 projects to both $M1_c$ and $M1_r$, whereas area 2 projects largely to $M1_r$. SII and PV project to both subfields, but more to $M1_r$. These differences in inputs suggest that $M1_r$ and $M1_c$ have somewhat different functional roles, with $M1_r$ being more influenced by higher-order somatosensory fields (area 2, PV, and SII).

M1 also receives inputs from ventral and dorsal premotor fields, the supplementary motor area (SMA) and limbic cortex ventral to SMA. Thus, premotor "planning" areas can affect the output of motor cortex. The callosal connections of MI are uneven, with the hand region having the most sparse callosal connections (Gould *et al.*, 1986). Subcortical projections of M1 in owl monkeys include dense, bilateral terminations in the putamen of the corpus striatum (Whitworth *et al.*, 1991).

B. The Supplementary Motor Area

Stimulation of cortex rostral to the hindlimb representation in M1 produces movements from a field known as the supplementary motor area (SMA). The first detailed microstimulation map of SMA was determined from owl monkeys (Gould *et al.*, 1986). Hindlimb, forelimb, face, and eye movements are evoked from SMA in a caudorostral sequence. Overall, the much smaller representation in SMA resembles that in M1 by being disjunctive or mosiaclike with given movements sometimes represented at several sites. The most rostral part of SMA is sometimes distinguished as the supplementary eye field, E-SMA.

Major inputs of SMA are from posterior parietal cortex, providing evidence that the SMA receives more processed sensory information than M1 and is a higher-order field more concerned with motor programs than producing given movements. SMA also has connections with more rostral parts of frontal cortex, premotor areas, and of course, M1 (Stepniewska *et al.*, 1991, 1993). The E-SMA is interconnected with the frontal eye field (Huerta *et al.*, 1987).

C. The Dorsal and Ventral Premotor Areas

Movements can be evoked from cortex just rostral to M1 in dorsal (PMD) and ventral (PMV) premotor fields. PMD and PMV are considered distinct from M1 for several reasons. First, in general, higher levels of stimulating current are needed to evoke movements in premotor cortex. Second, the movement patterns, though mosiaclike as in M1, are distinct from that in M1 (Preuss *et al.*, 1991). Most notably, the ventral area has a large representation of hand and digit movements that is quite distant from the hand region of M1 (see Gould *et al.*, 1986). Third, the architecture differs from that of M1 so that a sparse granular layer is evident and pyramidal cells are less pronounced in premotor cortex. The connections of PMD and PMV have not been fully studied, but both fields are interconnected with SMA and M1 (Stepniewska *et al.*, 1991).

D. The Frontal Eye Field

A small 2-mm oval of cortex just rostral to M1 and between PMD and PMV constitutes the frontal eye field (FEF). In FEF, stimulation with microelectrodes produces rapid contralateral deviations of the eyes (Gould *et al.*, 1986; Huerta *et al.*, 1987). The major cortical connections of the FEF are with higher-order visual areas of parietal and temporal cortex, the E-SMA, an adjoining frontal visual area, FV, more rostral frontal cortex, and adjoining parts of M1 related to eyelid movements (Huerta *et al.*, 1987).

In addition to FEF and E-SMA, cortex between these two fields, when stimulated with higher levels of current, can produce eye movements (Preuss *et al.*, 1991). Thus, there are several visuomotor fields in the frontal lobe of owl monkeys and other primates.

VI. ADULT PLASTICITY: THE MUTABILITY OF CORTICAL MAPS

The extensive and detailed information on the topographic organizations of sensory representations in the cortex of owl monkeys has served as a framework for another important line of research on how sensory representations are altered in adult monkeys as a result of changes in the significance of various inputs. As is apparent from Figs. 2, 4, and 6, each major sensory system includes a number of nuclei and cortical areas. Most nuclei and areas systematically represent a receptor surface and are arranged in a hierarchical manner so that higher areas depend on inputs from one or more lower structures for activation. Somewhat surprisingly, given our usual concept of the adult brain as having a relatively fixed organization, partial deactivations of cortical maps (e.g., removing all sensory inputs related to the hand) can be followed by extensive reorganization of the maps so that neurons formerly activated by inputs from the hand become activated by inputs from the face (see Kaas, 1991, for review). Such findings demonstrate that the internal organizations of sensory maps are dynamically maintained and subject to change. The research on this adult plasticity is considered important because it can help us understand how normal organization is maintained, and the plasticity could be important in functional compensations that improve performances on sensory-motor skills and perceptual abilities, allow recalibrations of sensory systems after receptor surface damage, and mediate recoveries after central nervous system damage such as that following strokes.

Most of the research on adult plasticity of sensory cortex has been on the somatosensory cortex of owl monkeys, where the anterior parietal fields are accessible on the dorsolateral surface of the brain. The early studies showed that areas 3b and 1 reorganize after section of the median nerve, which subserves the thumb half of the glabrous hand. Cutting the median nerve deactivates the input, but the input neurons survive and can regenerate. After such a cut, the deprived parts of the hand maps in areas 3b and 1 (and presumably other cortical fields) rapidly reorganize over days to weeks so that they fail to remain silent, but instead come to be activated by inputs from the back (dorsal surface) of the hand, and to a lesser extent by the lateral glabrous hand (Merzenich *et al.*, 1983a,b). We now know that most of the cortical changes are a result of alterations at earlier stages of the processing hierarchy, since reorganization is seen at the level of the ventroposterior nucleus of the thalamus (Garraghty and Kaas, 1991). Nevertheless, some of the reactivation may depend on changes that occur in cortex; the deprivation by nerve section is followed by a reduction in the expression of the inhibitory neurotransmitter, GABA, in the reorganized cortex, which would release previously subthreshold excitatory inputs (Garraghty *et al.*, 1991).

Other reorganizations of area 3b in owl monkeys have been demonstrated after digit amputations (Merzenich *et al.*, 1984), after partial lesions of area 3b (Jenkins and Merzenich, 1987), and as a consequence of repeated microstimulation of a given cortical site (Recanzone and Merzenich, 1988). The reorganizations appear to be highly reversible, since deactivation by nerve crush is followed by reorganization that is later replaced by a recovery of the original organization after regeneration of the crushed nerve (Wall *et al.*, 1983). In addition, reorganizations in area 3b of owl monkeys fail to create orderly maps out of disordered inputs resulting from cut and regenerated nerves (Wall *et al.*, 1986) or from nerves that have been forced to regenerate into the wrong skin field (Wall and Kaas, 1986), even after long recovery periods. Thus, there are limits to adult plasticity, and functional connection patterns formed in development have some advantage over those newly expressed in adults.

In other mammals, adult plasticity has been demonstrated in the motor, auditory, and visual systems (see Kaas, 1991, for review), and thus plasticity is a general feature of mammalian cortex. Because of the number of well-studied sensory and motor maps in owl monkeys, they are likely to be used in further studies of adult plasticity in both the somatosensory and other systems.

VII. CONCLUSIONS AND DIRECTIONS FOR THE FUTURE

1. Owl monkeys have been extremely useful in studies of the organization and connections of neocortex in primates, largely because they have relatively small brains with relatively few fissures. In addition, as the only nocturnal monkey, they offer advantages in studies of the visual system by having neurons with larger, easier to study receptive fields and visual areas that do not have extremely large representations of foveal or central vision, and thus patterns of visuotopic organization are easier to reveal.
2. The auditory, visual, somatosensory, and motor systems in owl monkeys appear to reflect a basic primate plan, in that similar organizations exist in other primates. Each system consists of a number of cortical and subcortical areas and nuclei, connected in a complex, hierarchical, distributed manner, with components of parallel and serial processing. The extent of species differences is not yet clear.
3. The sensory and motor areas are plastic even in adult mammals in that neurons are able to acquire new receptive fields and process information from changed sets of sensory afferents as a result of alterations in the effectiveness of inputs following injury or experience. These changes might help mediate recovery from injury and the learning of sensory-motor skills.
4. As a result of their extensive use in anatomical, histochemical, and electrophysiological studies of cortical organization, the organization of neocortex is better understood in owl monkeys than in most primates. The extensive understanding of the organization of neocortex in owl monkeys can serve as a framework for guiding further research and posing future experiments. Thus, it is likely the owl monkey will continue to be a major and important research animal for studies of brain organization.
5. Future research can be usefully directed toward: (a) exploring the poorly understood regions of cortex to see how they are subdivided into areas and how they relate to the currently proposed fields; (b) evaluating the significance of the interconnections between areas and fields by deactivating and selectively stimulating subsets of connections; (c) further evaluating the validity of current conclusions about subdivisions, connections, and processing sequences; and (d) determining the limits and extents of plasticity in cortical areas, especially the higher-order fields, and determining what plasticity means in terms of brain function.

REFERENCES

Allman, J. M., and Kaas, J. H. (1971a). A representation of the visual field in the posterior third of the middle temporal gyrus of the owl monkey (*Aotus trivirgatus*). *Brain Res.* **31,** 85–105.

Allman, J. M., and Kaas, J. H. (1971b). Representation of the visual field in striate and adjoining cortex of the owl monkey (*Aotus trivirgatus*). *Brain Res.* **35,** 89–106.

Allman, J. M., and Kaas, J. H. (1974a). The organization of the second visual area (VII) in the owl monkey: A second order transformation of the visual field. *Brain Res.* **76,** 247–265.

Allman, J. M., and Kaas, J. H. (1974b). A crescent-shaped cortical visual area surrounding the middle temporal area (MT) in the owl monkey (*Aotus trivirgatus*). *Brain Res.* **81,** 199–213.

Allman, J. M., and Kaas, J. H. (1975). The dorsomedial cortical visual area: A third tier area in the occipital lobe of the owl monkey. (*Aotus trivirgatus*). *Brain Res.* **100,** 473–487.

Allman, J. H., and Kaas, J. H. (1976). Representation of the visual field on the medial wall of occipital–parietal cortex in the owl monkey. *Science* **191,** 572–575.

Allman, J. M., Miezen, F., and McGuinness, E. (1986). Direction and velocity-specific responses beyond the classical receptive field in the middle temporal visual area (MT). *Perception* **14,** 105–126.

Baker, J. F., Petersen, S. E., Newsome, W. T., and Allman, J. M. (1981). Response properties in four extrastriate visual areas of the owl monkey (*Aotus trivirgatus*): A quantitative comparison of medial, dorsomedial, dorsolateral, and middle temporal areas. *J. Neurophysiol.* **45,** 397–416.

Brodmann, K. (1990). "Vergleichende Iokalizationlehre der grosshirnrinde." Barth, Leipzig.

Cusick, C. G., and Kaas, J. H. (1986). Interhemispheric connections of cortical, sensory and motor maps in primates. *In* "Two Hemispheres—One Brain" (F. Lepore, M. Ptito, and H. H. Jasper, eds.), pp. 83–102. Liss, New York.

Cusick, C. G., and Kaas, J. H. (1988). Cortical connections of area 18 and dorsolateral visual cortex in squirrel monkeys. *Visual Neurosci.* **1,** 211–237.

Cusick, C. G., Gould, H. J., III, and Kaas, J. H. (1984). Interhemispheric connections of visual cortex of owl monkeys (*Aotus trivirgatus*), marmosets (*Callithrix jacchus*) and galagos (*Galago crassicaudatus*). *J. Comp. Neurol.* **230,** 311–336.

Cusick, C. G., Steindler, D. A., and Kaas, J. H. (1985). Corticocortical and collateral thalamocortical connections of post central somatosensory cortical areas in squirrel monkeys. *Somatosensory Res.* **3,** 1–31.

Cusick, C. G., Wall, J. T., Felleman, D. J., and Kaas, J. H. (1989). Somatotopic organization of the lateral sulcus of owl monkeys: Area 3b, S-II, and a ventral somatosensory area. *J. Comp. Neurol.* **282,** 169–190.

Diamond, I. T., Conley, M., Ito, K., and Fitzpatrick, D. (1985). Laminar organization of geniculocortical projections in *Galago senegalensis* and *Aotus trivirgatus. J. Comp. Neurol.* **242,** 584–610.

Felleman, D. J., and Kaas, J. H. (1982). Receptive field properties of neurons in the striate cortex of the owl monkey (*Aotus trivirgatus*). *Assoc. Res. Vision and Ophthal. Abstr.* **4,** 118.

Felleman, D. J., and Kaas, J. H. (1984). Receptive field properties of neurons in the middle temporal visual area (MT) of owl monkeys. *J. Neurophysiol.* **52,** 488–513.

FitzPatrick, K. A., and Imig, T. J. (1978). Projections of auditory cortex upon the thalamus and midbrain in the owl monkey. *J. Comp. Neurol.* **177,** 537–556.

FitzPatrick, K. A., and Imig, T. J. (1980). Auditory cortico-cortical connections in the owl monkey. *J. Comp. Neurol.* **192,** 589–610.

Florence, S. L., and Kaas, J. H. (1992). Ocular dominance columns in area 17 of Old World macaque and Talapoin monkeys: Complete reconstructions and quantitative analyses. *Visual Neurosci.* **8,** 449–462.

Florence, S. L., Wall, J. T., and Kaas, J. H. (1991). Central projections from the skin of the hand in squirrel monkeys. *J. Comp. Neurol.* **311,** 563–578.

Garraghty, P. E., and Kaas, J. H. (1991). Functional reorganization in adult monkey thalamus after peripheral nerve injury. *Neuroreport* **2,** 747–750.
Garraghty, P. E., Pons, T. P., Sur, M., and Kaas, J. H. (1989). The arbors of axons terminating in middle cortical layers of somatosensory area 3b in owl monkeys. *Somatosensory Motor Res.* **6,** 401–411.
Garraghty, P. E., Florence, S. L., and Kaas, J. H. (1990a). Ablations of areas 3a and 3b of monkey somatosensory cortex abolish cutaneous responsivity in area 1. *Brain Res.* **528,** 165–169.
Garraghty, P. E., Pons, T. P., and Kaas, J. H. (1990b). Ablations of areas 3b (SI proper) and 3a of somatosensory cortex in marmosets deactivate the second and parietal ventral somatosensory areas. *Somatosensory Motor Res.* **7,** 125–135.
Garraghty, P. E., LaChica, E. A., and Kaas, J. H. (1991). Injury-induced reorganization of somatosensory cortex is accompanied by reductions in GABA staining. *Somatosensory Motor Res.* **8,** 347–354.
Gould, J. H., III, Cusick, C. G., Pons, T. P., and Kaas, J. H. (1986). The relationships of corpus callosum connections to electrical stimulation maps of motor, supplementary motor, and frontal eye fields in owl monkeys. *J. Comp. Neurol.* **247,** 297–325.
Graham, J., Wall, J. T., and Kaas, J. H. (1978). Cortical projections of the medial visual area in the owl monkey (*Aotus trivirgatus*). *Neurosci. Lett.* **15,** 109–114.
Graham, J., Lin, C.-S., and Kaas, J. H. (1979). Subcortical projections of six visual cortical areas in the owl monkey (*Aotus trivirgatus*). *J. Comp. Neurol.* **187,** 557–580.
Hassler, R. (1966). Comparative anatomy in the central visual systems in day and night active primates. *In* "Evolution of the Forebrain" (R. Hassler and H. Stephen, eds.), pp. 419–434. Thieme, Stuttgart.
Huerta, M. F., Krubitzer, L. A., and Kaas, J. H. (1987). The frontal eye field as defined by intracortical microstimulation in squirrel monkeys, owl monkeys, and macaque monkeys. II. Cortical connections. *J. Comp. Neurol.* **265,** 332–361.
Imig, T. J., Ruggero, M. A., Kitzes, L. M., Javel, E., and Brugge, J. F. (1977). Organization of auditory cortex in the owl monkey. (*Aotus trivirgatus*). *J. Comp. Neurol.* **171,** 111–128.
Jenkins, W. M., and Merzenich, M. M. (1987). Reorganization of neocortical representations after brain injury: A neurophysiological model of the basis of recovery from stroke. *Prog. Brain Res.* **1,** 249–266.
Kaas, J. H. (1983). What, if anything, is S-I? The organization of the "first somatosensory area" of cortex. *Physiol. Rev.* **63,** 206–231.
Kaas, J. H. (1986). The structural basis for information processing in the primate visual system. *In* "Visual Neuroscience" (J. D. Pettigrew, W. R. Levick, and K. J. Sanderson, eds.), pp. 315–340. Cambridge Univ. Press, Cambridge, UK.
Kaas, J. H. (1989). Why does the brain have so many visual areas? *J. Cogn. Neurosci.* **1,** 121–135.
Kaas, J. H. (1991). Plasticity of sensory and motor maps in adult mammals. *Annu. Rev. Neurosci.* **14,** 137–167.
Kaas, J. H., and Huerta, M. F. (1988). Subcortical visual system of primates. *In* 'Comparative Primate Biology" (H. P. Steklis, ed.), Vol. 4, pp. 327–391. Liss, New York.
Kaas, J. H., and Krubitzer, L. A. (1991). The organization of extrastriate visual cortex. *In* "Vision and Visual Dysfunction" (J. Cronly-Dillon, gen. ed.), Vol. 3, pp. 302–359. Macmillan, London.
Kaas, J. H., and Krubitzer, L. A. (1992). Area 17 lesions deactivate area MT in owl monkeys. *Visual Neurosci.* **9,** 399–407.
Kaas, J. H., and Morel, A. E. (1993). Connections of visual areas of the upper temporal lobe of owl monkeys: The MT crescent and dorsal and ventral subdivisions of FST. *J. Neurosci.* **13,** 534–546.
Kaas, J. H., and Pons, T. P. (1988). The somatosensory system of primates. *In* "Comparative Primate Biology" (H. P. Steklis, ed.), Vol. 4, pp. 421–468, Liss, New York.

Kaas, J. H., Guillery, R. W., and Allman, J. M. (1972). Some principles of organization in the dorsal lateral geniculate nucleus. *Brain, Behav. Evol.* **6,** 253–299.

Kaas, J. H., Lin, C.-S., and Casagrande, V. A. (1976). The relay of ipsilateral and contralateral retinal input from the lateral geniculate nucleus to striate cortex in the owl monkey: A transneuronal transport study. *Brain Res.* **106,** 371–378.

Kaas, J. H., and Lin, C.-S., and Wagor, E. (1977). Cortical projections of posterior parietal cortex in owl monkeys. *J. Comp. Neurol.* **177,** 387–408.

Kaas, J. H., Huerta, M. F., Weber, J. T., and Harting, J. K. (1978). Patterns of retinal terminations and laminar organization of the lateral geniculate nucleus of primates. *J. Comp. Neurol.* **182,** 517–554.

Kaas, J. H., Nelson, R. J., Sur, M., Lin, C.-S., and Merzenich, M. M. (1979). Multiple representations of the body within "S-I" of primates. *Science* **204,** 521–523.

Krubitzer, L. A., and Kaas, J. H. (1989). Cortical integration of parallel pathways in the visual system of primates. *Brain Res.* **478,** 161–165.

Krubitzer, L. A., and Kaas, J. H. (1990a). The organizations and connections of somatosensory cortex in marmosets. *J. Neurosci.* **10,** 952–974.

Krubitzer, L. A., and Kaas, J. H. (1990b). Cortical connections of MT in four species of primates: Areal, modular, and retinotopic patterns. *Visual Neurosci.* **5,** 165–204.

Krubitzer, L. A., and Kaas, J. H. (1990c). Convergence of processing channels in extrastriate cortex of monkeys. *Visual Neurosci.* **5,** 609–613.

Krubitzer, L. A., and Kaas, J. H. (1992). The somatosensory thalamus of monkeys: Cortical connections and a redefinition of nuclei in marmosets. *J. Comp. Neurol.* **319,** 123–140.

Krubitzer, L. A., and Kaas, J. H. (1993). The dorsomedial visual area (DM) of owl monkeys: Connections, myeloarchitecture, and homologies in other primates. *J. Comp. Neurol.* **334,** 497–528.

Lachica, E. A., Beck, P. D., and Casagrande, V. A. (1992). Parallel pathways in macaque monkey striate cortex: Anatomically defined columns in layer III. *Proc. Nat. Acad. Sci.* U.S.A. **89,** 3566–3570.

Lin, C.-S., and Kaas, J. H. (1977). Projections from cortical visual areas 17, 18, and MT onto the dorsal lateral geniculate nucleus in owl monkeys. *J. Comp. Neurol.* **173,** 457–474.

Lin, C.-S., and Kaas, J. H. (1979). The inferior pulvinar complex in owl monkeys: Architectonic subdivisions and patterns of input from the superior colliculus and subdivisions of visual cortex. *J. Comp. Neurol.* **187,** 655–678.

Lin, C.-S., Merzenich, M. M., Sur, M., and Kaas, J. H. (1979). Connections of areas 3b and 1 of the parietal somatosensory strip with ventroposterior nucleus in the owl monkey (*Aotus trivirgatus*). *J. Comp. Neurol.* **185,** 355–372.

Lin, C.-S., Weller, R. E., and Kaas, J. H. (1982). Cortical connections of striate cortex in the owl monkey. *J. Comp. Neurol.* **211,** 165–176.

Merzenich, M. M., and Brugge, J. F. (1973). Representation of the cochlear partition on the superior temporal plane of the macaque monkey. *Brain Res.* **50,** 275–296.

Merzenich, M. M., Kaas, J. H., Sur, M., and Lin, C.-S. (1978). Double representation of the body surface within cytoarchitectonic areas 3b and 1 in "S-I" in the owl monkey. (*Aotus trivirgatus*). *J. Comp. Neurol.* **181,** 41–74.

Merzenich, M. M., Kaas, J. H., Wall, J. T., Nelson, J. R., Sur, M., and Felleman, D. J. (1983a). Topographic reorganization of somatosensory cortical areas 3b and 1 in adult monkeys following restricted deafferentation. *Neuroscience* **8,** 33–55.

Merzenich, M. M., Kaas, J. H., Wall, J. T., Sur, M., Nelson, R. J., and Felleman, D. J. (1983b). Progression of change following median nerve section in the cortical representation of the hand in areas 3b and 1 in adult owl and squirrel monkeys. *Neuroscience* **10,** 639–665.

Merzenich, M. M., Nelson, R. J., Stryker, M. P., Cynader, M. S., Shoppman, A., and Zook, J. M. (1984). Somatosensory cortical map changes following digit amputation in adult monkeys. *J. Comp. Neurol.* **224,** 591–605.

Merzenich, M. M., Nelson, R. J., Kaas, J. H., Stryker, M. P., Jenkins, W. M., Zook, J. M., Cynader, M. S., and Schoppmann, A. (1987). Variability in hand surface representation in areas 3b and 1 in adult owl and squirrel monkeys. *J. Comp. Neurol.* **258,** 281–296.

Mishkin, M. (1979). Analogous neural models for tactual and visual learning. *Neuropsychologia* **17,** 139–151.

Morel, A., and Kaas, J. H. (1992). Subdivision and connections of auditory cortex in owl monkeys. *J. Comp. Neurol.* **316,** 1–36.

Newsome, W. T., and Allman J. M. (1980). Interhemispheric connections of visual cortex in the owl monkey, *Aotus trivirgatus,* and the bushbaby, *Galago senegalensis. J. Comp. Neurol.* **194,** 209–233.

Norden, J. J., and Kaas, J. H. (1978). The identification of relay neurons in the dorsal lateral geniculate nucleus of monkeys using horseradish peroxidase. *J. Comp. Neurol.* **182,** 707–726.

Nudo, R. J., and Masterton, R. B. (1990). Decending pathways to the spinal cord. III. Sites of origin of the corticospinal tract. *J. Comp. Neurol.* **296,** 559–583.

Petersen, S. E., Baker, J. F., and Allman, J. M. (1985). Direction-specific adaptation in area MT of the owl monkey. *Brain Res.* **346,** 146–150.

Petersen, S. E., Miezen, F. M., and Allman, J. M. (1988). Transient and sustained responses in four extrastriate visual areas of the owl monkey. *Exp. Brain Res.* **70,** 55–60.

Pollen, D. A., Nagler, M., Paugman, J., Kronauer, R., and Cavanaugh, P. (1984). Use of gabor elementary functions to probe receptive field substructure of posterior inferotemporal neurons in the owl monkeys. *Vision Res.* **24,** 233–241.

Pospichal, M. W., Florence, S. L., and Kaas, J. H. (1994). The postnatal development of geniculocortical axon arbors in owl monkeys. *Visual Neuroscience* **11,** 71–90.

Preuss, T. M., Stepniewska, I., and Kaas, J. H. (1991). Microstimulation studies of premotor and oculomotor cortex in owl monkeys. *Soc. Neurosci. Abstr.* **17,** 1022.

Recanzone, G. H., and Merzenich, M. M. (1988). Intracortical microstimulation in somatosensory cortex in adult rats and owl monkeys results in a large expansion of the cortical zone of representation of a specific cortical receptive field. *Soc. Neurosci. Abstr.* **14,** 223.

Sereno, M. I., McDonald, C. T., and Allman, J. M. (1986). Organization of extrastriate cortex in the owl monkey. *Soc. Neurosci. Abstr.* **12,** 1181.

Sherman, S. M., Wilson, J. R., Kaas, J. H., and Webb, S. V. (1976). X- and Y-cells in the dorsal lateral geniculate nucleus of the owl monkey (*Aotus trivirgatus*). *Science* **192,** 475–477.

Stepniewska, I., Preuss, T., M., and Kaas, J. H. (1993). Architectonics, somatotopic organization and ipsilateral cortical connections of the primary motor area (M1) of owl monkeys. *J. Comp. Neurol.* **329,** 1–34.

Stepniewska, I., Gaspar, P., and Kaas, J. H. (1991). Comparison of cortical projections to primary and supplementary motor areas in owl monkeys (*Aotus trivirgatus*). *Soc. Neurosci. Abstr.* **17,** 1020.

Sur, M. (1980). Receptive fields of neurons in areas 3b and 1 of somatosensory cortex in monkeys. *Brain Res.* **198,** 465–471.

Sur, M., Merzenich, M. M., and Kaas, J. H. (1980). Some quantitative features of the body surface representation in areas 3b and 1 in owl monkeys. Magnification, receptive field area and "hypercolumn" size. *J. Neurophysiol.* **80,** 295–311.

Sur, M., Wall, J. T., and Kaas, J. H. (1981). Modular segregation of functional cell classes within the postcentral somatosensory cortex of monkeys. *Science* **212,** 1059–1061.

Sur, M., Wall, J. T., and Kaas, J. H. (1984). Modular distributions of neurons with slowly adapting and rapidly adapting responses in areas 3b of somatosensory cortex in monkeys. *J. Neurophysiol.* **51,** 724–744.

Tootell, R. B. H., and Born, R. T. (1991). Architecture of primate area MT. *Soc. Neurosci. Abstr.* **17,** 524.

Tootell, R. B. H., Hamilton, S. L., and Silverman, M. S. (1985). Topography of cytochrome-oxidase activity in owl monkey cortex. *J. Neurosci.* **5,** 2786–2800.

Ungerleider, L. G., and Mishkin, M. (1982). Two cortical visual system. *In* "Analysis of Visual Behavior" (J. D. Ingle, M. a. Goodale, and R. J. W. Mansfield, eds.), pp. 549–586. MIT Press, Cambridge, MA.

Wagor, E., Lin, C.-S., and Kaas, J. H. (1975). Some cortical projections of the dorsomedial visual area (DM) of association cortex in the owl monkey. (*Aotus trivirgatus*). *J. Comp. Neurol.* **163,** 227–250.

Wall, J. T., and Kaas, J. H. (1986). Long-term cortical consequence of reinnervation errors after nerve regeneration in monkeys. *Brain Res.* **372,** 400–404.

Wall, J. T., Felleman, D. J., and Kaas, J. H. (1983). Recovery of normal topography in the somatosensory cortex of monkeys after nerve crush and regeneration. *Science* **221,** 771–773.

Wall, J. T., Kaas, J. H., Merzenich, M. M., Sur, M., Nelson, R. J., and Felleman, D. J. (1986). Functional reorganization in somatosensory cortical areas 3b and 1 of adult monkeys after median nerve repair: Possible relationships to sensory recovery in humans. *J. Neurosci.* **6,** 218–233.

Weller, R. E., and Kaas, J. H. (1985). Cortical projections of the dorsolateral visual area in owl monkeys: The prestriate relay to inferior temporal cortex. *J. Comp. Neurol.* **234,** 35–59.

Weller, R. E., and Kaas, J. H. (1987). Subdivisions and connections of inferior temporal cortex in owl monkeys. *J. Comp. Neurol.* **256,** 137–172.

Weller, R. E., Wall, J. T., and Kaas, J. H. (1984). Cortical connections of the middle temporal visual area (MT) and the superior temporal cortex in owl monkeys. *J. Comp. Neurol.* **228,** 81–104.

Weller, R. E., Steele, G. E., and Cusick, C. G. (1991). Cortical connections of dorsal cortex rostral to VII in squirrel monkeys. *J. Comp. Neurol.* **306,** 521–537.

Whitworth, R. H., LeDoux, M. S., and Gould, H. J. (1991). Topographic distribution of connections from the primary motor cortex to the corpus striatum in *Aotus trivirgatus*. *J. Comp. Neurol.* **307,** 177–188.

13

Parasites of the *Aotus* Monkey

Manuel Tantaleán
Tropical Medicine Institute "Daniel Alcides Carrión"
Universidad Nacional Mayor de San Marcos
Casilla 2670
Lima 100, Perú

Alfonso Gozalo
Nonhuman Primate Reproduction and Conservation Center
Tropical and High Altitude Research Veterinary Institute
Universidad Nacional Mayor de San Marcos
Casilla 621
Iquitos, Peru

In spite of the existence of abundant literature on parasites in neotropical primates, there is not much information on parasites in the *Aotus* monkey. Most of the reports on this species have been made on animals that have been in captivity for some time, admitting the possibility of their acquiring the infections during the same. According to our experience and studies performed in newly captured animals, the *Aotus* monkey is, in contrast with other neotropical monkeys like *Saimiri* and *Saguinus*, one of the primate species least affected by parasitic infections in its natural habitat. However, they are sensitive to a series of pathogens, of both human and animal origin, when they are exposed to them during captivity, probably facilitated by an immunodepression caused by stress. In this chapter we will review the parasitic species that we have found in the *Aotus* monkey as well as those reported by other researchers, including those reported in other species of neotropical primates that could affect the *Aotus* monkeys under given conditions.

I. FLAGELLATES

Several species of flagellates have been reported as parasitizing neotropical primates (Table I). In monkeys of the genus *Aotus* only one hemoflagellate (*Trypanosoma minasense*) has been observed, and the discovery of antibodies against *T. cruzi* is mentioned; however, the latter has not been observed in blood smears.

Trypanosoma cruzi is approximately 20 μm long. It takes on a U, C, or S shape, having a central nucleus and a large blepharoplast located near the pointed end of the body; it has a wavy membrane bordered by a flagellum that becomes free at the end of the body (Fig. 1). This protozoan is the only member of the Trypanosomatidae family that makes mastigote nests, reproducing itself in this stage. Some nests are found in various tissues of hot-blooded vertebrates. The infection of the vertebrate occurs through eroded skin, when metacyclic trypomastigotes found in the feces of the vector penetrate the wound left by the sting. The vectors consist of hematophagous insects of the Reduviidae family. In contrast with the other trypanosomal species, *T. cruzi* is extremely pathogenic.

Trypanosoma minasense is 29–46 μm long; the nucleus is usually located in or slightly before the middle of the body. The blepharoplast is small and anterior to the posterior end. The wavy membrane is well developed and the free flagellum is from one-sixth to one-third as long as the body (Fig. 2). *Trypanosoma minasense* is not pathogenic but it may be found in mixed infections with *T. cruzi*.

Hemoflagellates are infrequent in *Aotus* monkeys. Diagnosis is made by light microscope examination of a blood smear. Treatment is unknown. Preventive measures are based on controlling the vector insect.

Occasionally intestinal flagellates (*Trichomonas* spp. and *Pentatrichomonas* spp.) are observed in *Aotus* monkeys. These have no great significance in the animal's health unless the parasitic load is high, in which case a persistent watery diarrhea occurs.

Trichomonas hominis is characterized by the presence of four free flagella and a wavy membrane bordered by a flagellum that later becomes free. It

TABLE I Flagellated Parasite Species Reported in Neotropical Primates

Trypanosoma cruzi, T. sanmartini, T. minasense, T. rangeli
T. saimirii, T. diasi, T. lambrechti
Chilomastix mesnili, Chilomastix spp.
Trichomonas spp.
Pentatrichomonas hominis
Giardia lamblia, Giardia spp.

has a vesicular nucleus that is located in the anterior end. It lacks cystic shape. The trophozoite measures between 5 and 15 μm, and the host is infected orally.

Pentatrichomonas hominis has a pyriform shape and is 8–20 μm long by 3–14 μm wide. It usually has five anterior flagella but in some cases it can have four and sometimes only three. It is not pathogenic.

Diagnosis is made by observing the parasites under microscopic fecal examination. Treatment is based on the use of metronidazole (35–50 mg/kg/day BID for 10 days).

II. SARCODINA

We have not observed sarcodina parasites in *Aotus* monkeys. However, several species of intestinal amoebas are reported in other species of neotropical monkeys, there being a possibility of cross infection in situations of captivity (Table II). These parasites can produce enteric situations, whose severity depends on the species involved. Diagnosis is made by finding the cysts upon fecal examination under light microscope. For treatment, metronidazole (35–50 mg/kg/day BID for 5–10 days) or paromomycin (25–30 mg/kg/day BID for 5–10 days) is recommended.

FIGURE 1 *Trypanosoma cruzi.*

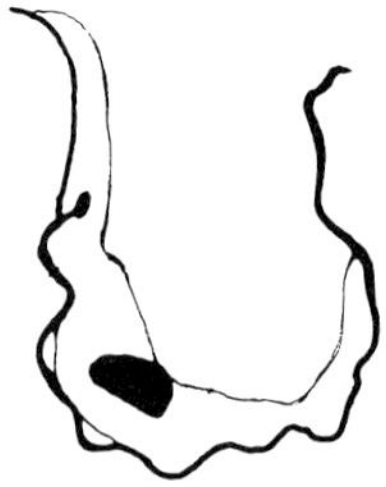

FIGURE 2 *Trypanosoma minasense.*

III. SPOROZOANS

Sporozoans are extremely rare in owl monkeys. Spontaneous infections with *Toxoplasma gondii* have been reported in *Aotus* monkeys and in other Cebidae species as well as in Callitrichidae. The definitive hosts are felines like the domestic cat, where sexual reproduction takes place, resulting in oocysts, which upon maturing infect various vertebrates like the monkey orally; but the infection can also be acquired if the host ingests poorly cooked meat from other intermediate hosts containing parasitic cysts or indirectly in a congenital manner. Extracellular trophozoites have a half-moon shape, with one widened end where the nucleus is located. They are called tachyzoites; by contrast, the intracellular ones vary in shape because they are almost rounded and are found in the cytoplasm of the invaded cells. Cysts are formed in various organs; they have a spherical shape and contain a great amount of bradyzoites. It is thought that primates may become infected both in their natural habitat as well as in captivity and the disease is reported as extremely pathogenic, with anorexia, neurological symptoms, and diarrhea being observed. Diagnosis is made by serology and/or observation of the parasite by histological section. For treatment, sulfadiazine, pyrimethamine, and clindamycin are recommended.

Plasmodium brasilianum is reported as being extremely pathogenic in the Cebidae and Callitrichidae, producing anemia, fever (quartan malaria), hepatosplenomegaly, depression, and death. This parasite is similar to *P. malariae* in humans and is thought to be a mutant variety of *P. malariae* that

TABLE II Parasite Species of the Sarcodina Group Reported in Neotropical Primates

Entamoeba histolytica, E. chattoni
Iodamoeba butschlii
Endolimax nana, Endolimax spp.

was introduced by explorers in Amazonia. Schizogony occurs every 72 hr. The schizonts are in the shape of a band and the merozoites number from 8 to 10. The gametocytes are round and small. The vector is a mosquito of the genus *Anopheles*. Diagnosis is made by observing the parasite in a blood smear under a light microscope. Recommended treatment consists of using chloroquine phosphate (10 mg/kg IM, followed by 5 mg/kg at 6 hr, then 5 mg/kg/day for 2 days) and primaquine (0.3 mg/kg/day for 14 days).

The sporozoan species reported in neotropical primates are listed in Table III.

IV. CILIATES

No ciliates have been reported in monkeys of the genus *Aotus*. In the literature, only one species of ciliates in neotropical monkeys is mentioned, *Balantidium coli*, affecting *Alouatta*, *Ateles* and *Cebus*.

Balantidium coli is the protozoan parasite of greatest size, the trophozoite measuring 40–70 μm and the cyst 50–55 μm. The trophozoite is ovoid and covered with cilia, but one end is reduced where the cytostome opens, the cytopyge on the other side. They have two nuclei; the largest or macronucleus is kidney shaped and the small one or micronucleus is hard to visualize. The infecting form is the cyst, which enters the organism orally. It may or may not be associated with a diarrheic condition. In severe cases it produces ulcerative colitis. Diagnosis is made by observing the cysts in a fresh fecal smear under a light microscope. For treatment, metronidazole (35–50 mg/kg/day BID for 5–10 days) is recommended.

V. TREMATODES

According to our experience and that reported in the literature, there are three species of trematodes found in monkeys of the genus *Aotus* in

TABLE III Sporozoan Parasites Reported in Neotropical Primates

Sarcocystis spp.
Cryptosporidium spp.
Isospora arctopitheci, *I. callimico*
Plasmodium brasilianum, *P. simium*
Toxoplasma gondii
Encephalitozoon cuniculi
Microsporum spp.
Pneumocystis carinii
Haemobartonella spp.

their natural habitat: *Athesmia heterolecithoides* (= *A. foxi*) and *Zonorchis goliath*, which live in the biliary ducts, and *Phaneropsolus orbicularis*, which parasitizes the small intestine.

Athesmia heterolecithoides (= *A. foxi*) is a worm with an elongated flat body and a cuticle with small papillae. They are from 7 to 9 mm long by 1.5 mm wide. The oral sucker is subterminal and the acetabulum is preequatorial. The genital pore is postbifurcal. The intestinal caeca are thin. The testicles are strongly lobulated, postacetabular, and preovarian. The ovary is oval and the uterus is mostly postacetabular. The vitellarium is unilateral, cecal, or intercecal. The eggs have an operculum and are approximately 30 μm long.

Zonorchis goliath has a long, thin body, 9.5–13 mm long by 2.5–3.5 mm wide, with a subterminal oral sucker and the acetabulum in the anterior part of the body. The intestinal caeca are thin and sinuous. The long testicles are postacetabular. The round ovary is posttesticular and preequatorial. The vitellogens are lateral. The uterus is located in the posterior half of the body, with an ascending bend to the postpharyngeal genital pore. The operculated eggs are 34–41 μm long by 22–26 μm wide.

Phaneropsolus orbicularis has an ovoid body and is 68–89 μm long by 50–60 μm wide. The cuticle is spiny. The oral sucker is located in an anterior cone and the acetabulum is in the middle of the body. The intestinal caeca are very short and transversal. The genital pore opens at the level of the pharynx. The spherical testicles are preacetabular and the almost spherical ovary is posttesticular. The uterus is mainly postacetabular. The eggs are operculated and are approximately 40 μm long.

The parasitized animals are clinically healthy and did not show any pathological condition. However, *A. heterolecithoides* and *Z. goliath* can produce mechanical obstruction of biliary ducts if they are present in great quantities. The infection is diagnosed by the discovery of the characteristic eggs in the feces or in the postmortem examination of animals from natural areas. The indicated treatment consists of the use of praziquantel (40 mg/kg).

TABLE IV Trematode Species Reported in Neotropical Primates

Athesmia heterolecithoides
Platynosomun amazonensis, *P. marmoseti*
Phaneropsolus orbicularis
Controrchis biliophilus
Zonorchis goliath
Echinostoma aphylactum
Schistosoma mansoni
Diplostomid mesocercariae
Neodiplostomum tamarini

Table IV lists the trematode species reported in New World primates.

VI. CESTODES

Although infections by cestodes are not frequent in *Aotus* monkeys in their natural habitat, they are frequent and very pathogenic in captive animals. We have observed that the owl monkey is susceptible to infection with the rat tapeworm (*Hymenolepis diminuta*).

Hymenolepis diminuta is approximately 70 cm long. It has a scolex without spines with four suckers (Fig. 3), and proglottids that are wider than they are long and with a unilateral genital pore. They have three testicles and

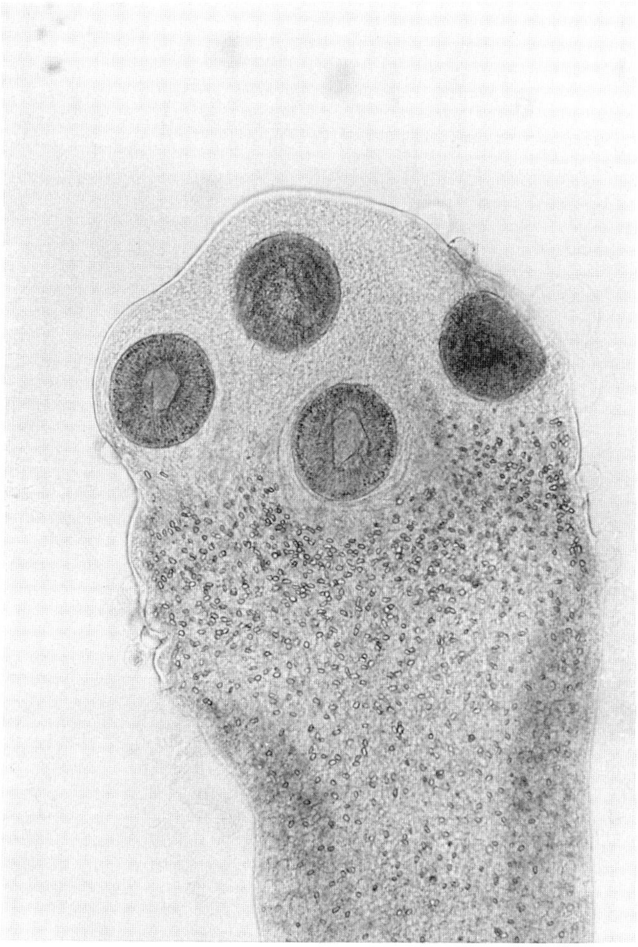

FIGURE 3 Anterior end of *Hymenolepis diminuta.*

the uterus is sacciform. The eggs are spherical, yellow in color, and with a hexacanth embryo. The infecting form is the cysticercoid, which is found in arthropods, mainly *Tribolium confusum*. The infection of the definitive host is oral. The curious thing is that the infected monkey eliminates the tapeworm's eggs in the feces with a weakly formed shell (Fig. 4).

The *Aotus* monkey is susceptible to intestinal cestode infections, producing a severe enteritis with loss of body fluids and destruction of the enteric mucosa, besides the damage produced by the deficient absorption of nutrients and the mechanical obstruction. The diagnosis is based on identifying the eggs in the feces under a light microscope. The treatment of this and other taeniasis is based on the use of niclosamide (150 mg/kg) or praziquantel (15–20 mg/kg) in a single dose.

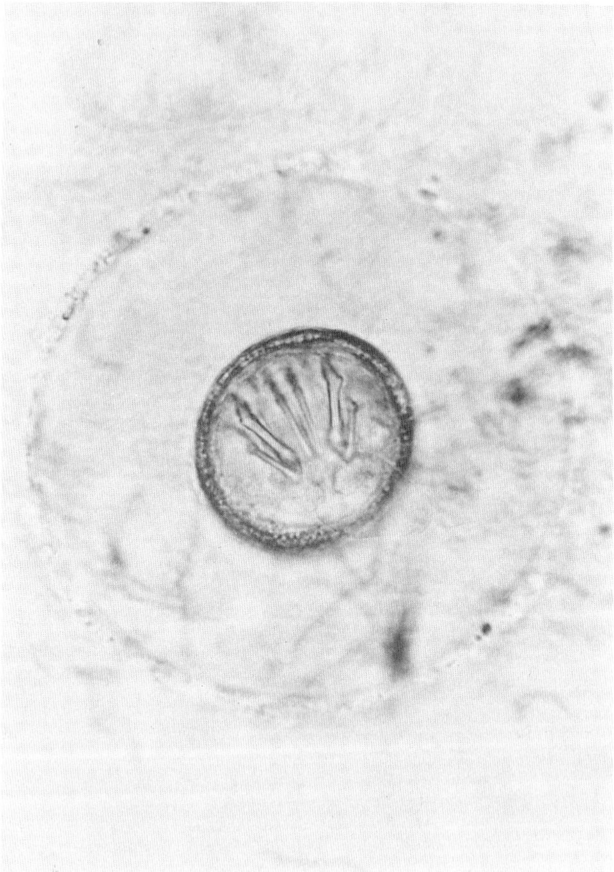

FIGURE 4 Egg of *Hymenolepis diminuta;* note the weakly formed shell.

We have also found a case of hydatidosis (*Echinococcus granulosus*) in the liver of a captive animal. The symptomatology is according to the location of the cyst and the area affected by it. Diagnosis is based on radiographic observation of the cyst or discovery during the autopsy. Recommended treatment consists of the surgical removal of the cyst.

Table V lists the cestode species reported in neotropical primates.

VII. NEMATODES

Nematodes are the parasitic group most frequently found in *Aotus* monkeys in their natural habitat, those of the genus *Trypanoxyuris* (*T. microon* = *T. interlabiata*) being the most frequent and least pathogenic at the intestinal level, followed by *Molineus torulosus* and occasionally *Primasubulura jacchi*.

Trypanoxyuris microon is characterized by having a circular cephalic plate and bilobulated lips, the dorsal one being the most developed with thickened edges. The males (Fig. 5) are 1.5–2.5 mm long and have simple lateral fins and a caudal sac with four pairs of pedunculated papillae, the first of which is more developed and the second is surrounded by a chitinous ring. The spicule is simple, and the tail is short and pointed. The female is 3.5–6.4 mm long and has double lateral fins. The eggs have a thick shell and are symmetrical. The life cycle is direct and infection occurs upon ingesting larval eggs (Fig. 6). *Trypanoxyuris* usually parasitizes the caecum and is well tolerated by the host. According to a study made on 128 field *Aotus nancymae*, 54.7% of the monkeys were parasitized by these oxyures.

When alive, *Molineus torulosus* is red. The cuticle has longitudinal grooves and the cephalic area has a cuticular widening. The cervical papillae are very small and the excreting pore is protruded. The male is 8–9 mm long

TABLE V Cestode Species Reported in Neotropical Primates

Atriotaenia megastoma
Hymenolepis spp., *H. nana*, *H. cebidarum*, *H. diminuta*
Diphyllobothrium erinaci (sparganosis)
Spirometra reptans (sparganosis)
Echinococcus granulosus (hidatidosis)
Paratriotaenia oedipomidatis
Vampirolepis spp.
Bertiella mucronata, *B. fallax*, *B. satyri*
Moniezia rugosa
Matheovataenia brasiliensis
Raillietina allouatae, *R. demerariensis*

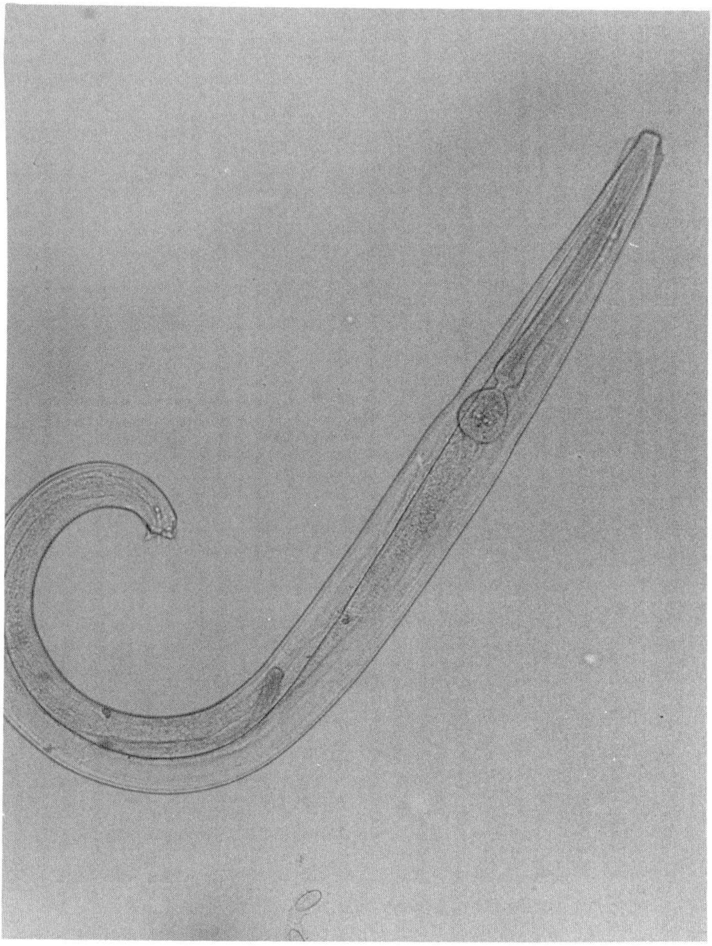

FIGURE 5 Male specimen of *Trypanoxyuris microon*.

and has a copulating sac with a well-defined dorsal lobe; the ventral ribs are together and pointed toward the front; the anterodorsal rib is short and the middle and posterolateral ones are together and reach the edge of the sac. The externodorsal rib is slightly sinuous and the dorsal one is double divided. The spicules are complex, quite chitinous, with two distal barbs joined by a membrane. It has a gubernaculum. The female is 10–12 mm long, has a truncated tail, and ends in a cuticular spine. The eggs have a thin shell and are 59–63 μm long by 37 μm wide (Fig. 7). *Molineus torulosus* also seems to be slightly pathogenic; it is generally found in the small intestine.

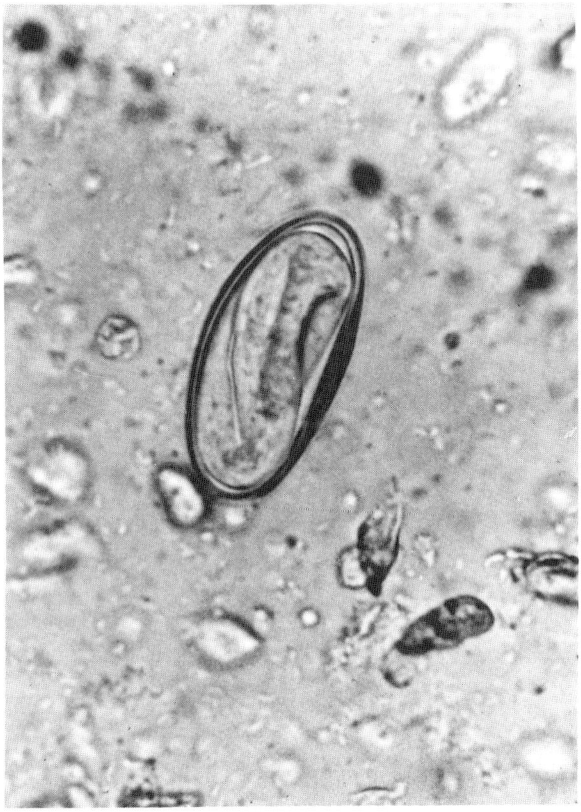

FIGURE 6 Egg of *Trypanoxyuris microon.*

When alive, *Primasubulura jacchi* is pink in color, with an elongated body with a well-attenuated posterior end. They have cephalic fins, a buccal capsule, and an esophagus with a posterior bulb that has chitinous valves (Fig. 8). The males (8–13 mm long) have equal spicules, quite chitinous and finned, with a gubernaculum. In the posterior area they have a caudal sucker without a chitinous edge and 11 pairs of papillae. The females are 13–20 mm long, and the eggs are 53 μm long. *Primasubulura jacchi* is a larger nematode and probably causes more intestinal irritation but there is no knowledge of it causing enteritis.

Diagnosis of these nematodes is based on the identification of the characteristic eggs in a fecal smear observed under the light microscope. Treatment consists of the use of pyrantel pamoate (probably a 1:1 mixture of pyrantel and oxantel pamoate, e.g., Quantrel, 11 mg/kg in one dose) for the oxyures and mebendazole (15 mg/kg/day for 3 days) or thiabendazole

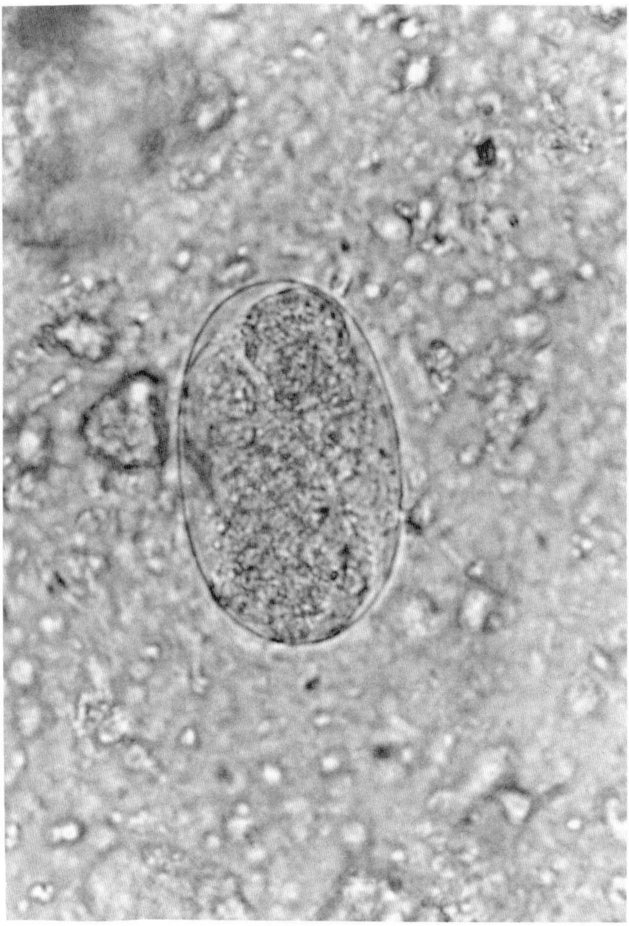

FIGURE 7 Egg of *Molineus torulosus*.

(50 mg/kg/day for 2 days) or ivermectine (200 mg/kg) in the case of the other nematodes.

Occasionally *Filariopsis barretoi* is observed. This parasite lives in the lung and eliminates larval eggs that hatch in the bronchial tract. The larvae are swallowed and eliminated with the feces. Occasionally an inflammatory reaction occurs around the adult parasite. However, it is not an important pathogen. This parasite has an elongated body but rounded extremities. The cuticle is grooved and the mouth has very small lips. The male is 40 mm long by 170 μm wide; it has a precloacal papilla; the spicules are short and banana shaped; and the gubernaculum is small (Fig. 9). The

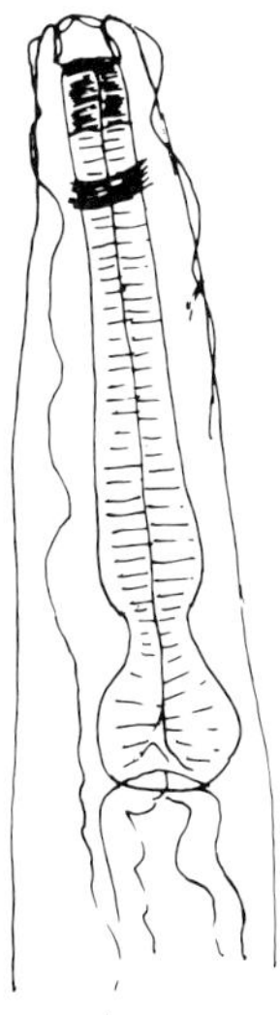

FIGURE 8 Anterior end of *Primasubulura jacchi.*

female is approximately 120 mm long and has a short, triangular tail, with a subterminal vulva near the anus. Diagnosis is based on recognizing the larva in the feces or the adults in the pulmonary parenchyma during autopsy. The recommended treatment consists of the use of Levamisole (10 mg/kg) in a single dose.

Frequently filariae are found in the abdominal cavity (*Dipetalonema gracile*) and in a lesser degree in the subcutaneous tissue (*Mansonella* spp. = *Tetrapetalonema*) in *Aotus* monkeys from their natural habitat.

Dipetalonema gracile has a long body; the females are 14–21 cm long by 400 μm wide and the males are 7–11 cm long, the esophagus having a

FIGURE 9 Caudal end of *Filariopsis barretoi* male.

muscular and glandular area. In the male, the caudal area is coiled four to five times. The haft of the left spicules is shorter than the right and the distal end of the right one is pleated. Their rugose area is interrupted at the medial line at the level of the tail and the postcloacal medial papillae are separated. The *vagina vera* of the female is very sinuous with a long, thin tail and simple caudal barb (Figs. 10 and 11).

Mansonella (*Tetrapetalonema*) spp. are worms of lesser size than *Dipetalonema*. They are elongated and the greater axis of the cephalic end is in the dorsoventral direction. The body is uniform in thickness for almost its whole length. The female has the vulva at the height of the middle of the esophagus; the tail is thin and has four terminal papillae. The male's cloacal papillae are grouped around the cloaca; the spicules are uneven, elongated, and simple in structure.

These parasites are slightly pathogenic, their infection generally being asymptomatic. We have found only one case of myocardiopathy associated with a severe infection with microfilaria, some of which were extravascularly located in the myocardium (Fig. 12). Diagnosis is based on recognizing the microfilaria in a blood smear or finding the adult parasites in the abdominal cavity or in the subcutaneous tissue during autopsy or surgery. The recommended treatment consists of using diethylcarbamazine 6–20 mg/kg daily for 6–15 days.

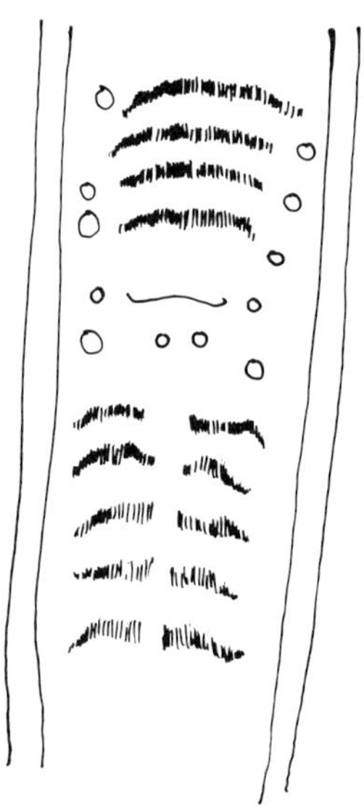

FIGURE 10 Cloacal area of *Dipetalonema gracile* male.

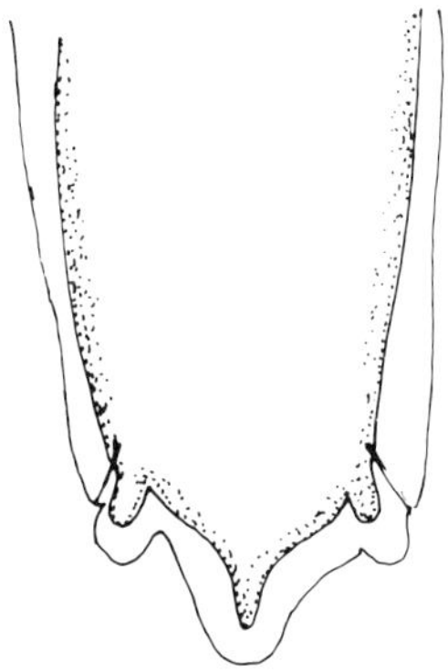

FIGURE 11 Caudal end of *Dipetalonema gracile* female.

Table VI lists the nematode species reported in neotropical primates.

VIII. ACANTHOCEPHALANS

In spite of the prevalence of *Prosthenorchis elegans* infections in other species of neotropical primates, it seems that owl monkeys possess a certain natural resistance to or are less exposed to infections with acanthocephalans, as can be deduced from the few occasions in which these parasites have been found in *Aotus* monkeys.

Prosthenorchis elegans males are 25–30 mm long and the females are 35–50 mm long. The body is cylindrical and has rugosities; at the height of the neck there is a collar formed by cuticular folds (Fig. 13). The proboscis is subspherical with 12 regular alternate rows of three hooks each, each hook having a barb. It has long lemnisci. The males have two testicles in tandem and the cement glands are compact and eight in number. The eggs have three coats and are 63–77 μm long by 42 μm wide. The infecting form is the cystacanth, which is found in cockroaches of the genus *Blatella*. The parasites of the genus *Prosthenorchis* are extremely pathogenic; they produce severe lesions of the intestinal mucosa and mechanical obstruction. Clinically, diarrhea, signs of peritonitis, and emaciation may be observed. Diagnosis is made by observing the characteristic eggs in the feces under the microscope. There is no effective treatment. Experiments have been made in the surgical removal of the parasite with variable results. Prevention is based on controlling the intermediate host (cockroaches).

Only two species of acanthocephalans are reported in neotropical monkeys, *Prosthenorchis elegans* and *P. spirula*.

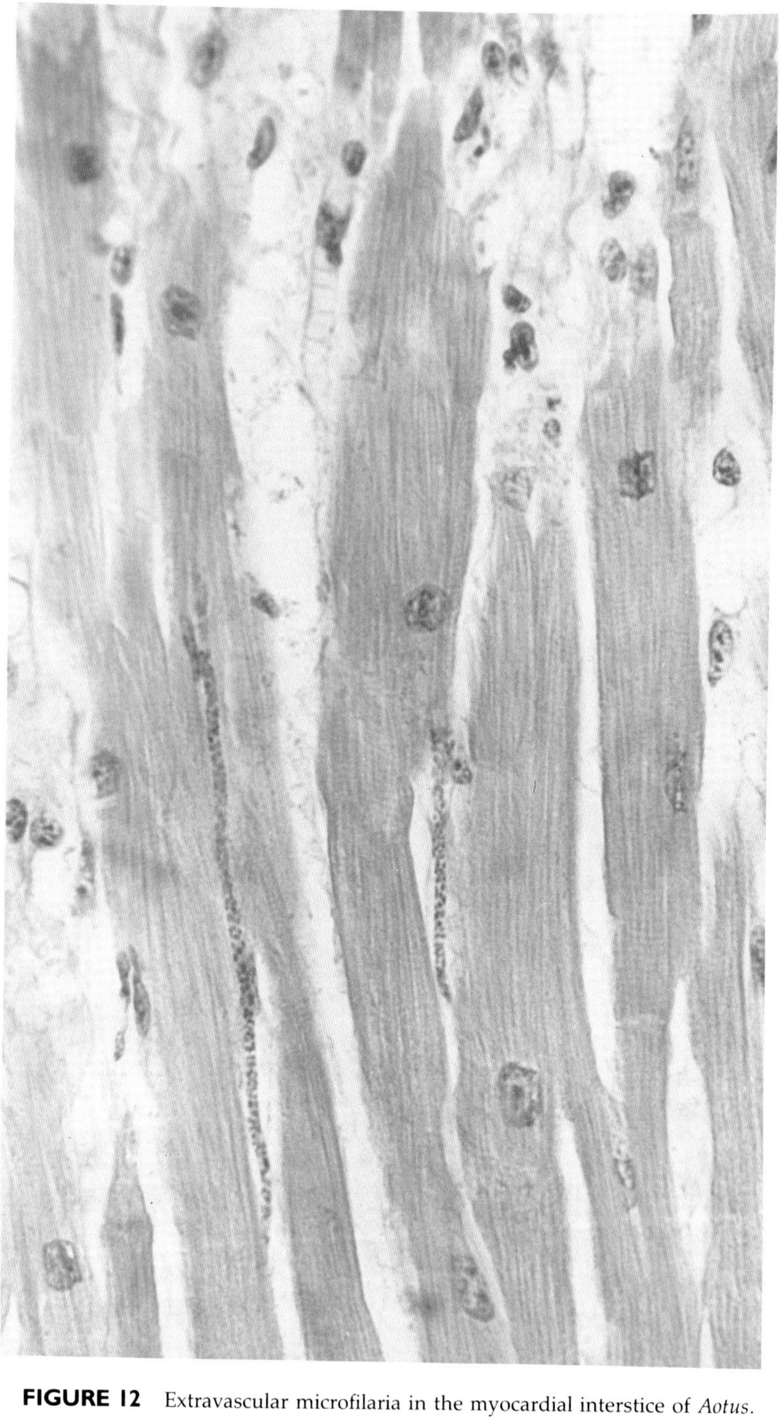

FIGURE 12 Extravascular microfilaria in the myocardial interstice of *Aotus*.

TABLE VI Nematode Species Reported in Neotropical Primates

Trypanoxyuris spp., *T. microon* (= *T. interlabiata*), *T. trypanoxyuris*, *T. atelis*, *T. duplicidens*, *T. lagothricis*, *T. minuta*, *T. sceleratus*, *T. brachytelesi*, *T. callithricis*, *T. tamarini*, *T. oedipi*, *T. goeldii*
Primasubulura jacchi
Molineus spp., *M. torulosus*, *M. vexillarius*, *M. elegans*
Filariopsis barretoi
Ancylostoma spp., *A. brasiliense*
Necator spp., *N. americanus*
Ascaris lumbricoides
Dipetalonema gracile, *D. caudispina*, *D. tenue*, *D. barbascalensis*
Mansonella spp. (= *Tetrapetalonema*), *T. obtusus*, *T. marmosetae*, *T. tamarinae*, *T. atelensis*, *T. parvum*, *T. vanhoofi*, *T. nicollei*
Strongyloides spp., *S. cebus*
Graphidiodes berlai
Longistriata dubia
Angiostrongylus costaricensis
Spirura guianensis
Trichospirura leptostoma
Pterygodermatites nycticebi
Gongylonema spp.
Physaloptera dilatata
Loa loa
Parlitomosa zakii
Trichuris spp., *T. trichiura*
Capillaria hepatica

IX. FLIES

Fly larvae of the genera *Cuterebra* and *Dermatobia* have been found only occasionally in *Aotus* monkeys from their natural habitat, there usually being only one specimen per monkey.

Cuterebra larvae are characterized by being coriaceous, with a more or less depressed body covered by black spines. The stigmatic plates have

FIGURE 13 Anterior end of *Prosthenorchis elegans*.

three laterally disposed coiled spiracles that converge in their middle part toward the inferior internal angle.

Dermatobia hominis also has coriaceous larvae with dark spines arranged in transverse rows, the last segment being retractile and smooth. The stigmatic plates are very close and the spiracles are slightly curved. The anterior end of these larvae has double rows of hooks similar to spines.

Apparently they do not cause major discomfort to the animal, although there is the possibility of secondary infections. *Cuterebra* spp., *Dermatobia hominis*, and *Alouttamyia* spp. have been reported in other species of Cebidae. Diagnosis is based on observing and feeling the parasite in the subcutaneous tissue. Treatment consists of surgical removal of the same and the topical application of antibiotics.

X. LICE

According to our experience, it is common to find lice that are indistinguishable from *Gyropus ovalis* in monkeys of the genus *Aotus* from their natural habitat. The literature mentions other ectoparasites (*Tetragyropus aotophilus* and *Aotiella aotophilus*).

Gyropus ovalis is a common parasite in wild guinea pigs (genus *Cavia*). It is 1.0–1.2 mm long. Its head is wider than it is long and the antennae are short. Each leg has a tarsal claw; the ones in the second and third pair are grooved. The abdomen is wide in the middle and has several rows of fine, short dorsal hairs and six pairs of spiracles (Fig. 14). These parasites are usually slightly pathogenic and the animals tolerate the infestation well. However, the presence of pruritus and alopecia has been reported. Diagnosis is made by observing the parasites in hair samples. The recommended treatment consists of applying pyrethrin powder for veterinary use.

Table VII lists the ectoparasites reported in neotropical primates.

XI. TICKS

Ticks are extremely rare in neotropical primates, with none being reported in *Aotus* monkeys. The literature mentions only one report of *Ixodes loricatus* in a spider monkey (*Ateles* spp.).

XII. ACARIDS

The literature mentions only two acarids in *Aotus* monkeys, *Rosalialges cruciformis* and *Alouattalges corbeti*. According to our experience, it is very

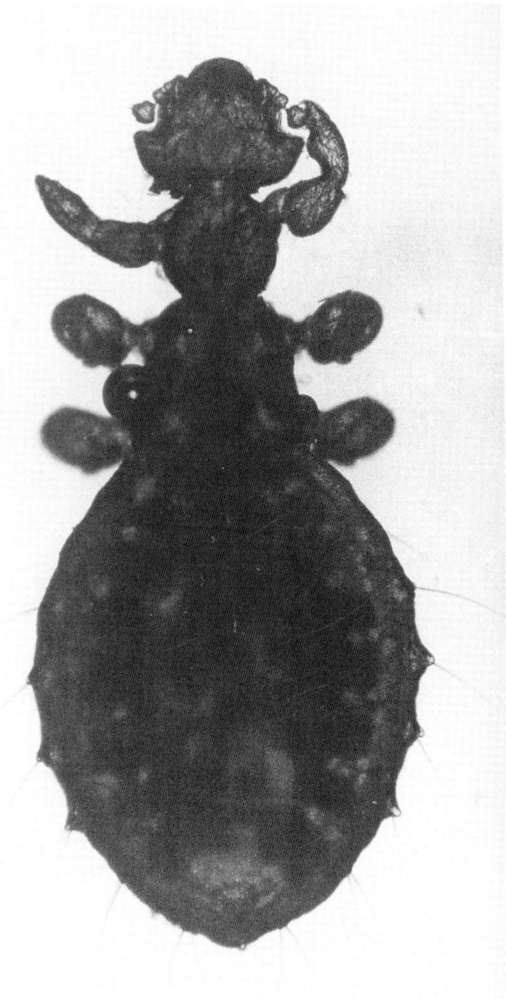

FIGURE 14 *Gyropus ovalis.*

uncommon to find acarids in *Aotus* monkeys. It seems that these parasites are slightly pathogenic in animals in their natural habitat. It is probable that infestations are exacerbated in animals subjected to the stress of captivity. Clinically, dermatitis may be observed. Diagnosis is made by observing the parasites in a lesion scrape under the light microscope. The recommended treatment consists of the topical application of ronnel.

Several species of acarids are reported in other species of neotropical monkeys (Table VIII).

TABLE VII Lice Reported in Neotropical Primates

Gyropus ovalis
Tetragyropus aotophilus
Gliricola pintoi
Pediculus humanus capitis, P. lobatus pseudohumanus
P. l. atelophilus
Harrisonia uncinata
Aotiella aotophilus
Cebidicola armatus, C. semirmatus
Trichodectes armatus, T. semirmatus

XIII. PENTASTOMIDES

Occasionally pentastomid nymphs of the *Porocephalus* spp. are found embedded in various organs, mainly the liver and the lungs, of *Aotus* monkeys.

Porocephalus crotali nymphal forms are of a white-cream color and have a cylindrical ringed body. They are 8–14 mm long and their hooks have an accessory lamina.

These infections are generally asymptomatic; the larva usually does not produce any inflammatory reaction and is covered by a fibrous membrane that is very resistant to cutting, which literally isolates it from the rest of the host, possibly avoiding in this manner the animal's immune response. Symptomatology is related to the location of the larva and the mechanical pressure it can exert on the underlying tissues. Diagnosis is made by finding

TABLE VIII Acarid Species Reported in Neotropical Primates

Sarcoptes scabiei
Demodex spp.
Prosarcoptes pitheci
Alouattalges corbeti
Fonsecalges saimirii
Listrocarpus cosgrovei, L. hapalei, L. saimirii, L. lagothrix
Pneumonyssoides stammeri
Audycoptes greeri, A. lawrenci
Rhyncoptes anastosi, R. cebi
Saimirioptes paradoxus
Rosalialges cruciformis
Mortelmansia brevis, M. longis, M. duboisi
Dunnalges lambrechti

TABLE IX Pentastomid Species Observed in Nonhuman Neotropical Primates

Porocephalus spp., *P. clavatus*, *P. crotali*
Armillifer armillatus
Linguatula serrata

the nymphs during autopsy. No treatment is known. The adult parasite lives in the respiratory tract of serpents, eliminating eggs that are swallowed with the mucus and are eliminated in the feces. Primates are infected by ingesting water or food contaminated with the feces of infected serpents.

Other pentastomid species have been reported in neotropical primates (Table IX).

GENERAL REFERENCES

Acha, P. N., and Szyfres, B. (1989). "Zoonoses and Communicable Diseases Common to Man and Animals," 2nd ed., Sci. Publ. No. 503. Pan Am. Health Organ., Washington, DC.

Adams, M. R., Lewis, J. C., and Bullock, B. C. (1984). Hemobartonellosis in squirrel monkeys (*Saimiri sciureus*) in a domestic breeding colony: Case report and preliminary study. *Lab. Anim. Sci.* **34**(1), 82–85.

Baker, J. R. (1972). Protozoa of tissue and blood. *In* "Pathology of Simian Primates. Part II. Infectious and Parasitic Diseases" (R. N. T-W-Fiennes, ed.), pp. 29–56. Karger, New York.

Blanchard, J. L., and Eberhard, M. L. (1986). Case report: Esophageal *Spirura* infection in a squirrel monkey (*Saimiri sciureus*). *Am. J. Primatol.* **10**(3), 279–282.

Brack, M. (1987). "Agents Transmissible from Simians to Man." Springer-Verlag, Berlin.

Brady, A. G., Pindak, F. F., Abee, C. R., and Gardner, W. A., Jr. (1988). Enteric trichomonads of squirrel monkeys (*Saimiri* sp.): Natural infestation and treatment. *Am. J. Primatol.* **14**(1), 65–71.

Burrows, R. B. (1972). Protozoa of the intestinal tract. *In* "Pathology of Simian Primates. Part II. Infectious and Parasitic Diseases" (R. N. T-W-Fiennes, ed.), pp. 2–28. Karger, New York.

Dunn, F. L., and Lambrecht, F. L. (1963). The hosts of *Plasmodium brasilianum*, Gonder and von Berenberg-Gossler, 1908. *J. Parasitol.* **49,** 316–319.

Dunn, F. L., Lambrecht, F. L., and due Plessis, R. (1963). Trypanosomes of South American monkeys and marmosets. *Am. J. Trop. Med. Hyg.* **12,** 524–534.

Fiennes, R. N. (1972). Ectoparasites and vectors. *In* "Pathology of Simian Primates. Part II. Infectious and Parasitic Diseases." (R. N. T-W-Fiennes, ed.), pp. 158–176. Karger, New York.

Flynn, R. J. (1973). "Parasites of Laboratory Animals." Iowa State Univ. Press, Ames.

Gozalo, A. (1993). Pentastomiasis in nonhuman primates and importance in public health. *Rev. Invest. Pecu. IVITA, UNMSM Perú* **6**(1), 50–51.

Gozalo, A., Aquino, R., and Montoya, E. (1990). Prevalence of gastrointestinal parasites in *Aotus nancymae* captured in Peruvian Amazonia. *In* "Primatology in Peru: La Primatología en el Perú," pp. 585–588. IVITA-OPS, Lima, Perú.

Hickey, T. E., Kelly, W. A., and Sitzman, J. E. (1983). Demodectic mange in a tamarin (*Saguinus geoffroyi*). *Lab. Anim. Sci.* **3,** 192–193.

King, N. W. (1976). Synopsis of the pathology of New World monkeys. *Sci. Publ. Pan Am. Health Organ.* **317,** 169–198.

Kuntz, R. E. (1972). Trematodes of the intestinal tract and biliary passages. *In* "Pathology of Simian Primates. Part II. Infectious and Parasitic Diseases" (R. N. T-W-Fiennes, ed.), pp. 104–123. Karger, New York.

Lok, J. B., and Kirkpatrick, C. E. (1987). Pentastomiasis in captive monkeys. *Lab. Anim. Sci.* **37,** 494–496.

Myers, B. J. (1972). Echinococcosis, coenurosis, cysticercosis, sparganosis, etc. *In* "Pathology of Simian Primates. Part II. Infectious and Parasitic Diseases" (R. N. T-W-Fiennes, ed.), pp. 124–143. Karger, New York.

Orihel, T. C., and Seibold, H. R. (1972). Nematodes of the bowel and tissues. *In* "Pathology of Simian Primates. Part II. Infectious and Parasitic Diseases" (R. N. T-W-Fiennes, ed.), pp. 76–103. Karger, New York.

Rosenblum, L. A., and Cooper, R. W. (1968). "The Squirrel Monkey." Academic Press, New York.

Seibold, H. R., and Wolf, R. W. (1971). Toxoplasmosis in *Aotus trivirgatus* and *Callicebus moloch*. *Lab. Anim. Sci.* **21,** 118.

Soulsby, E. J. L. (1987). "Parasitology and Parasitic Diseases in Domestic Animals," 7th ed. Nueva Editorial Interamericana S. A., México D. F.

Tantaleán, M., Gozalo, A., and Montoya, E. (1990). Notes on some helminth parasites from Peruvian monkeys. *Lab Primate Newsl.* **29**(2), 6–8.

Toft, J. D., II (1986). The pathoparasitology of nonhuman primates: A review. *In* "Primates: The Road to Self Sustaining Populations" (K. Benirschke, ed.), pp. 571–679. Springer-Verlag, New York.

Voller, A. (1972). Plasmodium and hepatocystis. *In* "Pathology of Simian Primates. Part II. Infectious and Parasitic Diseases" (R. N. T-W-Fiennes, ed.), pp. 57–75. Karger, New York.

Wolff, P. L. (1990). The parasites of New World primates: A review. *Proc. Am. Assoc. Zoo Vet.*, pp. 87–94.

Index